MODELING AND DIMENSIONING OF MOBILE NETWORKS

MODELING AND DIMENSIONING OF MOBILE NETWORKS

FROM GSM TO LTE

Maciej Stasiak
Poznań University of Technology, Poland

Mariusz Głąbowski
Poznań University of Technology, Poland

Arkadiusz Wiśniewski
Orange, Poland

Piotr Zwierzykowski
Poznań University of Technology, Poland

A John Wiley and Sons, Ltd., Publication

This edition first published 2011

Registered office
John Wiley & Sons Ltd, The Atrium, Southern Gate, Chichester, West Sussex, PO19 8SQ, United Kingdom

For details of our global editorial offices, for customer services and for information about how to apply for permission to reuse the copyright material in this book please see our website at www.wiley.com.

Library of Congress Cataloging-in-Publication Data

Stasiak, Maciej.
Modeling and dimensioning of mobile networks : from GSM to LTE / Maciej Stasiak, Mariusz Głąbowski, Arkadiusz Wiśniewski.
p. cm.
Includes bibliographical references and index.
ISBN 978-0-470-66586-2 (cloth)
1. Wireless communication systems. 2. Mobile communication systems. 3. Wireless metropolitan area networks. 4. Computer networks–Scalabiltiy. 5. Cell phone systems. I. Głąbowski, Mariusz. II. Wiśniewski, Arkadiusz. III. Title.
TK5103.2.S85 2011
004.6–dc22

2010026286

A catalogue record for this book is available from the British Library.

Print ISBN 9780470665862 (H/B)
ePDF ISBN: 9780470976043
oBook ISBN: 9780470976036
ePub ISBN: 9780470975992

Set in 10/12 Times by Thomson Digital, Noida, India
Printed and bound in Singapore by Markono Print Media Pte Ltd.

Contents

List of Figures

List of Tables

Preface

The first connection in the GSM network was set up in 1991 and this year marks the onset of the dynamic development of cellular telephony we are experiencing today. The unquestionable success of the GSM telephony has motivated further research in the field and the development new technologies for cellular telephony. Initially, it was assumed that cellular networks would also provide their users with multimedia services and would offer access to the Internet. Subsequent research eventually succeeded in working out a standard for the third-generation telephony (UMTS). However, UMTS network telephony has been developing at a much slower pace than its GSM predecessor, hampered by substantially high costs of rendering a network operational. Currently, services provided by the UMTS network are offered by most cellular network operators. The extension of a second-generation cellular network and the construction of a third-generation network must involve installation of new systems in the network that make it possible to increase the number of subscribers and to introduce new services, primarily those that have their quality parameters defined. Within this context, the need to design optimum network resources through a substantial reduction in the costs of network modernization is a topical and particularly important issue. The design process involves the determination of the network topology (the number of nodes, that of the structure, etc.), defining its resources (capacity) and preparing appropriate management and traffic-distribution strategies in such a way as to ensure that the total investment overlays for the creation of a network can be kept as low as possible. In order to design a network properly, and dimension its elements appropriately, knowledge of the characteristic methodologies of particular types of designed networks is indispensable. These methodologies, frequently very complex, stem from the problems and issues formulated in, for example, traffic theory, the theory of graphs, the theory of stochastic processes, combinatorics and integer programming. Knowledge of appropriate design and optimization methods facilitates effective functioning of cellular networks that are open to expansion and reconfiguration.

The creation of the GSM network, then the UMTS network, and, in the near future, the LTE network, has been, and still is a challenge for traffic theoreticians and engineers, which is particularly observable in 3G and 4G networks that service a mixture of traffic streams with multi-rate traffic. The analysis of present-day cellular networks servicing multi-rate traffic and guaranteeing a defined quality level of call service indicates a need for further research aimed at working out effective dimensioning methods and evaluation methods for traffic load-carrying capacity in such networks. The dimensioning process in 3G and 4G systems should allow designers to determine the capacity of particular elements of the system that will make it possible, given the assumed load of the system, to ensure the assumed level of GoS (Grade of Service) and QoS (Quality of Service). The dimensioning of a system also involves taking into consideration mutual dependencies of call-service processes in interfaces in the access part of the network (for example, in the radio interface and in the Iub interface in the UMTS system). Such an approach results in the creation of new traffic models that enable us to dimension and optimize 3G and 4G systems even when demands for resources are of a time-dependent (changeable) character.

The cellular network design process thus depends on the construction of the radio infrastructure part and the fixed part of the network, along with a determination of the capacity of individual elements of the system necessary to maintain complex quality parameters. It is therefore necessary to proceed with such activities as the evaluation of the capacity in radio interfaces and other interfaces in the access network of the system (for example, the Iub interface in the UMTS system). Determination of the capacity of radio interfaces in a network with soft and hard handover is by no means trivial. Because of the multi-service character of traffic, a response to this strong challenge is to work out appropriate methods for the evaluation of traffic capacities of the interfaces located in the "nonradio" part of the access network. The radio interfaces and other interfaces of the access network in 2G and 3G systems (as well as in the concept of the 4G system) are fundamental for the desired traffic effectiveness in the network. Due to the low capacity of the radio interface, which is also limited by interference from neighboring cells and by capacity as well as the organization (of resources) of joined "non radio" interfaces, operators take advantage of a number of mechanisms that allow them to increase the traffic effectiveness of the system such as compression (e.g. for services in HSDPA, HSUPA), reservation of resources for predefined classes of calls, introduction of thresholds making resources allocation conditional on the interference load at the level of the control function controlling access to resources, priorities for selected services or group of subscribers, optimum allocation of connections in a group of cells (implementation of retry handover and hard handover mechanisms), traffic overflow between neighboring cells and cells of other networks covering a given area of a given operator (for example, from 3G network cells to 2G cells).

This book aims to provide extensive information on modeling appropriate interfaces and groups of interfaces of 3G and 4G access networks that service a mixture of traffic characterized by different properties and different GoS and QoS requirements. In particular, the book presents the following analytical models of systems servicing multi-rate traffic in interfaces between the access network and implemented traffic-management mechanisms:

- prioritization models – with the possibility of introducing a hierarchy in the service process for particular call classes and pushing out calls with lower priority;
- reservation models – with the possibility of introducing a mechanism for the reservation of predefined classes of calls depending on the load of the system;
- threshold models – with the possibility of a change in service parameters (the number of allocated resources, service time) depending on predefined thresholds indicating the admissible load level in the interface;
- compression models – introducing a change in the parameters of serviced calls in a cell (compression of allocated resources) in the event of a lack of free resources for new calls;
- overflow models (with traffic overflow to neighboring cells) – multi-rate traffic overflows from cells (sectors) with the heaviest load or from other network of the operator;
- models for soft handover, softer handover and soft-softer handover connections in which the mobile station is connected to a number of base stations.

The models presented in the book form the basis of a coherent methodology for the dimensioning of the elements of the mobile network most sensitive to traffic load, namely the interfaces of the radio access network. This book provides the reader with extensive information on problems and issues in the traffic engineering, dimensioning and optimization of UMTS and LTE networks.

The authors have been involved in teaching and research investigations on the optimization and design of the UMTS network at the Faculty of Electronics and Communications of Poznań University of Technology for a number of years. Over the years, many original models worked out by the authors have been published in research journals and conference proceedings. The final version of the book has also been influenced by the authors' experience from a number of projects carried out for operators of cellular networks between 2003 and 2008.

Chapter 1 outlines the basic conceptual framework of a GSM system. The architecture of the system and its time structure with logical channels are discussed. Particular attention is given to data transmission technologies: HSCSD, GPRS and EDGE. The chapter also briefly delineates traffic-management mechanisms that are important for the GSM network, such as directed retry handover, traffic handover and queuing.

Chapter 2 briefly discusses the most important elements of the UMTS network. System architecture is presented as well as the basic operating principles of the WCDMA radio interface and channel and scrambling codes. Bearers and frame structure in the UMTS system are defined and explained. Special attention is given to a description of logical, transport, physical channels in the WCDMA radio interface. The chapter discusses essential methods for radio resource management in the UMTS system, including power control, handover control, call admission control, packet scheduler and load control. The HSPA technology (high-speed packet data transmission) for the uplink and the downlink is discussed and a brief presentation of the classification and categories of the most important services available in the UMTS network is provided.

Chapter 3 presents the concept of the evolution of multi-service systems towards the fourth generation system, known as the LTE system. Possible changes in the system architecture are discussed as well as a number of available proposals for new transmission techniques, including LTE MIMO. A classification of transport and physical logical channels potentially available for the LTE network is highlighted. The chapter also briefly presents a concept for resource management in this system.

Chapter 4 provides a discussion of basic issues and questions related to the analysis of single-rate systems. Basic concepts and properties of the call stream, service stream, Markov process and the birth and death process are defined and explained. The concepts of telecommunications traffic and traffic intensity are introduced and explained. The last section presents basic parameters for the quality of service and grade of service. Particular attention is given to the determination of blocking and the loss probability in telecommunications systems.

Chapter 5 familiarizes the reader with the most common models of the full availability group with single-rate traffic, which is known in engineering and the theory of traffic. A method for the determination of the occupancy distribution on the basis of the analysis of the birth and death process is presented and discussed. The chapter considers service of traffic streams generated both by an infinite number of traffic sources (Erlang model) and by a finite number of traffic sources (Engset model). All important parameters for full-availability groups, such as blocking probability and the loss probability, the occupancy probability of precisely determined channels and carried traffic, are defined and discussed. The differences between the Erlang and the Engest model are identified and commented on.

Chapter 6 formulates and discusses the problem of traffic overflow in single-service systems. The basics for the analysis of overflow systems and the concept of the primary group and the overflow group are presented. Special attention is given to a method for the determination of the parameters of overflow traffic: mean value, variance and the peakedness coefficient. Methods and algorithms for dimensioning of alternative groups, the equivalent random traffic method and the Fredericks-Hayward method are discussed in detail.

Chapter 7 analyzes of multi-dimensional, single-service and multi-service systems. Properties and characteristics multi-dimensional Markovian processes at the level of the so-called micro and macro states are presented and discussed. Particular attention is given to methods for the determination of the occupancy distribution in state-independent systems. Recursive and convolution algorithms for the determination of the occupancy distribution and other characteristics of the full-availability group carrying a mixture of different multi-rate traffic are presented and discussed. The following state-dependent systems are modeled: group with reservation, limited-availability group, threshold systems, systems with compression, and systems with priorities. The chapter also describes methods for modeling multi-rate systems for traffic generated by both a finite and infinite number of traffic sources.

Chapter 8 presents the concept of overflow traffic in multi-service systems. The occupancy distribution in the alternative group is determined. This is a mixture of multi-rate traffic characterized by different values of the peakedness coefficient. A method for the determination of the parameters of multi-rate traffic that overflows from primary groups, that is the average value and variance, is presented. The chapter also proposes a method for dimensioning alternative groups with overflow traffic from many primary groups servicing multi-rate traffic.

Chapter 9 presents an approach for dimensioning systems with virtual channel switching. The basics of modeling traffic sources with variable bit rates are given. Traffic source models of the type: ON/OFF IPP, ON/OFF IBP, MMPP and self-similar traffic sources are presented. The chapter also includes a presentation of example models for the determination of equivalent bandwidth and of the method for bandwidth discretization, which forms the base for dimensioning of broadband networks.

Chapter 10 discusses the issues concerning modeling of basic queuing systems. Classification and notation pertaining to queuing systems are discussed. Basic dependencies between the most important parameters for queuing systems are identified and defined. The most important part of the chapter demonstrates the parameters and characteristics of the following queuing systems: M/M/1, M/M/1/N, M/M/m, M/M/m/N, M/G/1, M/D/1 and the M/G/1 system with priorities without preemption. The chapter focuses on the M/G/R PS model used for modeling nodes in modern packet networks.

Chapter 11 is devoted to modeling and dimensioning radio interfaces in cellular networks. First, the method for bandwidth discretization in systems with soft capacity is discussed, with particular attention given to the method for determining the basic bandwidth unit (BBU) for the radio interface. Then, methods for modeling the radio interface with single-rate and multi-rate traffic are presented. The possibility of service in the interface with traffic from finite and infinite traffic sources is considered. The influence of interference upon the soft capacity of cell is taken into consideration through the application of the fixed-point methodology.

Chapter 12 discusses methods for modeling the Iub interface in the UMTS network on the basis of a model of the full-availability group with multi-rate traffic servicing traffic from a finite and infinite number of sources. The proposed model considers the possibility of introducing priorities for selected classes of calls. It also considers the feasibility of service-oriented solutions for traffic that undergoes compression, including the average throughput available for HSPA subscribers.

Chapter 13 sums up the book, providing a number of computational algorithms for selected elements of the radio access network, commonly used in engineering practice. The chapter discusses, for example, a model of a group of cells that services multi-rate traffic from finite and infinite traffic sources, handover model and traffic overflow between sectors of a given cell and between different cells. In the model of a group of cells the influence of interference on the soft capacity of cells has been taken into consideration through the application of the fixed-point methodology.

Acknowledgements

We would like to thank the first readers of our book: Professor Tadeusz Czachórski, Janusz Wiewióra and Tomasz Olszewski for offering valuable advice that helped shape the book in its final form, and our families for their patience, understanding and support.

Maciej Stasiak
Mariusz Głąbowski
Arkadiusz Wiśniewski
Piotr Zwierzykowski

Part I

Mobile Network Standards

1

Global System for Mobile Communications

1.1 Introduction

The Global System for Mobile Communications, or GSM,[1] is the so-called second-generation (2G) cellular mobile phone system. Its predecessors, first generation (1G) mobile phone systems, were analog systems such as the Nordic Mobile Telephone (NMT) system and the Advanced Mobile Phone System (AMPS) [1]. Designers of analog systems did not realize though, that cellular telephony would in a short time become a universal and popular service and thus the systems in question had a rather limited capacity. Moreover, most of them were incompatible with one another, which resulted in serious limitations because users of a given network were exclusively subscribers of a given operator.

The GSM standard was developed thanks to a European initiative aiming at creating a uniform, open cellular mobile phone system. Originally, the standard was to be applied and implemented in the 12 countries of the European Common Market. Accordingly, the European Conference of Postal and Telecommunications Administrations (CEPT) created the Groupe Spécial Mobile (GSM) in 1982 to develop a standard for a mobile telephone system that would operate in the 900 MHz bandwidth [2]. In 1987, a group of 15 operators from across Europe signed a memorandum of understanding in which they agreed to implement the GSM technology to develop a common cellular telephone system across Europe. In 1989 responsibility for GSM was transferred to the European Telecommunication Standard Institute, and the first phase of the GSM 900 specifications was completed one year later (GSM 900 Phase 1). Towards the end of phase 1 recommendations for the GSM 1800, operating in this band and aiming to service densely populated urban areas, were worked out. In October 1995, it was announced that work on the second phase of the GSM standard (GSM Phase 2) had been completed.

[1] Originally, the acronym GSM designated the Groupe Spécial Mobile, the name of the organization that was created to develop a standard for a mobile telephone system.

Modeling and Dimensioning of Mobile Networks: From GSM to LTE
Maciej Stasiak, Mariusz Głąbowski, Arkadiusz Wiśniewski and Piotr Zwierzykowski

Table 1.1 Frequency range and the number of available channels for the GSM 900 and 1800 MHz systems

Feature / Bandwidth	GSM 900	GSM 1800
Uplink (MHz)	890–915	1710–1785
Downlink (MHz)	935–960	1805–1880
Number of available channels	124	374

Work on the specifications for the GSM system continued in the following years (GSM Phase 2+). This included such specifications as the standard for the intelligent networks for the GSM CAMEL (Customized Application for the Mobile Network Enhanced Logic), fast data transmission with packet switching HSCSD (High Speed Circuit Switched Data), technology of packet transmission of data GPRS (General Packet Radio Service), and later also EDGE (Enhanced Data rates for GSM Evolution).

Since 1999, the leading role in the development and standardization of the GSM system has been taken by the Third Generation Partnership Project (3GPP) set up by regional standardization bodies working on new concepts in telecommunications systems in December 1998. Its scope of operation covers establishing norms and publishing technical reports on GPRS and EDGE data transmission as well as working out standards for the third-generation mobile networks [3].

One of the major objectives in designing the GSM system was to develop a digital system that would enable voice transmission, short text messaging (SMS) and data transmission. Moreover, the system was to handle international roaming. Table 1.1 shows the frequency bandwidths adopted in the GSM 900 and 1800 MHz systems for the uplink (from the mobile station to the base station) and for the downlink direction (from the base station to the mobile station). Full duplex transmission in both bandwidths is carried out based on separate frequency ranges. Each of the bands is divided into channels with a bandwidth of 200 KHz. For the GSM 900 system there are 124 available (separately for the uplink and for the downlink direction), and for the GSM 1800 there are 374 channels.

The designers of the GSM system decided to provide access to the radio link using frequency division multiple access (FDMA) and time division multiple access (TDMA) simultaneously (Figure 1.1). Each carrier frequency is then divided into eight time slots. In order to set up a connection in the GSM system it is necessary to assign each user a defined frequency channel and a time slot in which the signal can be transmitted or received.

1.2 System Architecture

In the GSM system, as shown in Figure 1.2, three basic subsystems can be distinguished [4]:

- base station subsystem (BSS);
- core network (CN);
- user equipment (UE).

The interfaces between particular elements of the system are defined. These interfaces determine rules for cooperation between devices.

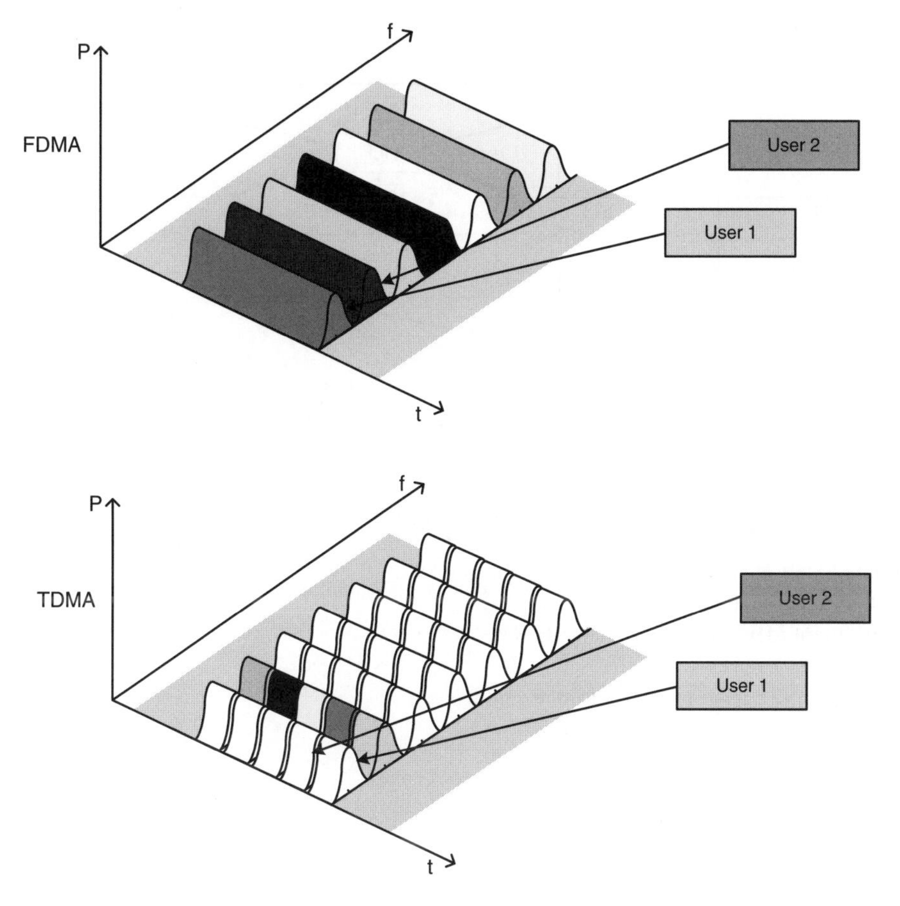

Figure 1.1 Radio link FDMA/TDMA access technology used in the GSM system.

The base station subsystem (BSS) includes a system of base stations and their controllers. Base stations provide optimum radio coverage of a given area included in the network through the radio interface in contact with mobile stations. Transmitted and received signals are also processed in base stations. The operation of the base station subsystem is controlled by the base station controller (BSC). This manages radio resources allocation to particular mobile stations, controls the setting-up of calls and the release of calls, gathers results of measurements carried out by base and mobile stations and, on the basis of the results obtained, decides on handover of connections between particular cells. The BSC controls and regulates the power, time alignment and time advance of signals transmitted by mobile stations. The exchange of information between the controller and base stations is executed by the Abis interface. Interface A enables the BSS system to be connected to the data link switching domain, whereas interface G_b enables it to be connected to the packet-switching domain [3, 5]. The main elements of the core network are:

- mobile switching center (MSC);
- visitor's location register (VLR);

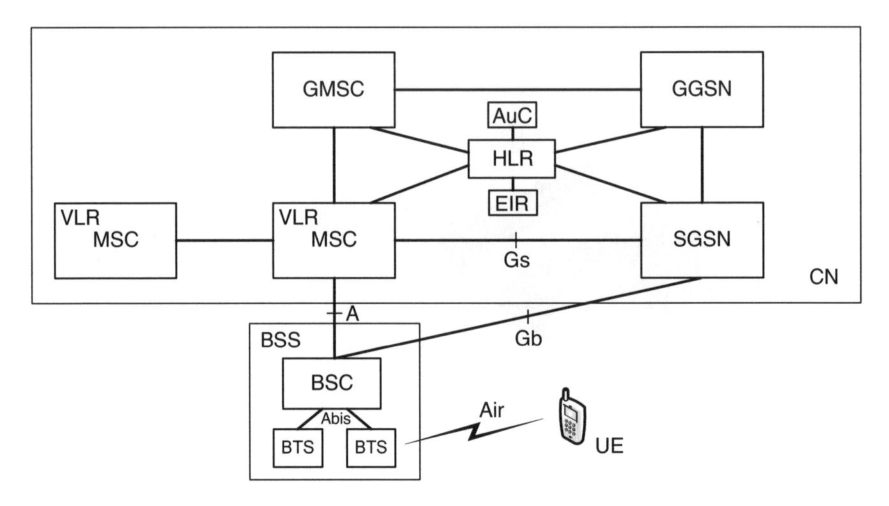

Figure 1.2 GSM system network structure.

- home location register (HLR);
- authentication center (AUC);
- equipment identification register (EIR);

and, within the packet network domain, nodes:

- serving GPRS support node (SGSN);
- gateway GPRS support node (GGSN).

The central element in circuit switching is the MSC. Its basic task is to control and regulate services provided by the system, circuit switching, and gathering billing information. Unlike switching centers used in global switched telephone networks, the MSC has to take into consideration the location or mobility of subscribers, which, in consequence, requires functions related to location registration (identification) of mobile stations. The VLR registry keeps information concerning mobile stations available in the area of one, or several, MSC switching centers. The information is collected at connection setup, login of a mobile station into the network, detection of a mobile station's initiation of using the channel, or during a handover mechanism initiation with a transition from one VLR register to another. The VLR register includes such data as: subscriber's location, the current status of the mobile station (ON/OFF), subscriber identity modules used by the network for authorization purposes, and a set of keys used in the authorization procedure [3].

The equivalent of the MSC switching center for packet switching is the SGSN. The node is responsible for providing appropriate services to individual mobile stations. It carries out procedures related to updating information on the location of the mobile station and stores location data. It also collects data used in authentication procedures and controls them as well as stores data on connections that have been set up. Using interface G_s the SGSN can be connected to the MSC, which allows it to carry on with recall and update procedures with reference to the current location of the mobile station within the area of a packet network and the area of a switching network [3].

The GGSN is an interface between the mobile packet network and external packet networks. The node converts incoming packets from nodes servicing SGSN to an appropriate protocol format for packet data used in public packet networks, packet data protocol (PDP). After the conversion, packets are transmitted to a destination network. In the case of incoming packets from external networks to the mobile network, conversion of the address in the PDP format to the GSM format follows. Then, packets are forwarded to an appropriate SGSN [1].

Setting up connections with channel-switching and packet traffic management requires cooperation between the MSC and SGSN with data bases shared by the whole of the system. The home location register (HLR) is a central database that contains details of each mobile phone subscriber authorized to use the GSM core network, and includes such data as: authorization of each of the subscriber for particular services, information that allows the user to be identified (unique keys used in authorization procedures) and information on the location of a given mobile station. The HLR closely cooperates with the AUC whose primary tasks are to generate sets of keys used in the encryption of transmission, identify the mobile station and the network, and to control and regulate the integrity of transmitted data. The EIR is a data base that keeps a list of numbers identifying a given mobile station—IMEI (*International Mobile Equipment Identity*). This provides the operator with a possibility to ban a given terminal from the network if its use is causing problems to the network or if they are stolen [3, 6].

1.3 Time Structure of the GSM System

In the GSM system each carrier frequency is divided into eight time slots. The physical radio channel between the base station and the mobile station operates using cyclic transmission in the same carrier frequency and the same time slot as a block of bits called a packet. Packet transmission is commenced every 4.615 ms and a single bit lasts 3.69 μs. It is possible to determine the transmission speed in the GSM system, which is 270.833 kbps. A typical packet, except the access packet, has 148 bits, thus its duration is about 546 μs. The duration of a single time slot is 577 μs, which allows for maintaining a steady interval between successively transmitted packets.

A basic frame in the GSM system is composed of eight successive time slots. Twenty-six frames make up a multiframe. A superframe is made up from 51 multiframes for the broadcast channel and the coupled control channel for the downlink direction, or 26 multiframes for speech channels. Duration for the superframe is 6.12 s. The highest level in the time hierarchy of the GSM system is occupied by the hyperframe, which is made up of 2048 superframes (Figure 1.3). Its duration time is 3 h, 28 min, 53 s, and 760 ms [1].

The GSM system has been designed in such a way so as to simplify the design of a mobile station maximally. To achieve that, TDMA technology has been used, with its ability to avoid the need for parallel signal transmission and reception despite transmission in full duplex. Transmission of the mobile station is delayed in relation to reception for the duration of three time slots (1.731 ms), as shown in Figure 1.4.

1.4 Logical Channels

As mentioned in Section 1.3, the physical channel is used for transmission of packets with varying time structure and hierarchical properties. Then the packets form particular logical channels.

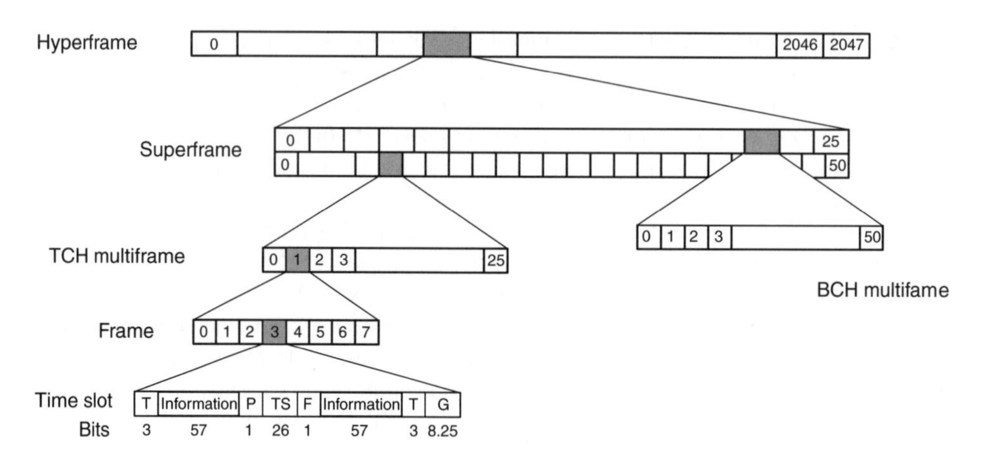

Figure 1.3 Time structure in the GSM system [1].

Logical channels, as shown in Figure 1.5, can be divided into two categories: the control channels and the traffic channels. Control channels are used to set up a connection in the radio network a for transmission of control data. Traffic channels are used to transmit user data.

In each cell of the GSM system for the downlink direction there is one carrier frequency, which, in the zero time slot, transmits system information vital for all mobile stations within the range of a given cell. In their matching coupled carrier frequency for the uplink, and also in the zero time slot, mobile stations can signal their readiness for a connection with the base station to be established. In the GSM system the following control channels can be distinguished:

- Frequency correction channel (FCCH) – used by the mobile station to tune to the carrier frequency, a frequency correction burst is transmitted on the channel by generating unmodulated sine waves.
- Synchronization channel (SCH) – transmits base station identity code (BSIC), which allows the mobile station to identify the base station and to convey synchronization information.
- Broadcast control channel (BCCH) – used for transmission of control information such as: radio channel frequency used by a given cell, neighbor cell list, information on the paging channel, configuration of logical channels in the base station.

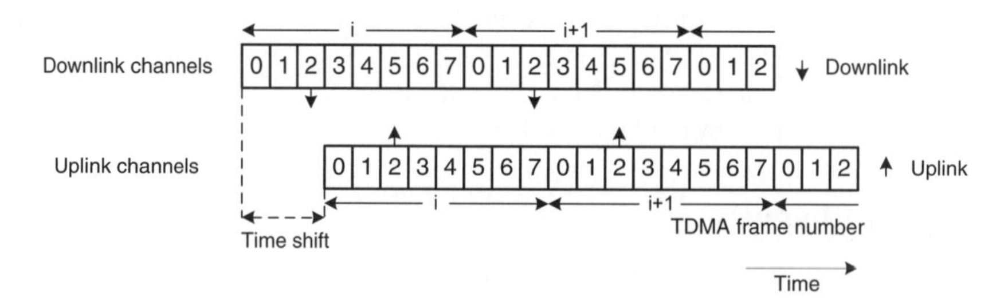

Figure 1.4 Time frame for reception and transmission in the mobile station.

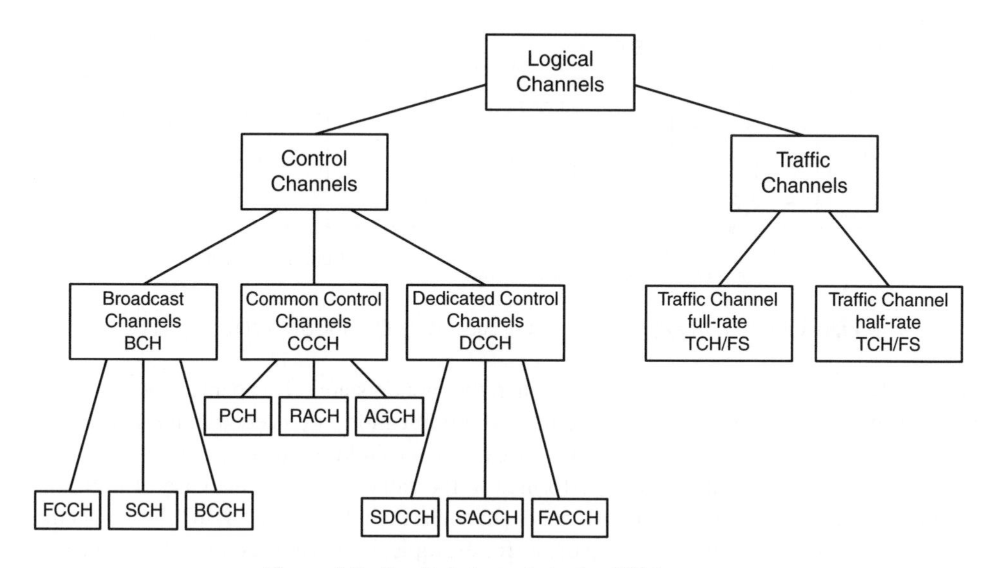

Figure 1.5 Logical channels in the GSM system.

- Paging channel (PCH) – with the PCH the base station initiates a connection with the mobile station.
- Random access channel (RACH) – used by mobile station for initial access to a system (with the RACH the mobile station initiates a connection with base station).
- Access grant channel (AGCH) – used by the base station to assign resources to a mobile station requesting access to the network.
- Stand alone dedicated control channel (SDCCH) – used to provide a reliable connection for signaling and SMS messages, for authentication, and to provide information on location update.
- Slow associated control channel (SACCH) – supports the SDCCH channel, used for sending network measurement reports and information related to power control procedures.
- Fast associated control channel (FACCH) – coupled with the speech channel, used for immediate transmission of information related to, for example, cell handover.

In the GSM system, speech signals are transmitted with traffic channels (TCHs). Speech can be transmitted at full rate, 13 kbps, or at half rate, 6.5 kbps. Half rate is mainly used by network operators in the case of heavy traffic in the radio interface. At the expense of lower quality, the capacity of a network can be increased by two different users sharing a given time slot.

1.5 High-Speed Circuit Switched Data (HSCSD)

High-speed circuit switched data is an additional feature of the GSM network and was introduced in phase 2. Its primary task is to enhance the capabilities of the GSM standard in data transmission. The GSM system architecture supporting the HSCSD is shown in Figure 1.6.

The HSCSD technology enables simultaneous application of several speech channels for a single data transmission link. A connection can be set up that makes use simultaneously of n channels (time slots) in the radio interface, where n takes on the values $n = 1, 2, \ldots, 8$.

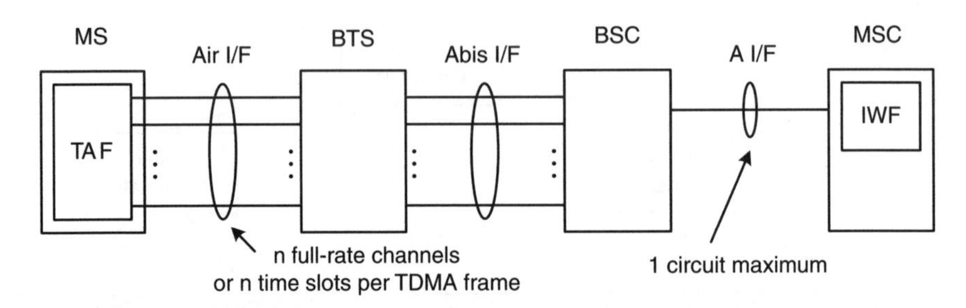

Figure 1.6 GSM system architecture supporting the HSCSD technology [7].

A HSCSD connection can be set up only when the mobile station is capable of using several radio channels simultaneously. Some additional modifications are needed in the BSS system that enable multiplexing a component data stream in one 64 kbps channel of interface A.

Transmission in the radio link is performed in the same way as in the case of traditional circuit-switching connections and all radio channels included in one HSCSD connection, from the viewpoint of network management (for example, in procedures related to connection switching, are treated as one logical channel [7, 8]).

An HSCSD connection can have a symmetrical configuration – the same number of speech channels is allocated for the uplink and the downlink direction – or a nonsymmetrical configuration. A nonsymmetrical configuration is chosen when the subscriber's requirements cannot be accomplished in a symmetrical configuration. First, the network attempts to meet the requirements related to data transmission speed for the downlink direction [1].

The maximum link transmission speed that can be achieved by the HSCSD technology depends on the number of channels used in the radio interface and on the applied coding. Due to the need to transmit all channels included in a HSCSD connection in one link 64 kbps in interface A, the transmission speed is limited to 57.6 kbps (Table 1.2).

1.6 Packet Transmission based on GPRS

Originally, the GSM system was designed to provide mainly speech services to users. Hence, the implementation of a technological improvement of a data transmission system

Table 1.2 Data rate performance in the radio interface in the HSCSD technology in relation to used number of channels and coding rate [8]

Data rate performance in radio interface (kbps)	TCH/F4.8	TCH/F9.6	TCH/F14.4
4.8 kbps	1	N/A	N/A
9.6 kbps	2	1	N/A
14.4 kbps	3	N/A	1
19.2 kbps	4	2	N/A
28.8 kbps	N/A	3	2
38.4 kbps	N/A	4	N/A
43.2 kbps	N/A	N/A	3
57.6 kbps	N/A	N/A	4

based on circuit switching – the HSCSD fast data transmission system – did not bring the expected results. Transmission with circuit switching does not match the packet and nonsymmetrical nature of data streams very well and thus the available resources were not used effectively and efficiently enough. This led to a decision that further development of data transmission within the GSM system should be continued and should employ GPRS packet technology.

The implementation of packet transmission in the GSM system, as opposed to HSCSD, required considerable changes in the structure of the system. New elements had to be included in the network – SGSN and GGSN nodes, whose functions are described in Section 1.2, and additional interfaces. The GPRS uses many of the resources of the GSM system but its main elements are connected with one another through a separate backbone network that employs the IP protocol in its operation [1, 4].

According to the assumptions given in specification [9], GPRS should allow the implementation of the following cost-effective approach, which is efficient from the network resources viewpoint, and support:

- pulse data transmission in which the time interval between individual moments of transmission is considerably higher than the average transmission delay;
- frequent transmission (several times per minute) of small amounts of data (bursty data transfer up to 500 octets);
- occasional transmission of large volumes of data.

The GPRS technology makes it possible to transmit data in several channels. Within one packet connection, the mobile station as well as the base station can make simultaneous use of eight time slots in a frame. All users of a packet service can share resources available for data transmission that are allocated, due to the asymmetry of traffic, separately for the uplink and for the downlink direction. The radio interface resources can be dynamically shared by a speech service and packet data transmission depending on the configuration of the network chosen by the operator [9].

Transmission with GPRS transmission made it necessary to introduce to the following logical channels to the radio interface, which are responsible for data packets and signaling information transmission [10]:

- Packet common control channel (PCCCH) – its functionality is implemented by the following channels:
  - Packet random access channel (PRACH) – this is used by the mobile station to initiate uplink transfer of user data or signaling information.
  - Packet paging channel (PPCH) – this is used to page a mobile station preceding downlink direction packet transfer. The channel can also be used for establishing speech connections (with circuit switched services).
  - Packet access grant channel (PAGCH) – this is used in the packet transfer establishment phase to send a resource assignment to a mobile station preceding packet transfer.
  - Packet notification channel (PNCH) – this is used to send point-to-multipoint multicast notification information to a group of mobile stations preceding multicast packet transfer.
- Packet broadcast control channel (PBCCH) – this is used to broadcast packet system information.

Table 1.3 Coding schemes in the GPRS system [10]

Coding scheme	Coding efficiency	Data rate (kbps)
CS-1	1/2	9.05
CS-2	2/3	13.4
CS-3	3/4	15.6
CS-4	1	21.4

- Packet data traffic channel (PDTCH) – this is a channel allocated for user data transfer. Several PDTCH channels can be allocated to a given mobile station. They can be allocated temporarily to one or more mobile stations.
- Packet associated control channel (PACCH) – this is used to transmit signaling information related to a given mobile station, such as information related to power control or packet reception acknowledgement messages.
- Packet timing advance control channel, uplink (PTCCH/U) – is used to ensure that the correct timing advance is maintained for each mobile station. In the uplink direction, the channel is used by a mobile station to send an access burst.
- Packet timing advance control channel, downlink (PTCCH/D) – this is used to send packets in the downlink direction to assess the needed timing advance in order to achieve frame synchronization.

Four coding schemes (CS) have been defined for the GPRS transmission: CS-1, CS-2, CS-3 and CS-4 (Table 1.3). Particular schemes are characterized by different user data transmission speed and by a various degree of error protection procedures. CS-1 coding, due to its high level of protection (error correction), is used for channels with highest interference and for signaling channels. CS-4 coding enables the fastest data transmission speed – 21.4 kbps for one channel, but has no protection. Hence, its application is limited to areas in the close vicinity of base stations with an appropriately high signal-to-noise ratio.

1.7 Packet Transmission based on EDGE

Voice transmission in the GSM system and the GPRS packet transmission uses of the Gaussian minimum shift keying (GMSK) modulation. This is a binary modulation and the initial speed of the modulator is 270.833 kbps. It is possible to achieve higher speed values with the same bandwidth of the radio channel thanks to the introduction of mechanisms such as multivalue modulations. This is why EDGE technology, like GMSK modulation, employs an eight-level phase shift keying (8PSK) modulation scheme, which allows a raw radio throughput increase of three times the raw radio throughput of GPRS.

In choosing the mode of modulation and coding for a specific transmission, the system can use one of nine coding schemes: MCS-1 – MCS-9 (Table 1.4). Each is characterized by a different data transmission speed at the cost of interference immunity. MCS-1 coding, employing GMSK modulation, is characterized by a coding efficiency of 0.53, which means that 47% of sent data is redundant and can be used in the receiver for error detection and correction. Transmission speed for one PDTCH channel with MCS-1 coding is 8.8 kbps. For

Table 1.4 Coding schemes in EDGE technology [10]

Coding scheme	Coding efficiency	Modulation	Data rate (kbps)
MCS-9	1.0	8PSK	59.2
MCS-8	0.92	8PSK	54.4
MCS-7	0.76	8PSK	44.8
MCS-6	0.49	8PSK	29.6 27.2
MCS-5	0.37	8PSK	22.4
MCS-4	1.0	GMSK	17.6
MCS-3	0.85	GMSK	14.8 13.6
MCS-2	0.66	GMSK	11.2
MCS-1	0.53	GMSK	8.8

MCS-9 coding scheme, which employs 8 PSK modulation, the coding efficiency is 1, which means that redundant data is not transmitted at all. The achieved link capacity in this case is 59.2 kbps for one radio channel [10]. Beside the new mode of modulation, EDGE also offers the possibility of controlling the link quality. This can be implemented based on the information on channel quality sent to the base station. On the basis of the information, the network chooses an appropriate combination of modulation and coding so that, with given parameters of the radio interface and the signal-to-noise value it can maximize user data transmission speed [1].

1.8 Traffic Management Mechanisms in Cellular Networks

In the GSM system, as in other cellular systems, varied traffic management mechanisms have been implemented that allow better use of available network resources and enhance the quality of service. Example mechanisms are described below.

1.8.1 Directed Retry Handover

Directed retry handover uses a particular property of the GSM system: overlapping of ranges from different cells. While establishing a connection in a given cell, a mobile station sends information on setting up a connection to the servicing base station. The base station checks if a given cell has appropriate resources to set up a connection. If the answer is positive, then the connection is established. If all speech channels are occupied, the base station provides the mobile station with information that enables it to identify neighboring cells. The mobile station checks the signal strength. If the signal strength is not good enough and does not match the criteria set up by the network operator, then a given call is blocked. Otherwise, the mobile station attempts to establish a connection through a neighboring cell.

1.8.2 Traffic Handover

In the GSM system a handover can be triggered by different events such as overload of a cell. This happens when a servicing cell has no, or few, voice channels available. The load can be equal to maximum capacity of the cell or appropriately lower. In the case when the subscriber is serviced by a cell whose load, in terms of the number of occupied frame slots, is greater or equal to the predefined load level and is within the range of another available cell, its connection can be handed over to a neighboring cell.

1.8.3 Queuing

A queuing system in the radio interface is a variety of mechanisms that allow a given call to wait in the case of lack of appropriate resources for a defined time span (several seconds) for channels required to set up a connection to be available. If appropriate resources are available within the defined time span, the call is accepted. Otherwise, the call is rejected.

References

[1] Wesołowski, K. (2002) *Mobile Communication Systems*, John Wiley & Sons, Ltd.
[2] Friedhelm, H. (ed.) (2002) *GSM and UMTS: The Creation of Global Mobile Communication*, John Wiley & Sons, Ltd.
[3] Kołakowski, J. and Cichocki, J. (2003) *UMTS. System telefonii komórkowej trzeciej generacji.* Wydawnictwo Komunikacji i Łączności.
[4] 3GPP. (2002) *TS 23.002 Network architecture v. 3.6.0*, 3GPP.
[5] Halonen, T., Romero, J., and Melero, J. (2003) *GSM, GPRS, and EDGE Performance: Evolution Towards 3G/UMTS*, John Wiley & Sons, Ltd.
[6] Holma, H. and Toskala, A. (2000) *WCDMA for UMTS. Radio Access For Third Generation Mobile Communications*, John Wiley & Sons, Ltd.
[7] 3GPP. (2000) *TS 23.034 High speed circuit switched data (HSCSD); stage 2 – technical aspects*, 3GPP.
[8] 3GPP. (2000) *TS 23.034 High speed circuit switched data (HSCSD); stage 1 – requirements*, 3GPP.
[9] 3GPP. (2003) *TS 23.60 General packet radio service (GPRS); service description; stage 2*, 3GPP.
[10] 3GPP. (2004) *TS 43.064 General packet radio service (GPRS); overall description of the GPRS radio interface; Stage 2 General packet radio service (GPRS); service description; stage 2*, 3GPP.

2

Universal Mobile Telecommunication System

2.1 Introduction

Work on 3G broadband systems commenced in 1985 and was carried out under the auspices of the International Telecommunication Union (ITU). It was decided that the new system, eventually named IMT-2000 (International Mobile Telecommunications), would provide the following standards [1–3]:

- Transmission rates:
  - transfer of up to 2 mbps inside of buildings and for slowly moving mobile stations (at the speed of less than 10 km/h);
  - 384 kbps for terminals moving with a speed of up to 120 km/h in built-up areas;
  - 144 kbps in non-built-up areas and in the case of fast moving terminals.
- Access to the Internet.
- Transmission in circuit switching operation and packet switching operation modes.
- Services in real time and multimedia and localization services.
- Simultaneous invocation of different services.
- Global roaming.
- Availability of services independently of the current localization of the subscriber and the radio interface used.
- High level of security of sent data.
- Smooth transition from the second generation systems to that of 3G.

Provision of standards for a system that would meet the above prerequisites was assigned to regional standardization organizations whose activity was mainly focused on working out a radio interface for the ground segment of the system. Eventually, three different standards for radio interface were agreed to be included in the IMT-2000 [3]:

- UMTS terrestrial radio access (UTRA) – broadband CDMA transmission working in frequency division duplex (FDD) and time division duplex (TDD) modes.

Modeling and Dimensioning of Mobile Networks: From GSM to LTE
Maciej Stasiak, Mariusz Głąbowski, Arkadiusz Wiśniewski and Piotr Zwierzykowski

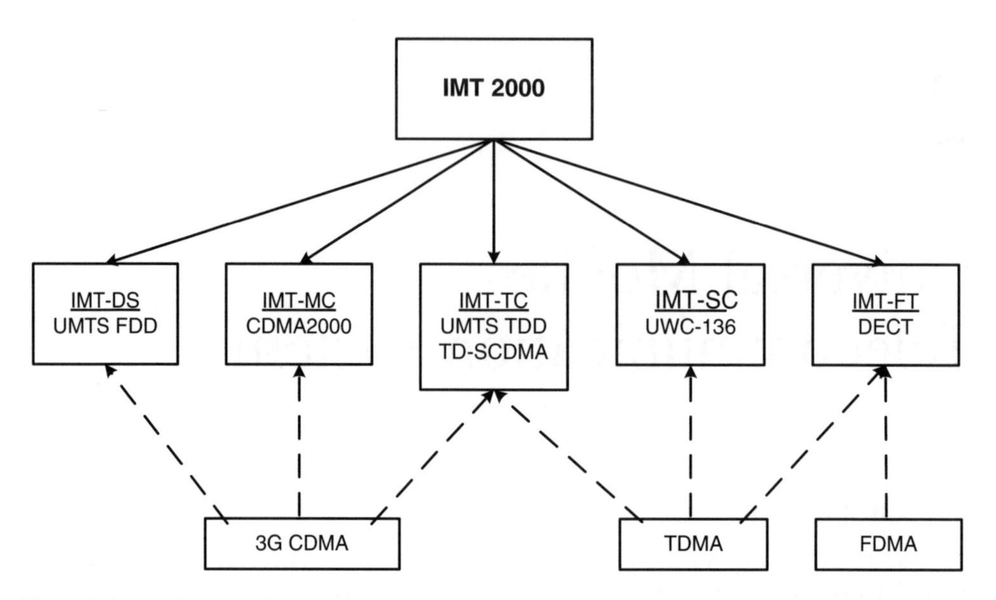

Figure 2.1 Radio transmission techniques in the terrestrial-based segment of the IMT-2000 system [2].

- Multicarrier CDMA (MC CDMA) – CDMA multitone transmission (the system is represented by the American cdma2000 system).
- Universal wireless communications (UWC 136) – standard based on convergence of the IS-136 and GSM EDGE standards. Figure 2.1 shows radio transmission techniques in the terrestrial-based segment of the IMT-2000 system.

Allocation plan for the use of radio frequency bands for the UMTS system in Europe includes the following frequencies: 1900–1980 MHz, 2010–2025 MHz, 2110–2170 MHz – Figure 2.2.

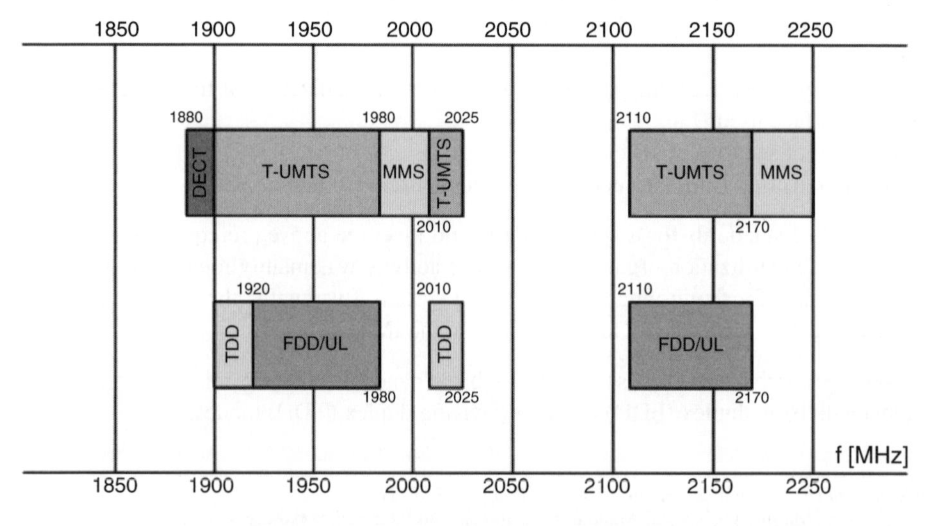

Figure 2.2 Frequency allocation plan for the UMTS system in Europe.

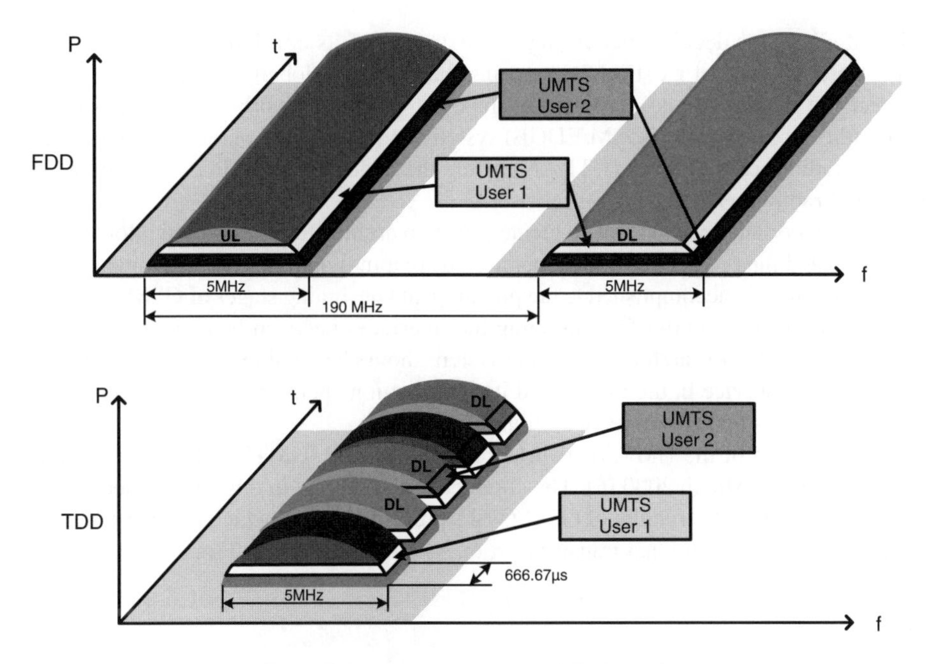

Figure 2.3 Duplex in FDD and TDD mode.

Multicarrier CDMA transmission techniques for integrated broadband have been accepted as an air interface standard for Europe. The system can operate in two modes: FDD and TDD duplex (Figure 2.3). The FDD mode provides separate 5 MHz channels, both for the link from the base station to the mobile subscriber (downlink), and from the subscriber to the base station (uplink). In the TDD mode, the 5 MHz channel is shared between the uplink and the downlink direction [4].

In the initial stage of work on the UMTS system, only the FDD mode was defined. The TDD mode was added later for the unpaired frequency ranges allotted by ITU for the UMTS system to be fully utilized. Due to potentially greater bandwidth capability and technological advancement of a system with frequency division duplex, operators of cellular networks decided to implement the above solution [4].

Since 1999, the 3GPP has been playing a leading role in the standardization of the UMTS system. Its interest encompasses globally applicable norms and standards and it publishes technical reports on third-generation systems and the development of GPRS and EDGE data transmission in the GSM system [2].

2.2 System Architecture

It was decided that the UMTS system should provide the users with many different types of services, including those that would only be developed in the future. Their implementation should be made easy and cost-effective and should be possible without costly development of the system. Thus, several decisions were taken at the planning and designing stage.

First of all, it was decided that during the first years of its development the system would use the core network of existing 2G systems that operate globally, for example GSM [5]. This decision implied the development of the existing core network of the GERAN – GSM EDGE radio access network (GSM/EDGE) system, and a new kind of radio network – UMTS terrestrial radio access network (UTRAN), using the wideband code division multiple access (WCDMA) radio interface.

Another important decision was to create a system architecture that would allow for its full modularity, and thus protect network operators from a monopoly of one equipment provider. This was not entirely accomplished at the planning and designing stages of GSM network, and it became possible only after fully defining the interfaces between its particular modules. It was also decided that the architecture of the system should be flexible enough to accommodate various types of service in the future, and its introduction should not entail costly expansion of the network or stoppages in its operation [1, 2].

The first version of the UMTS network architecture, R99 (UMTS Release 1999), was approved by 3GPP in March 2000 [6]. The core network structure in this version does not differ much from the network structure of GSM/GPRS presented in 1.2, which is consistent with the adopted strategy of evolutional transition from 2G network to 3G – Figure 2.4 [2]. UTRAN

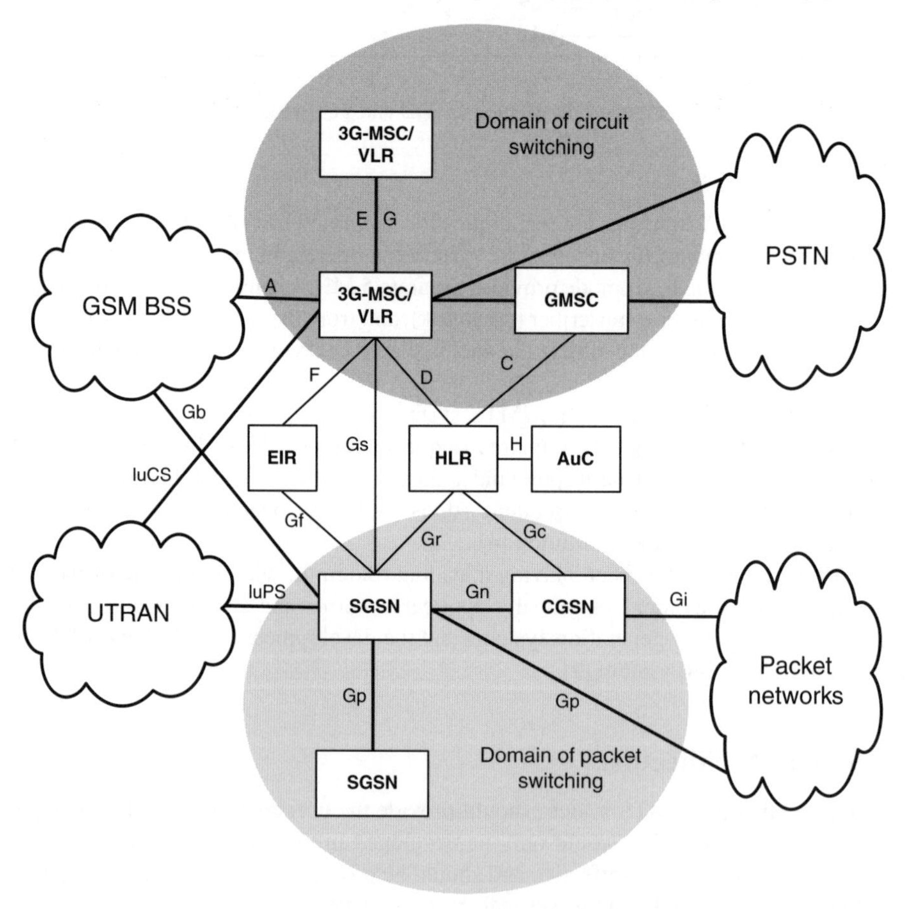

Figure 2.4 UMTS network architecture under version R99 [2].

radio network functions are carried out by two devices: base station Node B and the radio network controller (RNC). These elements are interconnected by the *Iub* interface.

The RNC radio interface controller is responsible for managing radio resources of UTRAN network. The major functions of the controller are:

- resource allocation to particular mobile stations;
- radio network admission control (AC);
- radio network load control (LC);
- power control (PC) – outer loop;
- switching connections between individual Node B;
- encryption of information in the radio link;
- allocation of channelization and scrambling codes;
- consolidation of measurement results taken by Node B and mobile stations.

The RNC controller is connected to the mobile switching center MSC via the interface IuCS and with the SGSN node via the interface IuPS. The Iur connection between particular RNCs has also been defined. A radio network controller can fulfill the following functions [4]:

- serving RNC (SRNC) – through the interface Iu this directs traffic to the core network, and in the case when it serves as a CRNC controller (Controlling RNC), it controls the operation of base stations supported by a given RNC;
- drift RNC (DRNC) – supports SRNC, is responsible for diversity effect[1]. It routes the data between Iub and Iur interface in the case that Node B's used by a mobile belong to different RNC.

In the UTRAN radio network Node B the following functions among others [4]:

- channel coding;
- data interleaving;
- signal spreading;
- data speed adjustment;
- modulation;
- power control – inner loop.

The UMTS network architecture in version R4 mainly introduces changes in the core structure. These changes are aimed at a transition towards networks operating exclusively on IP protocol. For this purpose, a division of signaling and switching functions is planned [7, 8]. And thus, in version R4 of the system, MSC will be replaced by a circuit switched media gateway (CSMGW), servicing traffic with channel switching, and MSC servers in which the VLRregister function has been integrated.

Transmission and processing of information generated by the user are carried out using media gateways. This is also where the conversion of information from networks into packet form takes place. Servers are responsible for controlling gateways, transmission and processing of information [2].

The system architecture in R5 and R6 versions is presented in Figure 2.5 and 2.6. The main changes in the architecture of the system involve the integration of HLR and AUC server in one single device called the home subscriber server (HSS) and the introduction of the IP

[1]The same data is transmitted by many RNC.

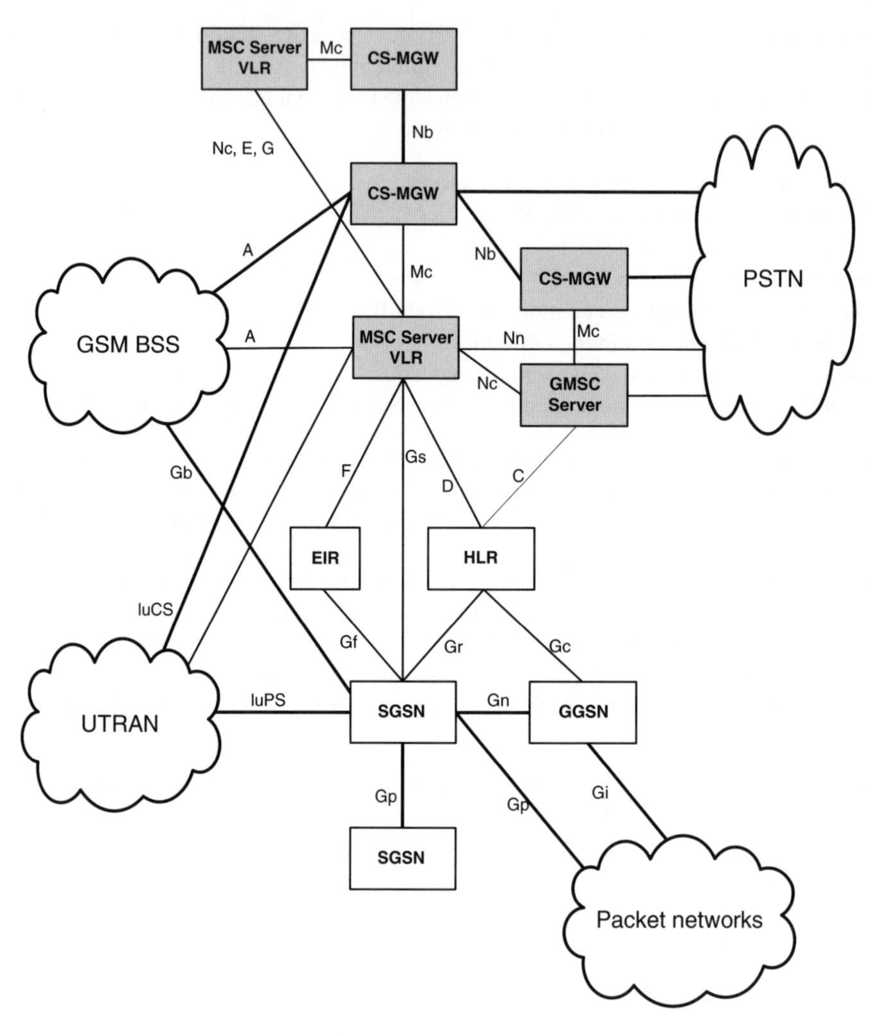

Figure 2.5 UMTS network architecture in R4 version [2].

multimedia subsystem (IMS), a device responsible for allowing access to services based on the IP protocol (for example, VoIP) [8].

2.3 Wideband Access with WCDMA Coding and Multiplexing – Essentials

As a radio link access method, the UMTS systems use wideband code division multiple access (WCDMA). According to Shannon's theorem, the throughput of a channel with bandwidth constrained to W Hz, in which the functional signal is degraded by additive white Gaussian noise with the power spectral density $N_0/2$ is [3]:

$$C_b = W \log_2 \left(1 + \frac{P_{av}}{WN_0}\right) \tag{2.1}$$

where P_{av} denotes the mean functional signal power.

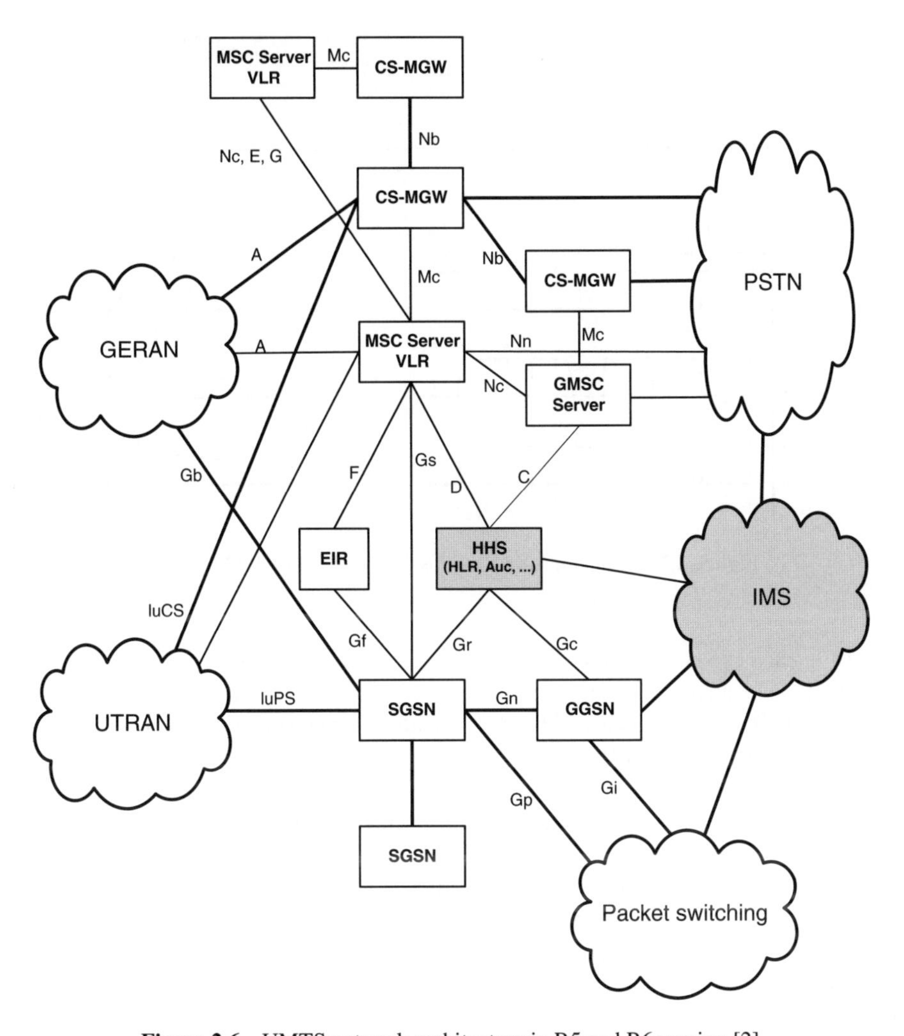

Figure 2.6 UMTS network architecture in R5 and R6 version [2].

From (2.1), the amount of information that can be transmitted by the channel reaches the boundary value called the link capacity of channel C_b when the transmitted signal has a Gaussian nature, and therefore is similar to noise. This means that an increase in the link capacity of a channel entails a broadening of the spectrum of the signal and a decrease of the signal/noise ratio below the noise level. It is this observation that has inspired designers of wireless telecommunications systems and has resulted in the UMTS system [3].

In the UMTS FDD radio interface a system with direct spectrum spreading by a pseudorandom sequence, direct sequence spread spectrum (DS-SS), was used [3]. During the modulation, information bits are spread in a wideband channel by a multiplication of the information signal and pseudorandom bits called "chips." The system utilizes chip at a rate of $R_{\text{chip}} = 3.84$ Mchps, spreading sequences with the length from 4 to 256 bits for the uplink and from four to 512 bits for the downlink direction. The length of the spreading sequence is separately

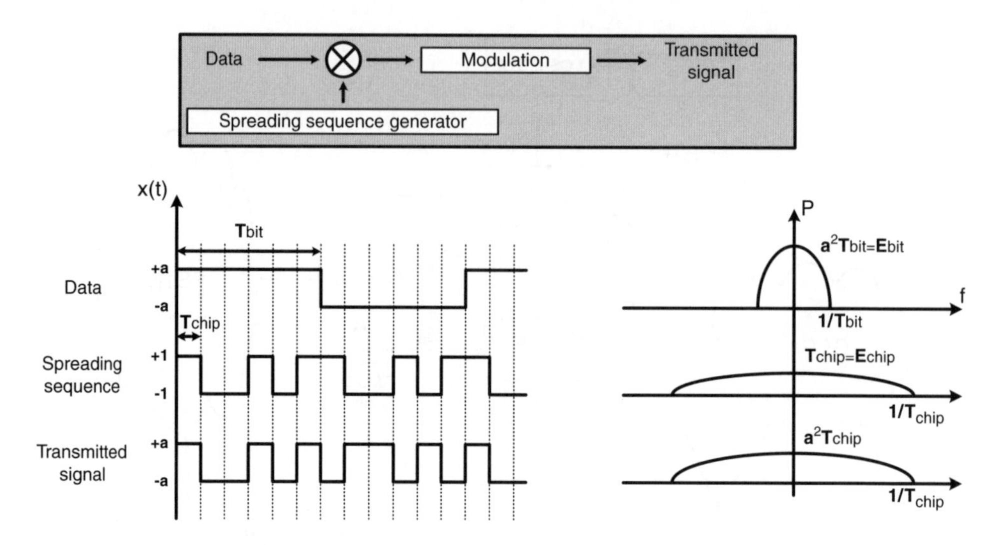

Figure 2.7 Spreading and transmission of the DS-CDMA signal.

selected for each physical channel in such a way as to obtain, as a result of the multiplication, the final sequence with an assumed link capacity of 3.84 Mchps. In the process, data from different physical channels can be spread with different spreading factors SF, according to the following dependence [2]:

$$T_{bit} = SF \times T_{chip} = 3.84 \text{ Mchps} \tag{2.2}$$

where T_{bit} denotes bit rate of the resulting signal and T_{chip} chip rate of the spreading signal.

The process of demodulation in the receiver is performed through a multiplication of the received signal with a phase-appropriate pseudorandom signal used for spreading.

The DS-CDMA signal-spreading process and signal despreading are presented in Figures 2.7 and 2.8, respectively.

2.3.1 Channelization Codes and Scrambling Codes

Wideband code division multiple access makes it possible to transmit different signals in the same frequency channel. Their reception is possible thanks to the application of two kinds of codes: channelization codes and scrambling codes. Channelization codes are used for retaining orthogonality [9] of signals coming from different physical channels, and in particular:

- in the uplink direction – to distinguish data channels and controlling channels coming from the same user;
- in the downlink direction – to distinguish calls carried out within the same cell (by different users).

Tables 2.1 and 2.2 show spreading factor for the downlink and the uplink direction and the corresponding DPDCH channel bit rate and the maximum user data rate.

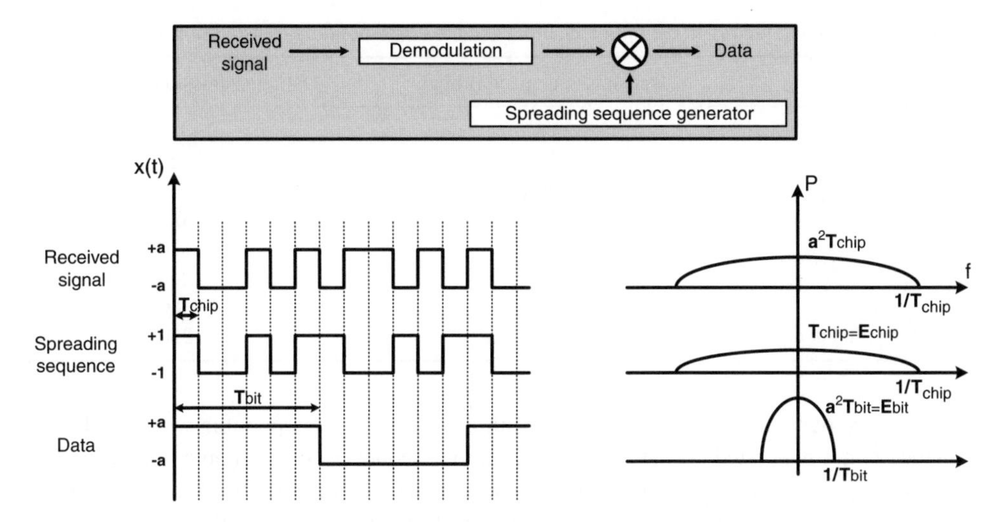

Figure 2.8 Despreading and reception of the DS-CDMA signal.

Scrambling codes are used:

- in the uplink direction for the identification of mobile stations;
- in the downlink direction for the identification of particular cells.

Channelization codes in the UMTS system are based on orthogonal variable spreading factor (OVSF) codes. The OVSF codes are defined with the code tree presented in Figure 2.9. In the code tree, each OVSF code is described as $c_{SF,k}$, where SF denotes the spreading factor of the code, and k is the number of the code for a given SF ($0 <= k <= SF - 1$). A given channelization code sequence modulates one bit of data signal. Different length of code sequences enables signals to be transmitted with different data rates in such a way as to maintain a steady chip rate 3.84 Mchps [3]. For each of the nodes of the tree derivative branch sequences are formed by a repetition of the sequence of the previous branch and by supplementing it with the same sequence in the upper branch or its negation in the lower branch.

Table 2.1 Spreading factors in the downlink channel and the corresponding transmission speeds

Spreading factor SF	Transmission speed (kbps)	Data transmission speed (kbps)
512	15	1–3
256	30	6–12
128	60	20–24
64	120	45
32	240	105
16	480	215
8	960	456
4	1920	936
4 with 3 parallel codes	5760	2300

Table 2.2 Spreading factors in the uplink channel and the corresponding transmission speeds

Spreading factor SF	Transmission speed (kbps)	Data transmission speed (kbps)
256	15	7.5
128	30	15
64	60	30
32	120	60
16	240	120
8	480	240
4	960	480
4 with 6 parallel codes	5760	230

Retaining orthogonality of code sequences with different length is possible when the longer sequence is not generated on the basis of the shorter sequence. For example, the sequences $C_{2,1}$ and $C_{4,1}$ are both orthogonal, whereas the sequences $C_{2,1}$ and $C_{4,3}$ do not satisfy the orthogonality condition. Each sector of the base station can transmit in the WCDMA radio interface traffic channels with 512 channelization codes.

The rate of the scrambling sequences in the UMTS system is equal to the chip rate of spreading sequences (3.84 Mchps). As the result of the scrambling process the rate of the coded stream does not increase the bandwidth. Each scrambling sequence consists of two binary code sequences of the same length that are subsequently treated as the real and imaginary parts of the complex scrambling sequence [2]. The standard defines two types of scrambling codes:

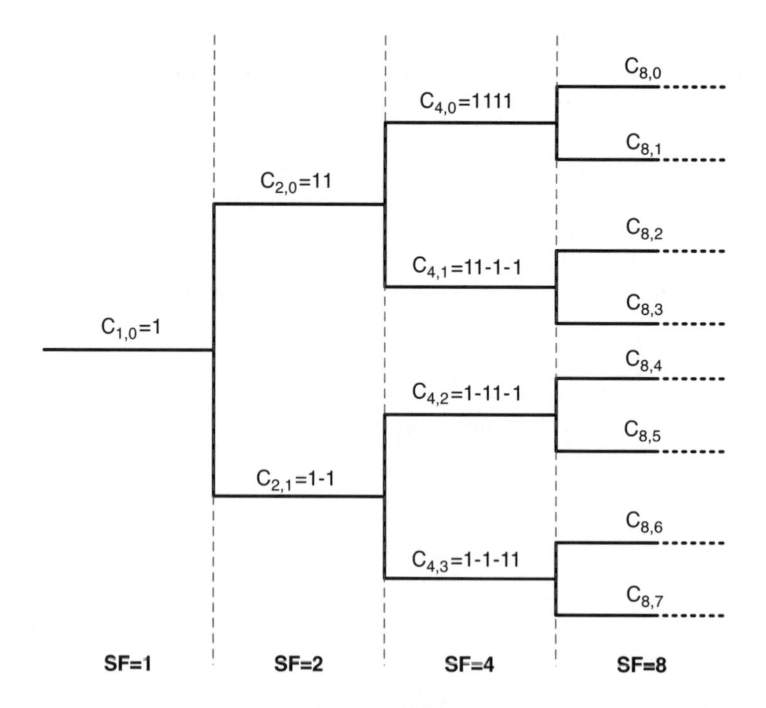

Figure 2.9 OVSF channelization code tree.

- short sequences $S(2)$ consisting of 256 symbols (repeated with the frequency of 15 KHz);
- long sequences (Gold codes) consisting of 38400 symbols (repeated every 10 ms).

2.3.2 Bearers in the UMTS System

A particular important role is played in the UMTS system by defined bearers of various kind that service transmission between different elements of the network. The bearers that service transmission in the UTRAN radio network, including the radio interface, are of particular significance. Transmission between a mobile station and the core network is carried out through the radio access bearer (RAB). The radio bearer (RB) in the radio interface is used for transmitting information between the RNC controller and the mobile station [2]. Bearers of the UMTS network are presented in Figure 2.10.

The layered architecture of UMTS bearers plays an important role in securing an appropriate quality of call. The UMTS system allows the user of the applications to negotiate characteristics of a particular bearer in such a way that it is appropriate for the transmitted information. Negotiations of the characteristics of the bearer is still possible when the call is on hold. Such negotiations can be initiated both by the subscriber and by the network – for example, prior to the initiation of the handover procedure.

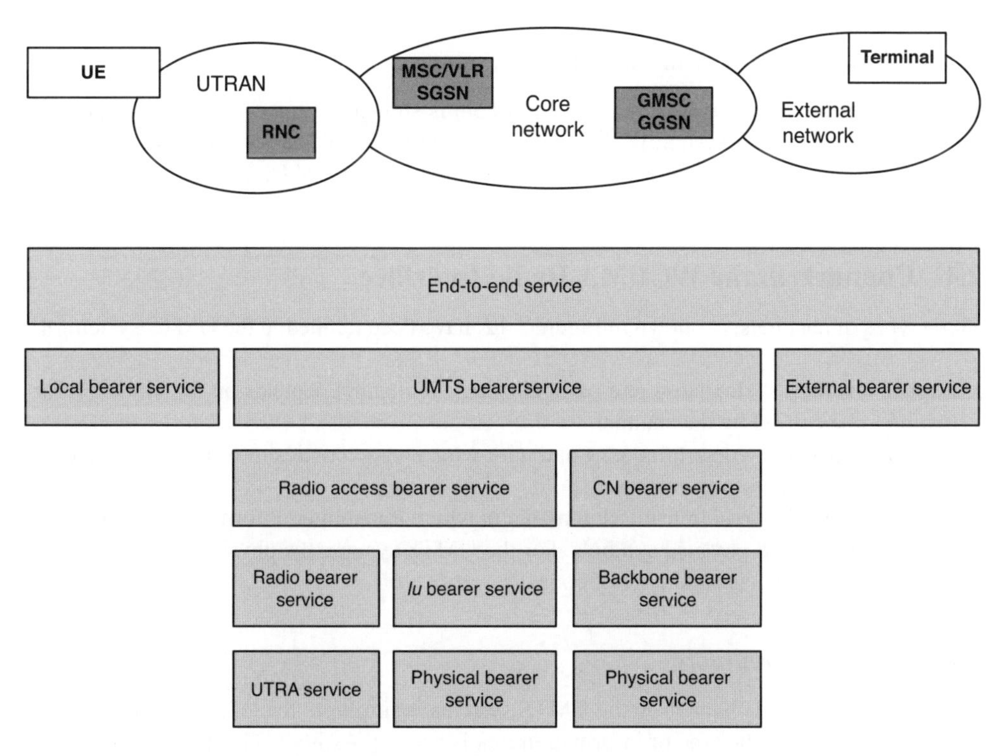

Figure 2.10 Bearers in the UMTS network.

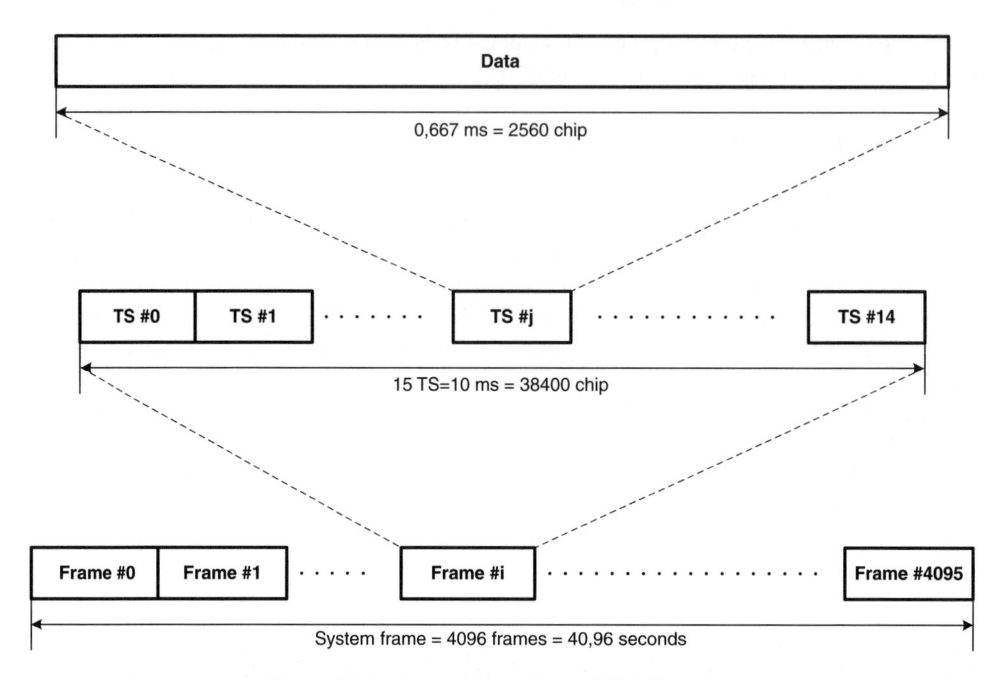

Figure 2.11 Frame format for the UMTS system.

2.3.3 Frame Structure in the UMTS System

The format of the frame used in the UMTS system is shown in Figure 2.11. The duration of the frame in the UMTS system is 10 ms. Each frame is divided into 15 slots with a length of 0.667 ms. For the chip rate of 3.84 Mchps adopted for the UMTS system, each time slot carries 2560 chips.

2.4 Channels in the WCDMA Radio Interface

Three types of channels, as shown in Figure 2.12, have been defined in the UMTS system [8]:

- Logical channels – allocation to a particular logical channel depends on the kind of information to be carried by the channel.
- Transport channels – determine how and with what characteristics information included in logical channels is to be transmitted.
- Physical channels – real transmission media in which transmitted information is mapped in the form of bits and physical symbols. For the UMTS system, the physical channel defines appropriate frequencies and a set of codes.

2.4.1 Logical Channels

The concept of logical channels in the UMTS systems is similar to that of the GSM. A logical channel is defined by the type of information that is carried. As in the GSM, logical channels are divided into two groups: control channels and data channels.

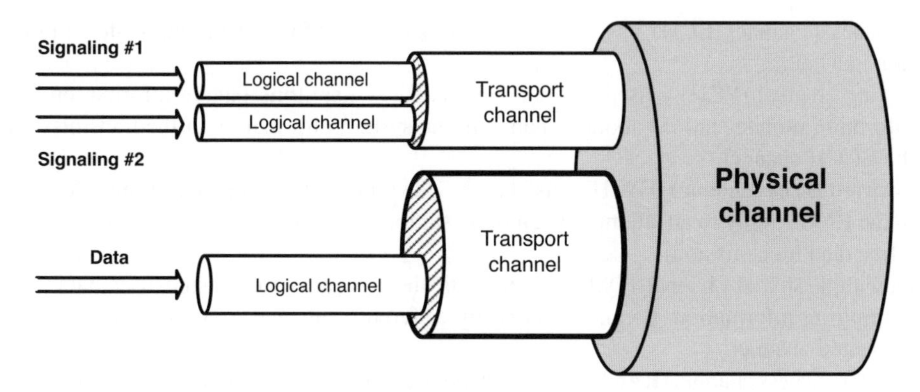

Figure 2.12 Channels in the UMTS system.

Data channels include [8, 10]:

- Dedicated traffic channel (DTCH) – used for transmission of user data information between the mobile station and the base station, a dedicated data channel to transmit and receive data both in the uplink and the downlink direction.
- Common traffic channel (CTCH) – a channel used for shared transmission in the downlink.

Control channels comprise the following channels [8, 10]:

- Dedicated traffic channel (DTCH) – used for transmission between the mobile station and the base station. It can operate both in the uplink and in the downlink.
- Broadcast control channel (BCCH) – used in the downlink direction for multicast transmission of system information.
- Paging control channel (PCCH) – used in the downlink direction for reception procedures in mobile stations.
- Dedicated control channel (DCCH) – used in the downlink and the uplink direction to transmit dedicated information between the base station and the mobile station.
- Common control channel (CCCH) – used in the downlink and the uplink direction to transmit controlling information between the network and the station.

2.4.2 Transport Channels

Information in logical channels is transmitted with the help of a set of transport blocks that are generated with a duration equal to the multiple of the length of the frame adopted for the radio interface (10 ms). Each set is allocated with a transport format (TF) that determines possible mapping, coding and interleaving of a given transport channel [2].

Transport channels can be divided into [8, 10]:

- Dedicated channel (DCH) – used in the uplink and the downlink direction for transmitting user data coming from higher layers of the network and controlling information.

- Broadcast channel (BCH) – used in the downlink direction for transmitting system information in a cell.
- Paging channel (PCH) – used in the downlink for transmitting paging information or for informing mobile stations about a change in the system information of the transmission in the BCCH channel.
- Forward access channel (FACH) – used in the downlink to transmit controlling information in the RNC's reply to an attempt by a mobile station to connect with the network (paging to a free data transmission).
- Downlink shared channel (DSCH) – used in the downlink to transmit user data and/or controlling information. It can be shared by different users and is coupled with the DCH dedicated channel.
- Random access channel (RACH) – used by mobile stations in the uplink for sending requests for resource allocations to the network. It can also be used for transmission of a small amount of user data in the uplink,
- Common packet channel (CPCH) – used in the uplink for packet transmission of a small amount of user data. It supports the internal controlling loop in the uplink.

After channel coding and interleaving, various transport channels can be multiplexed. A data stream obtained in the process is then allocated to a physical data channel [3].

2.4.3 Physical Channels

In the UMTS system each physical channel is defined by the carrier frequency, the spreading sequence and the components of the signal (in the uplink the inphase I and quadrature Q components of a signal are realized by different physical channels). Each connection is allocated one dedicated control channel and up to six dedicated physical data channels [3].

We can distinguish the following physical channels [8, 10]:

- Synchronization channel (SCH) – used in the downlink for transferring information enabling synchronization of a mobile station defined in the cell frame structure and time slots.
- Common pilot channel (CPICH) – used in the downlink in synchronization procedures.
- Primary/secondary common control physical channel (P-CCPCH/S-CCPCH) – used in the downlink for transmitting information directed to all mobile stations, such as scrambling codes used in the downlink.
- Dedicated physical data channel (DPDCH) – used in the uplink and the downlink to control the transmitting power of the mobile station and to send power control commands controlling transmission in the CPCH channel.
- Physical downlink shared channel (PDSH) – used to carry information transmitted in the shared DSCH channel.
- Page indicator channel (PICH) – used in the downlink to inform the mobile station that paging messages are in the paging channel.
- Physical random access channel (PRACH) – used in the uplink to indicate radio resource allocation demands by the mobile station.
- Physical common packet channel (PCPCH) – used in the uplink to carry packet data, shared by mobile stations within range of a cell.

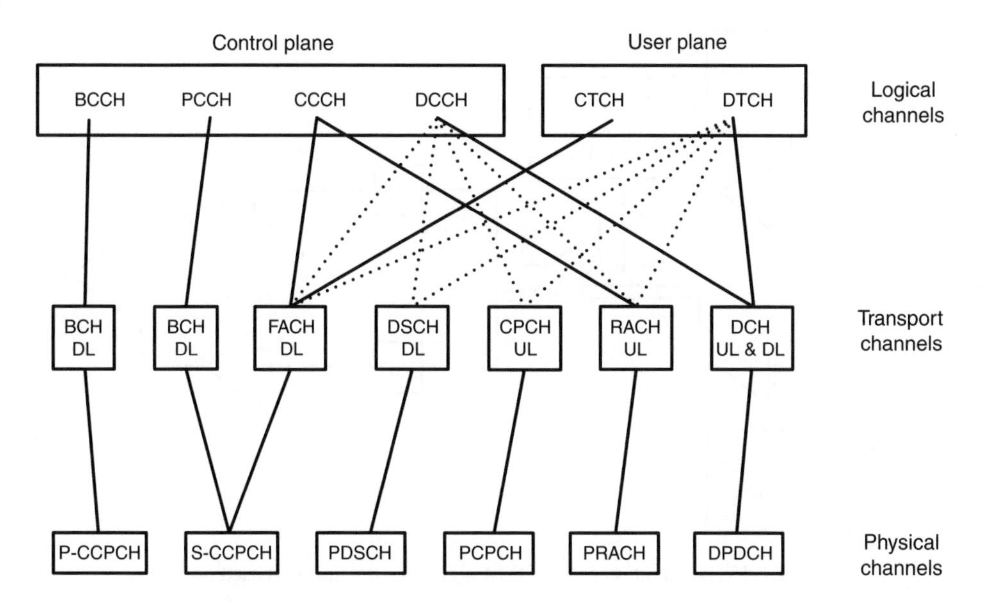

Figure 2.13 Mapping of logical, transport and physical channels.

- Acquisition indicator channel (AICH) – used in the downlink to inform the mobile station about the access demand sent in the RACH channel being accepted by the network.

Figure 2.13 presents the mapping of logical channels into transport and physical channels.

2.5 Modulation

2.5.1 Modulation in the Downlink

In the process of modulation for the downlink, QPSK modulation is used (Figure 2.14). During the modulation, dedicated data channels and control channels are appropriately multiplexed, and then serially broken down into two parallel streams that can be interpreted as a signal combined with the real part being the inphase *I* component and the imaginary component

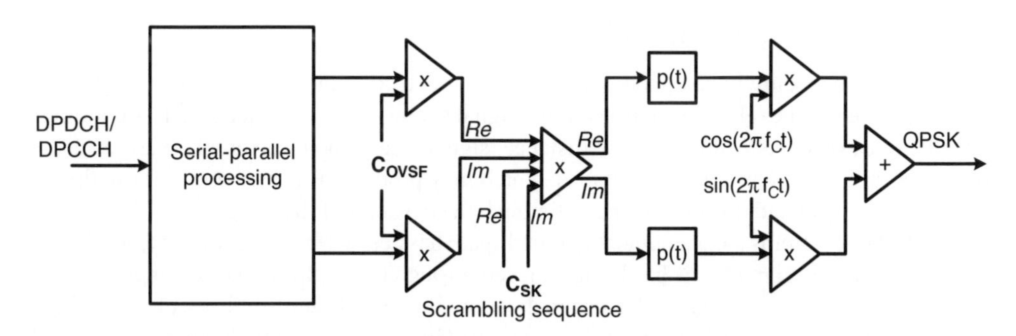

Figure 2.14 Generation of WCDMA signal for the downlink (based on [3]).

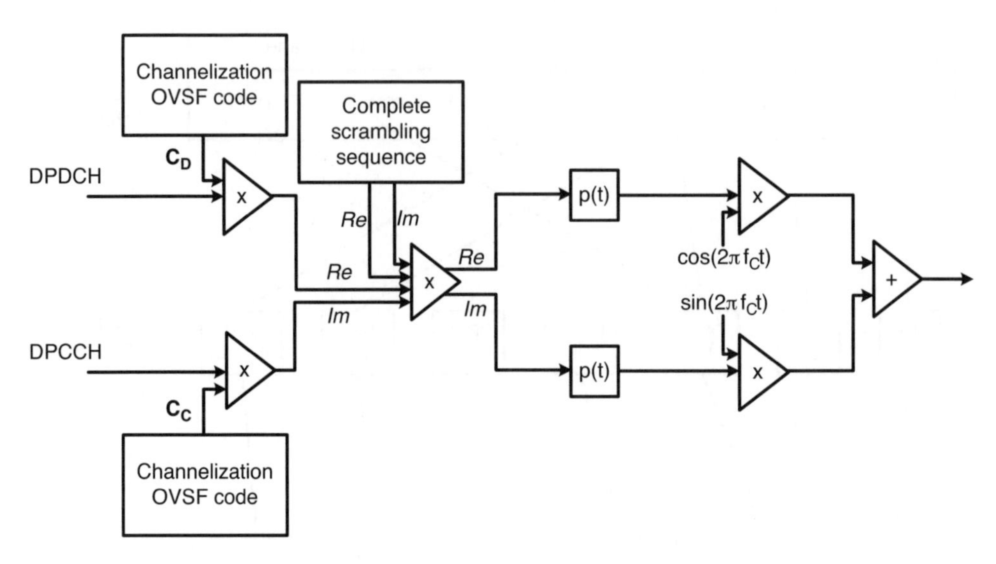

Figure 2.15 Generation of WCDMA signal for the uplink (based on [3]).

being the quadrature component Q of the coupled combined data signal. Binary signals of each of the components are spread with the application of the same OVSF codes. Subsequently, the inphase component and the quadrature component are scrambled with the application of a mutual pseudorandom sequence with a length of 10 ms. The mutual pseudorandom sequence is formed with the application of two appropriately shifted and truncated Gold codes [3].

2.5.2 Modulation in the Uplink

The modulation process for the uplink (Figure 2.15) is performed in a similar way as the modulation process for the downlink. The difference is that the inphase component I and the quadrature component Q are transmitted by different physical channels. These are respectively DPDCH and DPCCH. Both data streams are spread with the application of different mutually orthogonal codes, which two BPSK signals to be distinguished in the receiver.

2.6 Signal Reception Techniques

Multipath is an essential phenomenon used in signal detection in wideband CDMA systems. This is effected by the reflection of transmitted signal against different objects such as buildings or hills. Because of reflections and the distance, signals reach the receiver with various delay times and different phases and amplitudes, which may produce distortion and deformation in the received signal and, in consequence, diminish the capacity of the system [2].

In WCDMA systems the phenomenon of multipath has been used for the improvement of the quality of received signal. This has been achieved through the application of the multichannel RAKE receiver, in which particular channels allow the reception of replicas of transmitted signals (Figure 2.16). Signals that have been separated in individual channels are then processed and summed, which ensures better quality of reception than in the case of a single-channel

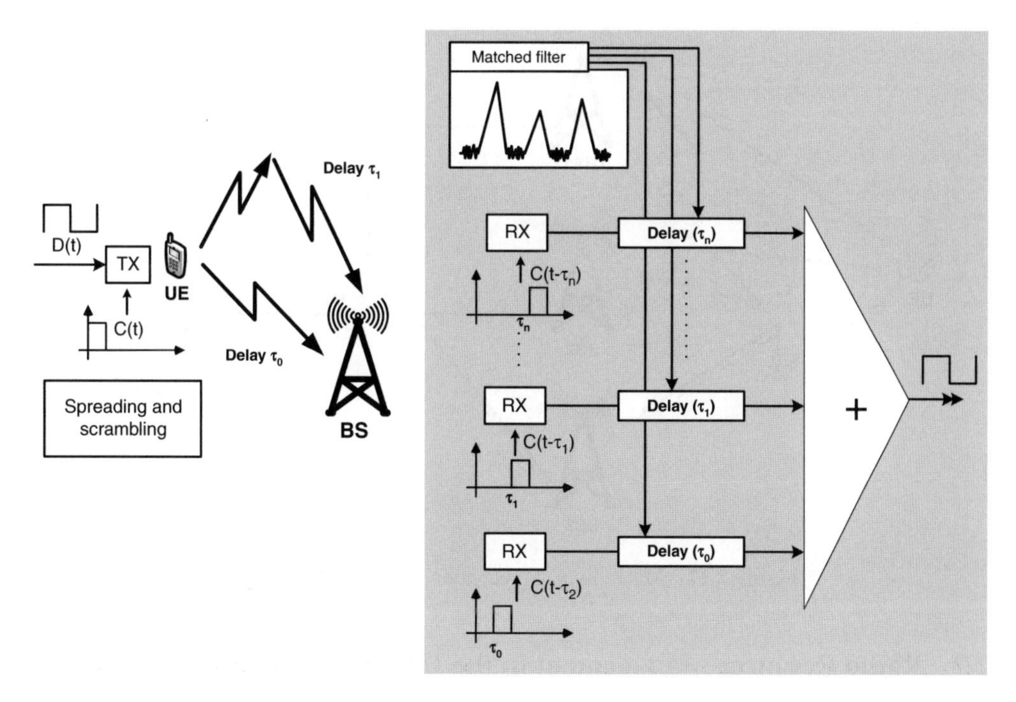

Figure 2.16 RAKE receiver operation.

receiver. After transferring the received high-frequency signal to the basic bandwidth, the signal is applied on the input of particular channels and on the input of the channel allocation system of the receiver. The channel allocation system determines the number of multipath components and the selection of the components with maximum concentrations, and allocates individual components to channels of the receiver. The input signal is correlated with the predefined code sequence and thus on the block output of the correlator the components I and Q of the original signal appear. The phase of the signal thus obtained is shifted relative to the distance covered and reflections in the radio channel [2].

The signal from the output of the correlation system proceeds to the channel estimation system, which, on the basis of the transmitted pilot sequence, reconstructs the phase of the signal. The value of the phase is then sent to the phase correction system, which changes the phase of the components I and Q of the processed signal. The next step of processing the received signal compensates for signal time delays in individual channels resulting from different length of propagation paths. The final stage includes the summing up of the signals from the outputs of individual channels. As a result of the operation, a signal is obtained whose amplitude is higher than that of any of the signals received in individual channels [2].

Another essential method of signal reception used in the UMTS system is macrodiversity, presented in Figure 2.17. Soft handovers enable the mobile station simultaneously to maintain connections with several base stations. In this way, the transmitted signal can be received by several base stations and summed up in appropriate RNC.

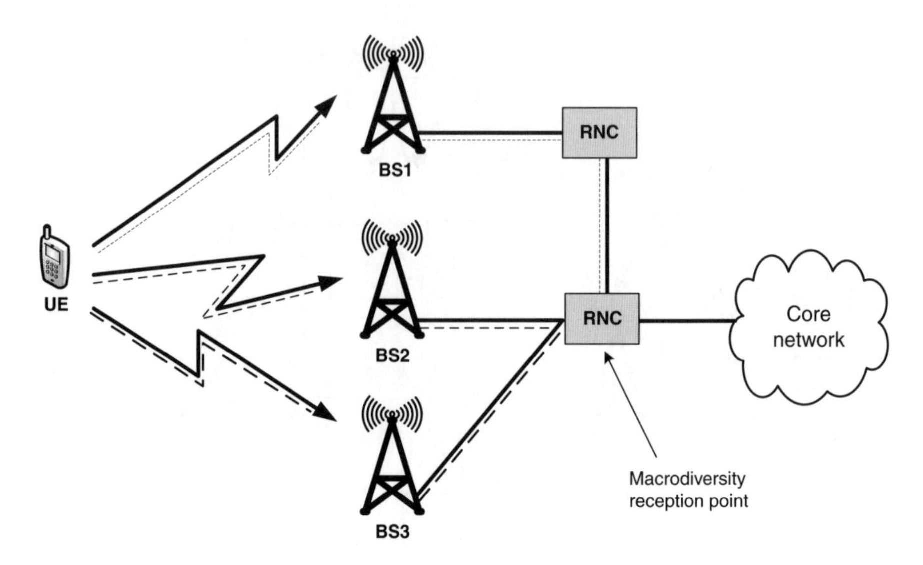

Figure 2.17 Macrodiversity reception in the UMTS system.

2.7 Radio Resource Management in the UMTS System

2.7.1 Power Control

Power control procedures enable the signal power transmitted in the radio channel to be adjusted to the changing distance between the transmitter and the receiver of the signal. This is implemented by a change in the intensity of the signal transmitted by, for example, a mobile station in a given value interval in such a way that, for a given moment for a given length of a link and a given interference level in the channel, this is the lowest applicable level that ensures a desired effect in the quality of transmission to be obtained.

Power control in the UMTS system, due to the radio channel being shared by all users, is of particular importance. Decoding of the received signal is possible only when the ratio of the energy per one bit E_b to the spectral density of noise N_0 is appropriate. A value that is too low E_b/N_0 will result in a situation when the receiver will not be able to decode the received signal, while a value of the energy per one bit in relation to noise that is too high will result in interference for other users of the same radio channel.

For the uplink, the lack of power control results in the near-far effect. This is a situation in which the received signal in the base station of the terminals at the edge of a cell, and thus those that have greater fading effect of the propagation path, encounters interference from the signals from terminals in the vicinity of the base station.

In the case of the downlink, the near-far problem is not that serious. Still, power control is also indispensable because of the necessity of sharing the available power in the base station by all users [5]. Power control in the UMTS system is performed by two mechanisms:

- open-loop power control;
- closed-loop power control, in which we can distinguish mechanisms of inner and outer power control loop.

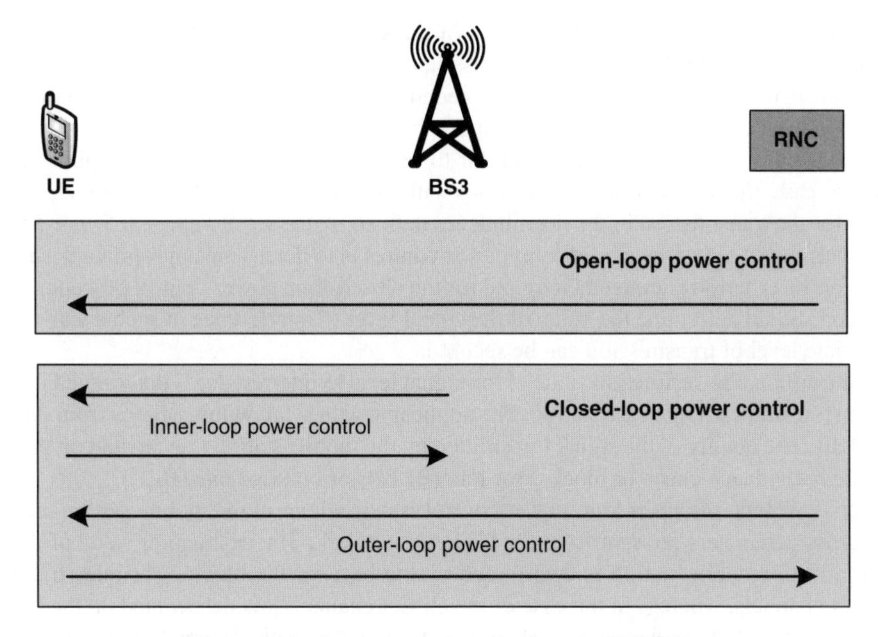

Figure 2.18 Power control mechanisms in WCDMA.

Open-loop power control is the capability of the transmitter to set its output power to an initial value suitable for the receiver. The mobile station in the uplink adjusts transmitting power based on the level of the CPICH signal received in idle modes. The mobile station also receives parameters related to the admissible power level transmitted by the cell in the BCCH channel. The mobile station evaluates the propagation loss in the radio channel and evaluates the power level with which transmission in the PRACH channel can be initiated. The transmitted power level is inversely proportional to the power of the received pilot signal. This mechanism is presented in Figure 2.18.

In the case of no reply signal returned from the base station in answer to a signal transmitted in the PRACH channel demanding radio resource allocation, the mobile station repeats the transmitted signal with a greater power intensity.

In the case of the downlink, the power value in the channels in which the transmission to a mobile station is initiated is determined on the level of signal measurement in the downlink taken and transmitted to the network by the mobile station, but also on the basis of the parameters that characterize a given connection such as the spreading factor and the required value E_b/N_0 [6].

The reason for power control is to control the signal power in the radio channel and to limit the interference level at the receiver. Closed-loop power control is the capability of the transmitter to adjust its output power in accordance with the transmission power command (TPC) symbols received in the uplink and, the speed of its operation should be appropriately high.

The base station sends a directive message to a terminal to increase the power by 1, 2 or 3 dB with the frequency of 1.5 KHz. A decision to either an increase or decrease in the power level is based on the ratio of the signal value to noise (SIR). The signal received by the base station is compared with a predefined value of the signal. If the signal power transmitted by a terminal exceeds the TPC command, the base station demands that the mobile station decrease the power. If the received signal is below predefined thresholds, the mobile station receives a directive message to increase the transmitted power level [5].

The algorithm for the downlink is similar to that for the algorithm for the uplink. In this particular case, the base station changes roles with the mobile station. Steps at which a power change for the transmission in the downlink are defined by the set of values: 0.5, 1, 1.5, 2 dB.

The purpose of using the closed-loop power control is to determine appropriate destination values for the criterion parameters required for the closed-loop power-control procedures. The open loop should determine the value of the signal level to interference in such a way that the appropriate level of transmission can be secured.

For the uplink, the destination value of the signal level to interference is determined in RNC, and then transmitted to the base station. The applicable values fall within a range from −8.2 dB to 17.3 dB. The quality of the signal transmitted by the mobile station is evaluated on the basis of cyclic redundancy check or block error rate (BLER) or bit error rate (BER).

In the downlink, the open-loop power control is carried out in the mobile station although some of its parameters are sent from the UTRAN network. The destination value of the SIR criterion in the mobile station is determined on the basis of the block error rate (BLER) in transport channels, whereas in the case of the CPCH channel, it is determined on the basis of the bit error ratio (BER) in the DPCCH channel for the downlink. The mobile station receives destination values BLER and BER from the network [2].

2.7.2 Handover Control

To allow for the provision of mobility of subscribers of cellular communications services it is necessary to create appropriate mechanisms to secure uninterrupted service for nomadic use of the mobile station between various cells of a network. In the UMTS system such a mechanism is provided by hard handover. Hard handover occurs each time the subscriber moves across a servicing cell boundary or when the connection to the current base station is broken and a new connection is immediately made to the target base station. The reason for the application of hard handover can be, for example, low quality of a connection due to interference, too low signal level or capacity-related problems.

Hard handover is usually initiated without the participation of a terminal: the mobile station does not need to perform any action. A hard handover process can occur between base stations or sectors of the same station within the range of different frequency channels. One of the scenarios of the hard-handover process is the process of transferring an ongoing call between 2G and 3G networks. Another way of transferring connections is the so-called soft handover, which can take on several variations [4]:

- soft handover;
- softer handover;
- soft-softer handover.

In the case of the soft handover the mobile station communicates with two or more sectors that belong to different base stations (Figure 2.19).

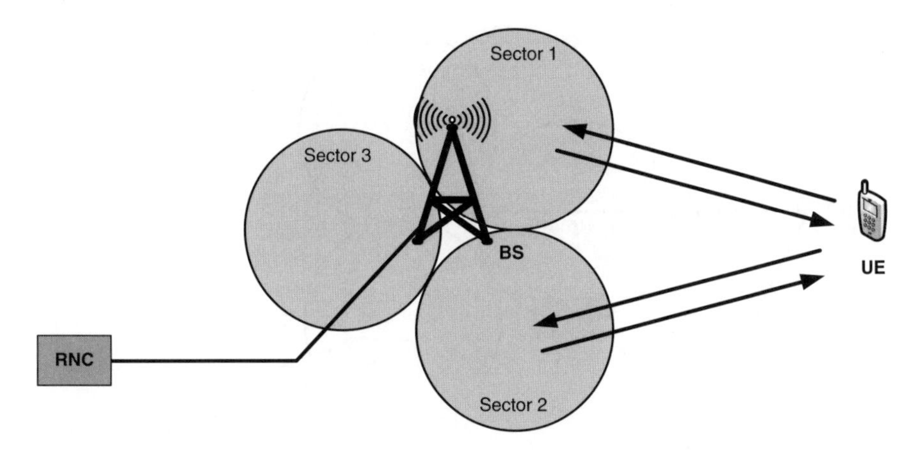

Figure 2.19 Operation of the soft handover.

It is possible to manage a transmission between a mobile station and several base stations that occurs while a soft handover process is conducted due to the phenomenon of macrodiversity. A soft handover may be necessary in a situation in which the channel in the source cell is retained and used for a while in parallel with the channel in the target cell. In this case, a connection to the other cell is established before the connection to the current cell is broken. It is estimated that the soft handover is conducted in about 20–40% of all connections [10].

In the case of the softer handover, the mobile station communicates with two or more sectors of the same base station. This is presented in Figure 2.20.

Similarly to the soft handover, the advantage of this procedure is a possibility of a reception and multiplexing of signals from a number of sectors of a base station. It is estimated that the softer handover is conducted in about 5–15% of connections [10].

A soft-softer handover is a combination of the above methods of soft handover and is conducted when the base station communicates with a number of sectors of the same base station and at least one sector of another base station.

2.7.3 Call Admission Control

In the UMTS system the load of the radio interface is related to the range and the quality of service offered by a network. With the increase in the load of the radio interface, the range of the network decreases. In order to ensure appropriate quality of service to subscribers it is necessary to control access to the radio network unconditionally and not to allow it to exceed the boundary values for the load of the radio interface predefined in the designing stage.

Admission control is carried out by the RNC of the base stations in which information related to the load of the cells connected to the controller is stored. Prior to setting up a new connection, the call admission control module checks if setting up of the call will not reduce below the predefined value, the range of a base station, and that it will not result in a lowering of the quality of the ongoing call. Admission control algorithms are also initiated when there is a modification of the bearer for an existing connection.

The procedures for the increment of load in the radio interface are performed both for the uplink and the downlink, and any change in the parameters of the bearer is possible only

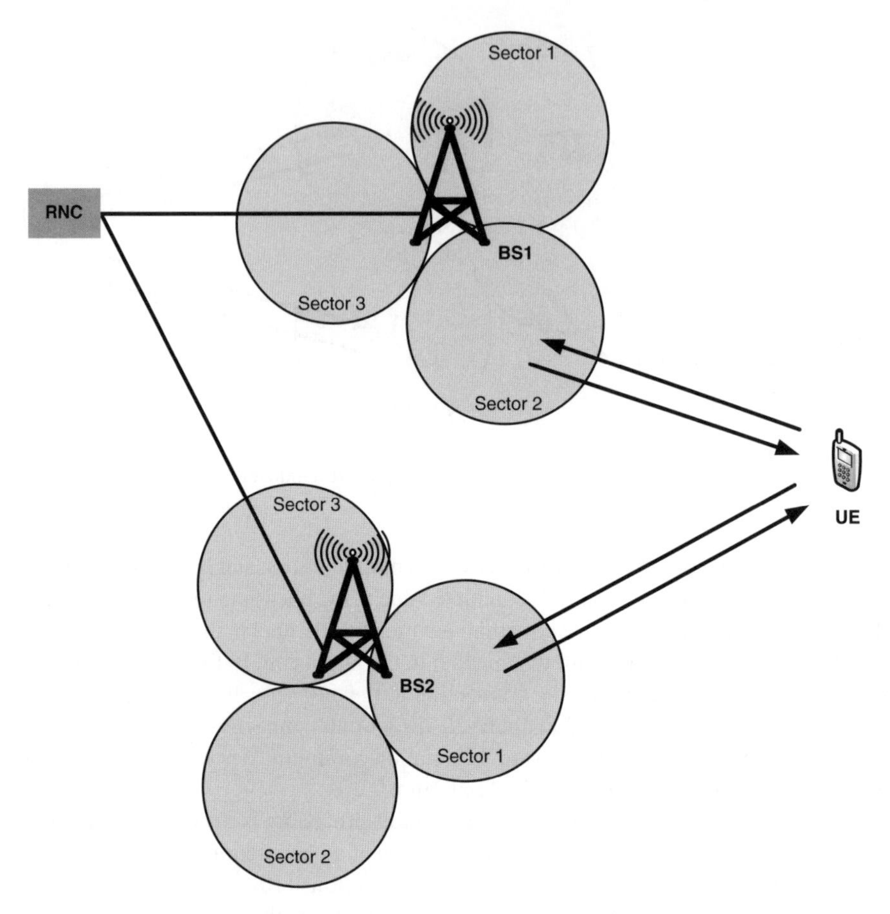

Figure 2.20 Operation of the softer handover.

when the call admission control module allows for its modification both in the link from the subscriber to the base station and from the base station to the subscriber [4].

Exemplary call admission control algorithms in a network with WCDMA radio interface will be presented and discussed later in this chapter.

2.7.3.1 Call Admission Control based on the Increase in the Load Factor η

The method for the call admission control based on the increase in the load of the link η uses the value of the load factor L_j that is introduced to a system by a user of a service of class j.

A new call of class j is serviced when the following conditions are satisfied [4]:

- for the uplink

$$\eta_{\mathrm{UL}} + L_j < \eta_{\mathrm{UL\,max}} \tag{2.3}$$

- for the downlink

$$\eta_{DL} + L_j < \eta_{DL\,max} \tag{2.4}$$

where η_{UL} and η_{DL} are the current values of the load prior to setting up a connection for the uplink and the downlink, respectively; L_j is the load factor introduced by a user of class j, and $\eta_{UL\,max}$ and $\eta_{DL\,max}$ are the maximum loads for respectively the uplink and the downlink.[2]

2.7.3.2 Call Admission Control based on the Received Interference Level and Transmitted Power of the Base Station

In this method a new call of class j is admitted for service when the conditions related to the maximum interference level for the uplink are satisfied [4]:

$$I_{total} + \Delta I < I_{max} \tag{2.5}$$

where I_{max} is the maximum admissible interference level for the uplink at the designing stage that will not cause a termination of a connection or lowering of the quality of currently serviced calls, I_{total} is the interference level prior to setting up of a new call, and ΔI is the estimated increment of the interference level in the uplink that will result from a new call.

Call admission control for the downlink allows for setting up a new connection when the transmitted power from the base station does not exceed the admissible values [4]:

$$P_{total} + \Delta P < P_{max} \tag{2.6}$$

where P_{max}, is the maximum admissible power level predefined for the downlink, P_{total} is the power level currently emitted by the power station, while ΔP is the estimated power increment needed for the service of a new call.

2.7.4 Packet Scheduler

The UMTS system allows for the use of common and dedicated channels as well as those shared by many users for packet transmission. The purpose of using packet resource allocation – the packet scheduler – is to allocate existing resources to the users of the system, which, in practice, means an allocation to a transport channel with appropriate characteristics. In addition, this module monitors the load of the network and, if necessary, initiates action aimed at its reduction. There are different resource-allocation algorithms for the following types of packets streams [2]:

- real-time packet streams used to carry out speech and streaming services with relatively high requirements related to link capacity and delays;
- the packet streams used by interactive services that are characterized by transmission of information in blocks and a greater tolerance of delays.

[2] A detailed presentation of the loads for the uplink and the downlink and the influence of calls of particular classes on the load of WCDMA radio interface is given in Chapter 11.

2.7.5 Load Control

The load control module in the WCDMA radio interface is responsible for the stable operation of the system and the control of used resources that prevents any overload of the system. When the values of the load of the radio interface exceed predefined values, the load control module should restore the defined load in the system as quickly as possible. In order to achieve that the load control can undertake the following action [4]:

- stop carrying out commands sent by mobile stations related to increasing power in the downlink;
- lower the admissible value (for a given service) of the energy ratio per one bit E_b to noise spectral density N_0 for the uplink;
- lower the transmission speed of packet connections;
- transfer the connection to another WCDMA;
- transfer the connection to a GSM network;
- lower the transmission speed for services in real time, e.g. ongoing AMR calls;
- terminate in a controlled way part of existing connections, including connections in real time.

It is assumed that the latter method of lowering load in the WCDMA radio interface is initiated only in exceptional cases when other action fails to achieve the desired results (in lowering load sufficiently).

2.8 High-Speed Packet Data Transmission

High-speed packet data transmission (HSDPA) in the downlink has been included in the system specification by 3GPP in version 5. Its introduction is aimed at increasing transmission speed in the downlink and at shortening delays in the network.

The equivalent of HSDPA for the uplink is the HSUPA fast packet data transmission in the uplink which became part of the UMTS system in its version 6 [10].

2.8.1 High-Speed Downlink Packet Access (HSDPA)

It is assumed that successive version of high-speed downlink packet access (HSDPA) will allow data transmission with speed of 1.8 Mbps, 3.6 Mbps, 7.2 Mbps and 14.4 Mbps. To make it possible, new solutions have been worked out in relation to the organization and management of transport and physical channels. The following channels have been introduced in the system [2, 10]:

- High speed downlink shared channel (HS-DSCH) – a channel shared by many mobile stations, used for transmitting user's data from higher layers of the network and controlling information. The channel is an extension of the DCH channel for high-speed data transmission.
- Physical channels:
  - high speed physical downlink shared channel (HS-PDSCH) – used for data transmission with the constant spreading factor equal to 16;

- shared control channel (HS-SCCH) – used to inform the mobile station about a planned transmission in the HS-DSCH channel;
- high speed dedicated physical control channel (HS-DPCCH) – used in the uplink to confirm transmitted data and to send the channel quality indicator.

In addition to new channels, the HSDPA technology introduces the following new mechanisms:

- Adaptive modulation and AMC coding – apart from the QPSK modulation, HSDPA permits the application, with a low level of interference, of 16 quadrature amplitude modulation 16 QAM. Modulation and coding schemes can be changed depending on the quality of the signal and the load of the radio link.
- High-speed packet transmission from the level of Node B – the HS-DSCH channel is shared by different users of the system to fully make use of available resources of the radio link depending on propagation conditions and the level of interference. On the basis of the signal-level indicator CQI in the downlink sent by mobile stations, the base station decides which user will be sent data, as appropriate.
- High-speed retransmission from the level of the Node B hybrid automatic repeat request (HARQ) – HSDPA technology includes the function of retransmission in the physical layer. The function is located in Node B thanks to which the process of retransmission that does not get RNC involved is carried out much faster. In addition, HARQ introduces the concept of incremental redundancy. In the case of receiving wrong data by a mobile station, the data is stored and reused by the decoder to restructure the received signal after a retransmission of redundant data to the mobile station (Figure 2.21). The base station sends incremental redundant data if the previous transmission has made it impossible to decode the received information.
- Multicode transmission – HSDPA makes for multicode transmission possible. The base station can transmit a signal to a mobile station using simultaneously up to 15 channelization codes with a spreading factor of 16.

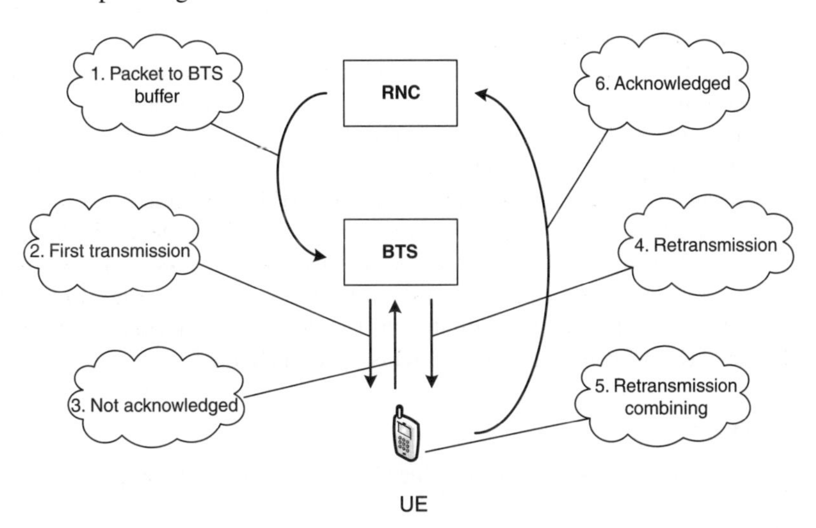

Figure 2.21 Operation of high-speed retransmission from Node B (HSDPA/HSUPA).

Table 2.3 A comparison of the properties of DCH channels (R99), HS-DSCH (HSDPA) and E-DSH (HSUPA)

Feature	DCH	HSDPA (HS-DSCH)	HSUPA (E-DCH)
Variable spreading factor	Yes	No	Yes
Fast power control	Yes	No	Yes
Adaptive modulation	No	Yes	No
BTS based scheduling	No	Yes	Yes
Fast L1 HARQ	No	Yes	Yes
Soft handover	Yes	No	Yes
TTI length (ms)	80, 40, 20, 10	2	10, 2

2.8.2 High-Speed Uplink Packet Access (HSUPA)

High-speed uplink packet access (HSUPA) is the counterpart of HSDPA for the uplink. It enables data transmission from the subscriber to the base station with the speed of 5.76 mbps. The HSUPA technology uses high-speed retransmission from the HARQ level of a mobile station with incremental redundancy, allows a transmission time interval (TTI) between subsequent transmissions to be used and introduces a new type of enhanced dedicated channel (E-DCH). The E-DCH, unlike the HS-DSCH used in HSDPA technology, is not a shared channel but a dedicated one. This means that each mobile station sets up, with the servicing Node B, its own E-DCH channel. Moreover, HSUPA does not use adaptive modulation. As in the R99 version of the UMTS system, BPSK modulation is used.

High-speed HARQ retransmission for HSUPA operates in a similar way to HSDPA. The base station informs the mobile station if it has received data packets or not. Erroneously received packets by the base station are immediately retransmitted by the mobile station. Having received them, Node B, also using the previously received signal, tries to recreate data sent by the mobile station. The retransmission is then repeated until the packets sent by the mobile station have been received properly or the number of admissible retransmissions has run out. Unlike HSDPA, the procedure for high-speed packet access in HSUPA operates differently. In HSDPA, the HS-DSCH channel is shared by all participants serviced by a given cell. Due to the above, the base station can allocate (for a short time though) all resources to just one mobile station, when other mobile stations do not receive demanded data. In HSUPA the E-DCH channel is a dedicated channel, which results in a situation in which cosharing is not possible. Because of this, high-speed transmission in HSUPA operates in a similar way to the function of packet scheduler for the R99 version of the UMTS system. The RNC informs all mobile stations about the maximum power they can use for transmission. If the interference level approaches the value that can cause instability in the system, the level of admissible transmission power allocated to all mobile stations is reduced. A comparison of the properties of DCH channels (R99), HS-DSCH and E-DSH is shown in Table 2.3.

2.9 Services

Four following service classes have been defined in the UMTS system [8]:

- conversational;
- streaming;

Table 2.4 Service classes of the UMTS system and their basic parameters [8]

Feature Service	Speech	Streaming	Interactive	Background
Maximum transmission speed (kbps)	< 2048	< 2048	< 2048	< 2048
Guaranteed transmission speed (kbps)	< 2048	< 2048	N/A	N/A
Symmetry	yes	no	no	no
Delay	100–250 ms	250 s	N/A	N/A

- interactive;
- background classes.

The main factors influencing the division of particular services in classes include their sensitivity to delays during transmission, transmission speed and acceptable bit error rate.

An example of a conversational service is a traditional telephone call in which two or more users participate. As a rule, transmission occurs alternately and traffic, generated in two directions, is symmetrical. The maximum permissible transmission delay is conditioned by human perception and it has been proven experimentally that it should not exceed 250 ms [8]. Videotelephony is an extension to the traditional phone service. In the case of this service the requirements related to the bit error rate are far higher and transmission between the participants is carried out non-stop (video component).

A *streaming* service, often described as a "server–user" service type, is used, for example, for watching television programmes offered by some of web sites. Transmission is effected with a delay of several seconds and the transmitted signal is buffered with the user. This mechanism ensures that any delays in transmission are imperceptible for the end user. The permissible delay values for this service class reach several hundred seconds. Traffic in the applications that use the streaming class is regular and asymmetric.

Interactive services are used by all applications to enable the user to obtain data from a particular location in the network. Examples of interactive services include web sites browsing, localization services, games, data downloading, or access to a given server.

The most common *background* service is electronic mail or short text messages. Such transmissions can be performed "in the background;" their delays are insignificant and can amount to many minutes.

Table 2.4 presents the basic information related to the requirements for particular classes of service.

References

[1] Braithwaite, C. and Scott, M. (2004) *UMTS Network Planning and Development: Design and Implementation of the 3G CDMA Infrastructure*. Newnes.

[2] Kołakowski, J. and Cichocki, J. (2003) *UMTS. System telefonii komórkowej trzeciej generacji*, Wydawnictwo Komunikacji i Łączności.

[3] Wesołowski, K. (2002) *Mobile Communication Systems*, John Wiley & Sons, Ltd.

[4] Holma, H. and Toskala, A. (2000) *WCDMA for UMTS. Radio Access For Third Generation Mobile Communications*, John Wiley & Sons, Ltd.

[5] Prasad, R., Mohr, W., and Konhauser, W. (2000) *Third Generation Mobile Communication Systems*, Artech House Universal Personal Communications Library.

[6] 3GPP (2002). *TS 23.002 Network architecture v. 3.6.0*. 3GPP.
[7] 3GPP (2003). *TS 23.002 Network architecture v. 4.7.0*. 3GPP.
[8] Kaaranen, H., Ahtiainen, A., Laitinen, L., Naghian, S., and Niemi, V. (2001) *UMTS Networks. Architecture, Mobility and Services*, John Wiley & Sons, Ltd.
[9] Faruque, S. (1996) *Cellular Mobile Systems Engineering*, Artech House, Inc.
[10] Laiho, J., Wacker, A., and Novosad, T. (2002) *Radio Network Planning and Optimization for UMTS*. John Wiley & Sons, Ltd.

3

Long-Term Evolution

3.1 Introduction

The steady increase in the number of subscribers to mobile telecommunication systems in recent years, paralleled by a dramatic increase in the average call time and the amount of data transmitted per user, have given rise to a need to work out a new system for mobile telephony that would satisfy the requirements of subscribers. Furthermore, a number of new services have emerged in recent years, including wireless transmission of TV signals to portable terminals. Well-established services such as video-telephony, video-streaming and localization have experienced an ever-growing interest among subscribers. A particular feature of these services is a high diversification of requirements concerning the transmission rate, tolerable transmission delay and acceptable bit error ratio. The International Telecommunications Union has responded to this situation by defining the requirements for a fourth generation system [1]. One of the systems for which the standardization process is already much advanced – in line with the requirements set up by ITU-R for 4G systems – is the Third Generation Partnership Project Long-Term Evolution (3GPP LTE).

The most important requirements identified for the LTE systems can be itemized in the following way [2]:

- Cost reduction in network data transmission (per bit):
  - improvement of spectrum efficiency;
  - cost reduction in backhaul data transmission.
- Reduction in setup time and round trip delay.
- Improvement in functioning of quality of service (QoS) mechanisms for various services.
- Focus on services utilizing the IP protocol.
- Broadening of multimedia multicasting services for selected groups of users (enhanced MBMS).
- Increase in the transmission rate to over 100 Mbps in the downlink and 50 Mbps in the uplink direction.
- Flexibility in the use of existing and new spectral resources.
- Possibility of carrier allocation with different bandwidth, ranging from 1.25 to 20 MHz.

Modeling and Dimensioning of Mobile Networks: From GSM to LTE
Maciej Stasiak, Mariusz Głąbowski, Arkadiusz Wiśniewski and Piotr Zwierzykowski

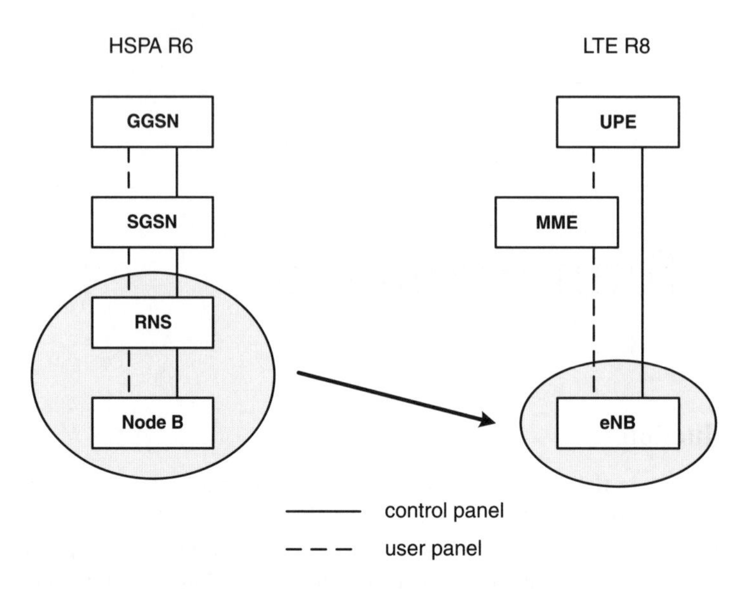

Figure 3.1 Network architecture evolution.

3.2 System Architecture

The system architecture for LTE stems from the system architecture of the previous generation, namely from UMTS. When the LTE system architecture was proposed it was assumed that the new system would have to satisfy even higher requirements concerning network signal transmission delay and that the costs of its construction and subsequent maintenance should be adequately lower. Hence, a decision was made to simplify the architecture of the system.

Figure 3.1 shows the evolution of the architecture from the 3GPP Release 6 version to Release 8 LTE [2]. Release 8 core network is also often referred to as evolved packet core (EPC). In Release 8 all radio protocols, subscriber mobility management, compression of headers and packet retransmissions have been located in the base station labeled E-UTRAN Node B (eNB), which also includes all algorithms and functions that in the architecture of the 3GPP Release 6 version were located in the radio network controller.

Evolved packet core LTE architecture is presented in Figure 3.2. The following elements can be distinguished:

- eNB – E-UTRAN Node B;
- eGW – access gateway;
- MME – mobility management entity;
- UPE – user plane entity.

The tasks to be assigned to MME include management and context storage for the control of users' terminals, authorization procedures for terminals, and mobility management schemes for terminals. The tasks of the UPE unit encompass management and subscriber terminal context storage. The unit also performs functions such as encoding, securing integrity of data blocks, and automatic repeat request (ARQ) to reduce data block discard. Individual data blocks

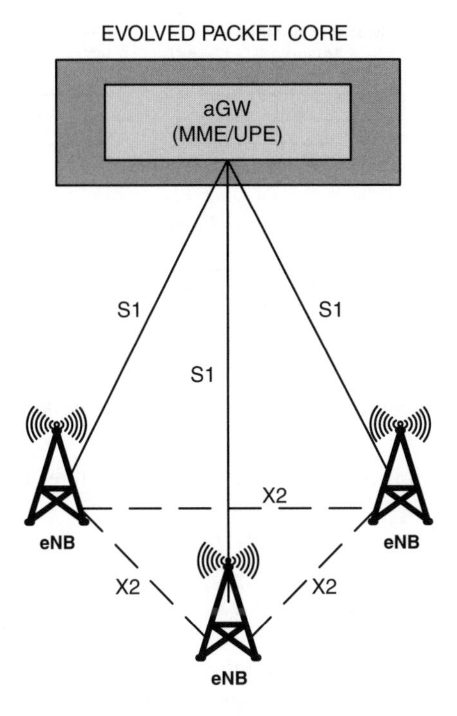

Figure 3.2 Long-term evolution network architecture.

located in network nodes (eGW) communicate with one another and with external networks. The blocks cooperate with base stations (eNB), which carry on all functions related to radio transmission [2].

3.3 Transmission Techniques in the LTE System

One of the key elements of LTE is the use of orthogonal frequency division multiplexing (OFDM) as the signal bearer and the associated access schemes, orthogonal frequency division multiple access (OFDMA) and single carrier frequency division multiple access (SC-FDMA).

The actual implementation of the technology will be different in the downlink and the uplink direction as a result of the different requirements between the two directions and the equipment at either end. However OFDM was chosen as the signal-bearer format because it is very resilient to interference. It is also a modulation format that is very suitable for carrying high data rates – one of the key requirements for LTE.

3.3.1 Long-Term Evolution OFDMA in the Downlink Direction

The principle of the OFDMA is based on the use of narrow, orthogonal subcarriers. In LTE the sub-carriers spacing is 15 KHz regardless of the total transmission bandwidth. The transmission is done after a fast fourier transform (FFT) block, which is used to change between the time and frequency domain representation of the signal.

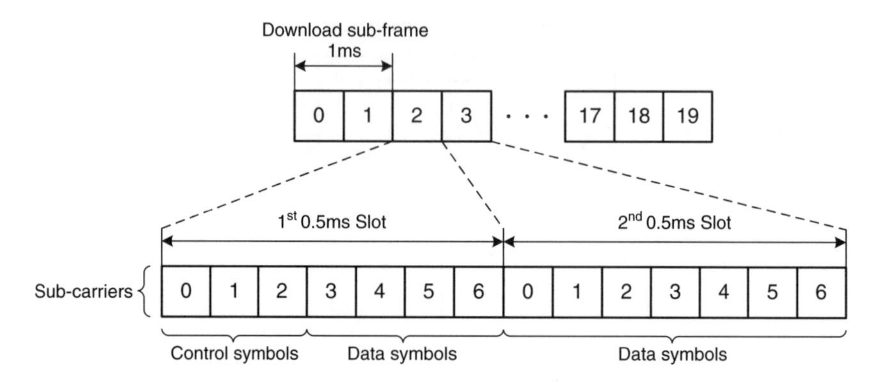

Figure 3.3 Long-term evolution frame structure in the downlink direction.

Within the OFDM signal it is possible to choose between three types of modulation:

- QPSK;
- 16QAM;
- 64QAM.

The duration of the LTE system frame, Figure 3.3, is 10 ms and the system consists of ten sub-frames with two slots each. Within one slot, seven OFDM symbols are transmitted, with a short cyclic prefix, and six OFDM symbols, when a longer cyclic prefix is used.

The OFDM signal includes N^{BW} subcarriers. The signal on a single subcarrier in one OFDM symbol is thus of fundamental importance. There are therefore $6N^{BW}$ or $7N^{BW}$ such elements depending on the length of the cyclic prefix. The total resources of a single slot are divided into the so-called physical resource blocks, each being composed of 12 subsequent subcarriers allocated in a given time slot. The physical resource block is the basic unit for radio resource allocation (Figure 3.4).

Figure 3.5 shows a block diagram of the transmitter and the receiver of the signal in the downlink direction in the case where single antennas are used. A data block from the n-th interval of the modulation can result from multiplexing several streams generated by users. Then it is arranged in the radio resource allocation block and the data is mapped in symbols from the constellation of elementary symbols (QPSK, 16-QAM, 64-QAM), which are then attributed to appropriate subcarriers. Signal samples in the time domain are performed with the help of M-point IFFT transformation. A cyclic prefix is added at the head of block of samples thus generated. This prefix is a sub-block of the sequence of samples copied from the rear of the block (trailer). The signal constructed in this way, with the inphase and quadrature components (the real and the imaginary part of the complex block of IFFT samples respectively), is then converted into analog form, converted to radio band and enhanced to be emitted through the antenna. On the receiver side the dual processes take place. So, after signal conversion to base band and conversion to a block of digital samples, the cyclic prefix is removed. Then, the subcarrier signals are correlated with reference signals using the FFT method. The sample block from the FFT output, corresponding to the frequency domain, is correlated on the basis of the estimated characteristics of the transmission channel. The samples obtained in this way from individual subcarriers are then mapped in the constellation points that indicate the positions

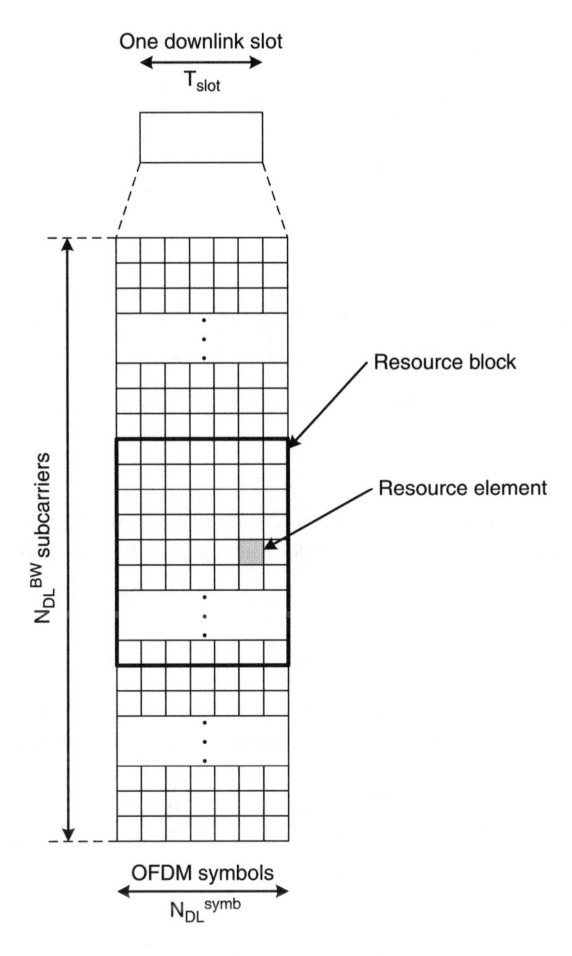

Figure 3.4 The downlink time–frequency resource grid.

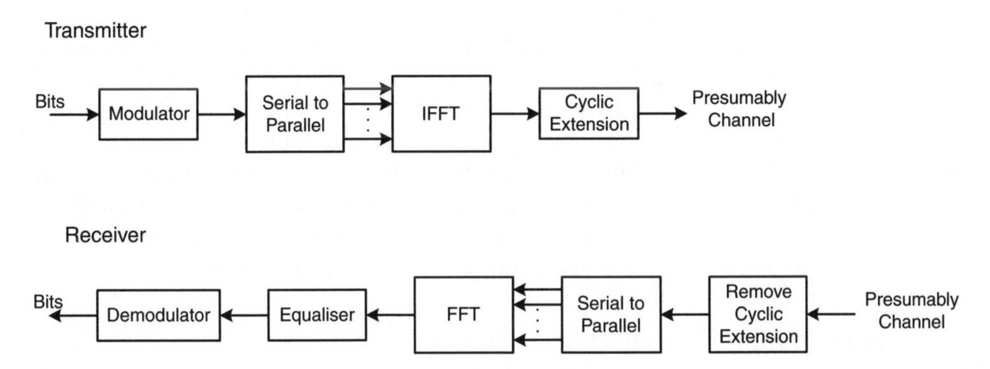

Figure 3.5 Block diagram of the radio transmitter and LTE receiver in the downlink operating in the OFDMA mode.

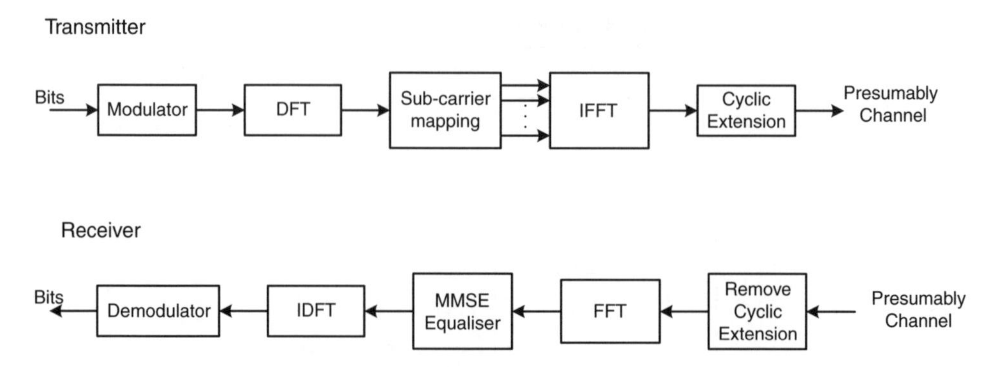

Figure 3.6 Block diagram of the radio transmitter and the LTE receiver in the uplink operating in the SC-FDMA mode.

of the symbols mapped in the constellation. Eventually, the binary block is determined on the basis of the symbols. The block includes the final decision on the properly received signal block.

3.3.2 Long-Term Evolution SC-FDMA in the Uplink

For the LTE uplink, a new concept is used for the access technique. Although still using a form of OFDMA technology, the implementation is called single carrier frequency division multiple access.

A fundamental problem for the OFDM transmission, from the mobile station perspective, is limited power resources derived from battery energy. Hence, efficiency of radio blocks is of fundamental significance, in particular that of the power enhancer. The OFDM signal is characterized by a high value of peak power ratio to average power, which implies the need for a high degree of linearity in the power enhancer and a decrease in the average power of emitted signal as compared to signals for which the peak power ratio to the average ratio is insignificant. In the uplink SC-FDMA transmission is therefore used. The diagram of the transmitter and the SC-FDMA receiver is presented in Figure 3.6

The binary data are first mapped into constellation points. The constellation is chosen according to the quality of the modulation channel (QPSK, 16-lub 64-QAM). The symbols are then arranged in blocks with the length N. Such a block is treated as a sequence of samples in the time domain and undergoes a frequency transformation according to the DFT algorithm (FFT). The obtained frequency samples are then mapped in selected subcarriers of SC-FDMA modulator. The block of samples thus obtained within the frequency domain is then transformed into the time form by the block that performs the IFFT algorithm. The block of samples in time is preceded with a cyclic prefix and the whole of the block is filtered so that spectral properties of the signal are shaped. Dual operations are performed in the receiver.

3.3.3 Long-Term Evolution MIMO

Multiple input multiple output (MIMO) is another major LTE technology innovations used to improve the performance of the system. It uses multiple antennas at both the transmitter

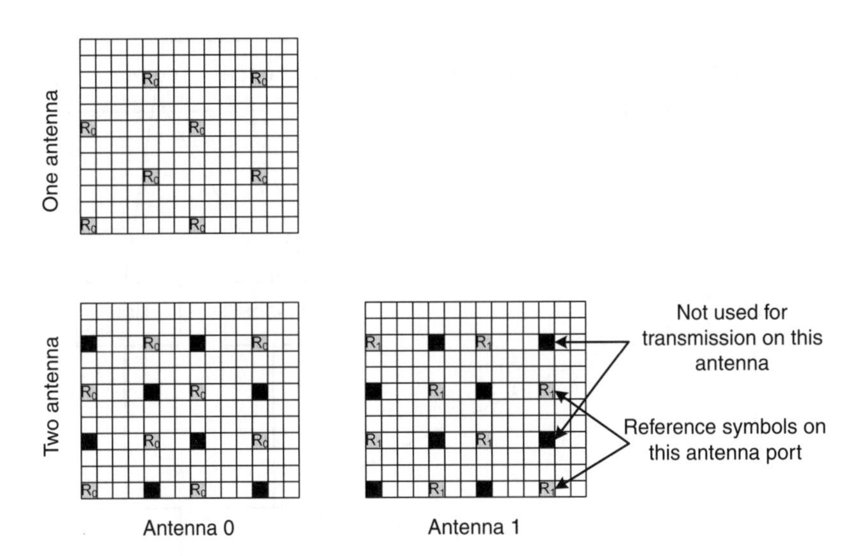

Figure 3.7 Distribution of reference signals in the transmission with one and two radio antennas.

and receiver to improve communication performance. It offers significant increases in data throughput and link range without using additional bandwidth or transmission power. This is achieved by higher spectral efficiency (more bits per second per hertz of bandwidth) and link reliability or diversity (reduced fading).

Although MIMO adds complexity to the system in terms of processing and the number of antennas required, it enables high data rates to be achieved along with much improved spectral efficiency. As a result, MIMO has been included as an integral part of LTE. The basic concept of MIMO uses the multipath signal propagation that is present in all terrestrial communications.

For the downlink direction, a configuration of two transmit antennas at the base station and two receive antennas on the mobile terminal is used as baseline, although configurations with four antennas are also being considered.

Figure 3.7 presents a configuration of reference symbols within two subsequent physical resource blocks included in a subframe for a single antenna transmission (SISO) and for two-antenna transmission. A similar configuration is also expected with the case of the application of four transmitting antennas.

3.4 Channels in the Radio Interface of the LTE System

Like UMTS, channels in the radio interface of the LTE system can be divided into three types (Figure 3.8) [2]:

- logical channels;
- transport channels;
- physical channels.

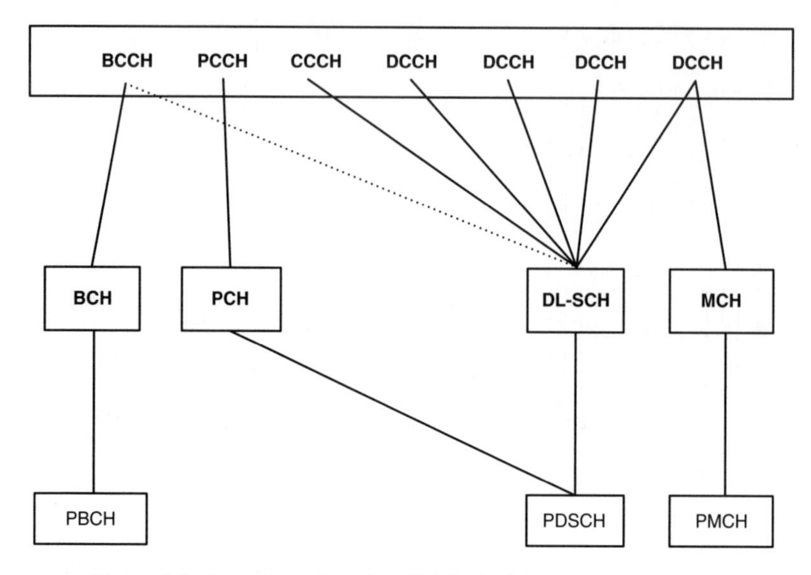

Figure 3.8 Mapping of the downlink logical and transport channels.

3.4.1 Long-Term Evolution Logical Channels

The concept of logical channels in the LTE systems is similar to that of the GSM and UMTS. LTE logical channels are divided into two groups: control channels and traffic channels.

The control channels are:

- the broadcast control channel (BCCH) – a downlink channel providing system information to all UE connected to the eNB;
- the paging control channel (PCCH) – the downlink channel used for transferring paging information;
- the common control channel (CCCH) – a uplink channel used for random access information, e.g. for actions including setting up a connection;
- the multicast control channel (MCCH) – a point-to-multipoint channel used for transferring MBMS information;
- the dedicated control channel (DCCH) – a point-to-point bi-directional channel used for exchanging control information, for example, for controlling actions including power control, handover.

The traffic channels are:

- a dedicated traffic channel (DTCH) – a bidirectional channel used for the transmission of user data;
- a multicast traffic channel (MTCH) – a downlink point-to-multipoint channel used for the transmission of MBMS data.

3.4.2 *Long-Term Evolution Transport Channels*

The LTE transport channels vary between the uplink and the downlink directions as each has different requirements and operates in a different manner. Physical layer transport channels offer information transfer to medium access control (MAC) and higher layers.

Downlink:

- the broadcast channel (BCH) – the LTE transport channel maps to broadcast control channel (BCCH);
- the downlink shared channel (DL-SCH) – this transport channel is the main channel for downlink data transfer and it is used by many logical channels;
- the paging channel (PCH) – this is used to convey the PCCH channel;
- the multicast channel (MCH) – is used to transmit MCCH information to set up multicast transmissions.

Uplink:

- the uplink shared channel (UL-SCH) – this channel is used for uplink data transfer;
- the random access channel (RACH) – this channel is used for random access requirements.

3.4.3 *Long-Term Evolution Physical Channels*

The LTE physical channels vary between the uplink and the downlink direction as each has different requirements and operates in a different manner.

Downlink:

- the physical broadcast channel (PBCH) – this carries system information for UEs in a coverage area;
- the physical control format indicator channel (PCFICH) – the purpose of this channel is to indicate dynamically how many OFDMA symbols are reserved for control information;
- the physical downlink control channel (PDCCH) – this channel is used by the eNB to carry control and scheduling information to the UEs;
- the physical hybrid ARQ indicator channel (PHICH) – this channel is used to report the hybrid ARQ status;
- the physical downlink shared channel (PDSCH) – this channel is used for unicast and paging functions;
- the physical multicast channel (PMCH) – this channel carries system information for multicast purposes;
- the physical control format indicator channel (PCFICH) – this channel provides information to enable the UEs to decode the PDSCH.

Uplink:

- the physical uplink control channel (PUCCH) – this sends hybrid ARQ acknowledgement;
- the physical uplink shared channel (PUSCH) – this physical channel found on the LTE uplink is the uplink counterpart of PDSCH;
- the physical random access channel (PRACH) – this uplink physical channel is used for random access functions.

3.5 Radio Resource Management in LTE

Radio resource management (RRM) aims at efficient usage of all available algorithms and techniques offered by the system and supports appropriate level of quality of service (QoS). Some examples of mechanisms for radio resource management in LTE are presented below.

3.5.1 *Admission Control*

The admission control (AC) algorithm in eNB decides whether or not a new call in the cell can be serviced. Admission control takes into account available resources in a cell, call priorities, quality requirements for a new call as well as current quality parameters in the cell. A new call can only be serviced when the expected quality of currently serviced calls with the same, or higher, priorities will be retained and the assumed QoS parameters for the new call will be satisfied [2].

Admission control algorithms have not yet been specified by 3GPP. Possible solutions with regard to the issue are to be decided by hardware suppliers.

3.5.2 *Frequency Domain Packet Scheduling*

Frequency domain packet scheduling (FDPS) uses frequency selective power variations on either the desired signal (frequency selective fading) or interference (fading or due to fractional other cell load) by only scheduling users on the physical resource blocks with high channel quality. Thanks to this solution, transmission in the radio channel is executed only when it is most efficient. The working principle for the frequency domain packet scheduling is presented in Figure 3.9.

3.5.3 *Interference Management and Power Settings*

The LTE system offers a number of mechanisms that underlie control of interference between neighboring cells. Version 8 of the system (Release 8) uses the distribution of frequencies

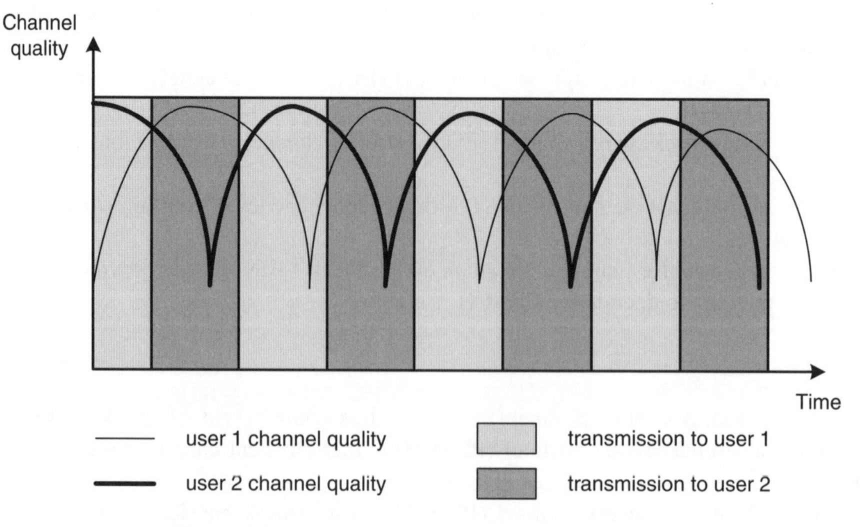

Figure 3.9 Frequency domain scheduling principle.

between neighboring cells and the possibility of power control in the radio channel. The X2 interface supports information transmission about inter cell interference between eNB base stations. In LTE two methods of interference level management in the radio interface can be distinguished [2]:

- The reactive method, based on monitoring the network quality. If interference is detected that is too high the system decreases the level of transmitted power in the radio channel or modifies the packet transmission process.
- The proactive method, based on an exchange of information between eNBs regarding how they plan to schedule packets to its users in the future. This allows neighboring eNBs involved in packet transmission to take into account the current load of the radio interface effected by neighboring cells. This communication is performed by means of the X2 interface.

Interference management in 3GPP Release 8 aims primarily at a constant improvement of the quality of shared data channels in the uplink and in the downlink direction (PDSCH and PUSCH).

3.5.4 Discontinuous Transmission and Reception (DTX/DRX)

Radio resources management mechanism in the LTE system offers discontinuous signal transmission and reception (DTX/DRX). Discontinuous transmission/reception in LTE allows a mobile station (MS) to enter the power-saving mode and to cease monitoring the PDCCH channel in a given sub frame. This operation can be effected based on activity requirements for a given mobile station in the uplink and in the downlink direction. For traffic in which a cycle pattern can be predicted with high probability, discontinuous reception can last 20 or 40 ms. This functionality lowers energy consumption in the mobile station and the level of interference in the radio channel. However, it should not be forgotten that too aggressive parameterization of the function, such as maximization of the periods of discontinuous transmission and reception, can eventually reduce the quality of service offered [2].

References

[1] *ITU-R M. 1645. (2003) Framework and overall objectives of the future development of IMT-2000 and systems beyond IMT-2000*, ITU.

[2] Holma, H. and Toskala, A. (2009) *LTE for UMTS – OFDMA and SC-FDMA Based Radio Access*, John Wiley & Sons, Ltd.

Part II

Teletraffic Engineering for Mobile Networks

Part II

Teletraffic Engineering for Mobile Networks

4

Basic Definitions and Terminology

4.1 Introduction

This chapter is a presentation of basic concepts related to traffic theory, which is a mathematical description of the network and telecommunication systems in probabilistic categories. The aim of traffic theory is to construct analytical models of telecommunication networks and systems that enable their proper design and effective management. Basic principles of traffic theory were formulated by a Danish mathematician Agner Krarup Erlang (1879–1929) at the beginning of the twentieth century. From work originally devoted to telecommunication applications, a new branch of probability theory emerged, called queuing theory. Nowadays, traffic theory is classified as part of technical cybernetics, a branch of science that has been developing very rapidly over the past few decades and has been using an increasing range of mathematical methodologies such as probability theory, algebra, graph theory, stochastic processes theory and Markov process theory.

4.2 Call Stream

The call stream is formed by concatenating calls arriving in random (or otherwise) time moments. Call streams are most commonly described by the following parameters:

- $\lambda(t)$ – call intensity – average number of calls (arrivals) during a time interval of length t;
- $P_k(t)$ – probability that there are exactly k arrivals during a time interval of length t;
- $f(t)$ – time distribution between successive calls.

4.2.1 Poisson Stream and its Properties

To describe calls occurring in communications systems, the so-called Poisson stream is most frequently used [1–4], which has the following properties:

Modeling and Dimensioning of Mobile Networks: From GSM to LTE
Maciej Stasiak, Mariusz Głąbowski, Arkadiusz Wiśniewski and Piotr Zwierzykowski

- *stationarity* – a stream is stationary if its intensity does not depend on time: $\lambda(t) = \lambda = \text{const}$; this means that the average number of arrival calls within a time interval remains unchanged;
- *memorylessness* – a stream has a memoryless property if the number of calls within any chosen time interval t_1 does not have any effect upon the number of calls in any other, randomly chosen, interval t_2; this means that the successively arriving calls are not mutually interdependent;
- *orderliness (singularity)* – a stream is singular if within an infinitely small time interval Δt, one call at most can arrive; the probability of the arrival of more than one call is omitted.

A Poisson call stream that is characterized by the above properties is often called *the simple stream*. The name stems from the simplicity of its mathematical description shown below.

4.2.2 Mathematical Model of Poisson Stream

4.2.2.1 Elementary Probabilities

In the Poisson stream, the probability of arrival (or the lack of it) of one call within the interval Δt is called the elementary probability and is directly proportional to the length of the time interval Δt:

$$P_1(\Delta t) = \lambda \Delta t + \theta(\Delta t) \tag{4.1}$$

$$P_0(\Delta t) = 1 - \lambda \Delta t + \theta(\Delta t) \tag{4.2}$$

where $\theta(\Delta t)$ is an infinitely small value in comparison with Δt:

$$\lim_{\Delta t \to 0} \frac{\theta(\Delta t)}{\Delta t} = 0 \tag{4.3}$$

The probability of the occurrence of more than one call is omittable:

$$P_{i>1}(\Delta t) = \theta(\Delta t) \tag{4.4}$$

4.2.2.2 Probability of the Occurrence of *k* Calls in Time Interval *t*

The occurrence of k calls in the time interval $t + \Delta t$ is possible only when $k - i$ calls $(0 \leq i \leq k)$ arrive in the time interval t, and, appropriately, i calls within the time interval Δt. As the Poisson stream is memoryless, the arrival of $k - i$ calls within the time interval t and i calls within the time interval Δt is not dependable. Therefore:

$$P_k(t + \Delta t) = P_{k-i}(t) P_i(\Delta t) \tag{4.5}$$

Taking all possible variants of the number of arrivals into consideration, and in keeping with the law of total probability, the following equation can be written:

$$P_k(t + \Delta t) = \sum_{i=0}^{k} P_{k-i}(t) P_i(\Delta t) \tag{4.6}$$

In Equation (4.6), the probabilities $P_i(\Delta t)$ will take on finite values only for $i = 0$ and $i = 1$. After taking into consideration Equations (4.1) and (4.2), Equation (4.6) can be rewritten in the following form:

$$P_k(t + \Delta t) = (1 - \lambda \Delta t) P_k(t) + (\lambda \Delta t) P_{k-1}(t) \tag{4.7}$$

With the limit for $\Delta t \to 0$, the following system of differential equations is obtained:

$$\frac{\mathrm{d}P_k(t)}{\mathrm{d}t} = -\lambda P_k(t) + \lambda P_{k-1}(t) \tag{4.8}$$

For $k = 0$, the solution of Equation (4.8) leads to the following formula:

$$P_0(t) = e^{-\lambda t} \tag{4.9}$$

After taking into account the result (4.9), the solution of Equation (4.8) for $k = 1$ takes on the following form:

$$P_1(t) = \lambda t e^{-\lambda t} \tag{4.10}$$

A continuation of the considerations for the successive values k on the basis of mathematical induction, leads eventually to the Poisson formula that determines the probability of arrival of k calls in time interval t:

$$P_k(t) = \frac{(\lambda t)^k}{k!} e^{-\lambda t} \tag{4.11}$$

4.2.2.3 Cumulative Distribution Function and Probability Density Function of Time Distribution between Calls

The cumulative distribution function of time distribution between successive calls $F(t)$ is, according to the definition, equal to the probability of an event in which time T between calls will be shorter than the given time t. This probability is equivalent to the probability of an event in which, in time interval t, one or more calls arrive. Thus, on the basis of Equation (4.11), we get:

$$F(t) = P(T < t) = \sum_{k=1}^{\infty} P_k(t) = 1 - P_0(t) = 1 - e^{-\lambda t} \tag{4.12}$$

The density probability function of time distribution between calls is, therefore, an exponential function:

$$f(t) = \frac{\mathrm{d}F(t)}{\mathrm{d}t} = \lambda e^{-\lambda t} \tag{4.13}$$

After taking into consideration Equation (4.13), the mean value m_T and variance σ_T^2 of the time between successive calls in the Poisson stream take on the following values:

$$m_T = \int_0^{\infty} t f(t) \mathrm{d}t = 1/\lambda \tag{4.14}$$

$$\sigma_T^2 = \int_0^{\infty} t^2 f(t) \mathrm{d}t - m_T^2 = 1/\lambda^2 \tag{4.15}$$

4.2.2.4 Poisson Stream Parameter

Let $\pi_1(t)$ denotes the probability of an arrival of at least one call within the time interval with the length t:

$$\pi_1(t) = \sum_{k=1}^{\infty} P_k(t) = 1 - P_0(t) \tag{4.16}$$

The stream parameter $\Lambda(t)$ at time moment t is the limit of the ratio of the probability of an arrival, within time interval $(t, t + \Delta t)$, of at least one call, to the length of the interval, with $\Delta t \to 0$:

$$\Lambda(t) = \lim_{\Delta t \to 0} \frac{\pi_1(t + \Delta t)}{\Delta t} \tag{4.17}$$

Taking into consideration the property of memorylessness and singularity and taking into account Equations (4.1) and (4.4), we can obtain the parameter value $\Lambda(t)$ for a Poisson stream:

$$\begin{aligned} \Lambda(t) &= \lim_{\Delta t \to 0} \frac{\pi_1(t + \Delta t)}{\Delta t} = \lim_{\Delta t \to 0} \frac{\pi_1(\Delta t)}{\Delta t} = \lim_{\Delta t \to 0} \frac{P_1(\Delta t)}{\Delta t} = \\ &= \lim_{\Delta t \to 0} \frac{\lambda \Delta t}{\Delta t} = \lambda \end{aligned} \tag{4.18}$$

The result (4.18) shows that, in the case of Poisson stream, the stream parameter and its intensity – defined as the average number of calls within a time interval – are equivalent. In traffic theory this result is generalized proving that in any stationary and singular stream the intensity is always equal to the stream parameter [3, 4].

4.2.2.5 The Memorylessness of the Poisson Stream

The exponential time distribution between successive calls is a necessary and sufficient condition for the "Poissonesque character" of a call stream [3]. Moreover, the exponential time distribution between calls shows a particular property that can be defined in the following way: the unconditional probability of an event in which the time τ between calls is exactly the same as the conditional probability determined with the assumption that during an earlier observation of the system in time t no call arrived. The above means that if the time between calls is of period t, then the knowledge of the fact has no influence upon the distribution of the remaining part of the time interval between calls. This distribution will be identical to the distribution of the whole of the interval.

Let us assume that the time interval between calls has the length t. We find the conditional probability $P(T > \tau | T > t)$ of the event that the interval will last for at least τ. On the basis of Bayes' theorem we can write:

$$P(T > t + \tau) = P(T > t)P(T > \tau | T > t) \tag{4.19}$$

After taking into account the cumulative distribution function (4.12), Equation (4.19) will be transformed into the following form:

$$e^{-\lambda(t+\tau)} = e^{-\lambda t} P(T > \tau | T > t) \tag{4.20}$$

To satisfy Equation (4.20), the conditional probability $P(T > \tau \,|T > t)$ "must" take on the following value:

$$P(T > \tau \,|T > t) = e^{-\lambda\tau} = P(T > \tau) \tag{4.21}$$

It results from Equation (4.21) that the conditional probability $P(T > \tau \,|T > t)$ is the same as the unconditional probability $P(T > \tau)$.

Equation (4.21) defines the memoryless property of time distribution between calls. The exponential distribution and the geometric distribution are the only distributions with such a property. It can therefore be proved [3] that the memoryless property of the exponential time distribution between calls is equivalent to the memoryless property in a Poisson stream. The memorylessness makes it possible to simplify considerably the mathematical analysis of service processes of systems to which the Poisson call stream is offered.

4.2.2.6 Comments

The Poisson stream plays an important role in traffic theory, similar to that of the Gaussian distribution in the probability theory [3]. This results from the fact that the sum of Poisson streams is also a Poisson stream with the intensity equal to the sum of the component streams. Moreover, the sum of a sufficiently great number of independent stationary and singular streams with a comparable intensity results in the creation of a stream with characteristics similar to the Poisson stream. If the number of component streams is infinitely great, then we get the Poisson stream. In particular, if each of the streams comes from a different traffic source, then the Poisson stream can be presented as a superposition of the streams generated by an infinite number of sources with the intensity tending towards zero.

4.3 Service Stream

4.3.1 Definition

A service stream is a set of moments of completion of the service of calls. Generally, the properties and characteristics of the service stream are dependable on the call stream, service quality parameters and the time distribution of a service. With "no-loss" service of the call stream and a constant service time h ($h =$ const), the properties and parameters of the service stream are the same as those of the call stream. The relation between these streams is expressed by the time shift h between the moments of call arrivals and the completion of its service.

4.3.2 Mathematical Model of Service Stream

4.3.2.1 Service Time

In traffic theory there are different service (holding) time distributions taken into consideration. Most frequently, it is assumed that the service time is constant (deterministic) or has an exponential distribution. This assumption of the exponential distribution of the service time was adopted by Erlang [2, 5].

4.3.2.2 Elementary Probability

With the above assumption about the exponential distribution of the service time, the elementary probability, i.e. the probability of completing (or not completing) a service of one call within a time interval Δt is expressed by equations analogous to (4.1) and (4.2):

$$P_1(\Delta t) = \mu \Delta t + \theta(\Delta t) \tag{4.22}$$

$$P_0(\Delta t) = 1 - \mu \Delta t + \theta(\Delta t) \tag{4.23}$$

where $\theta(\Delta t)$ is an infinitely small value in comparison to Δt (Equation (4.3)), while μ is the so-called service intensity, which is expressed by the dependence:

$$\mu = 1/h \tag{4.24}$$

where h is the mean value of the service time.

4.3.2.3 Cumulative Distribution Function and Probability Density Function of Service Time

In keeping with the assumption about exponential service time, the cumulative distribution function of the service time distribution $F(t)$ is equal, by definition, to the probability of an event, in which the duration of the service time T of a call will be shorter than the given time t, and can be written in the following form:

$$F(t) = P(T < t) = 1 - e^{-\mu t} \tag{4.25}$$

After differentiation of the distribution function (4.25), we obtain the exponential density function of the probability of the service time:

$$f(t) = \frac{\mathrm{d}F(t)}{\mathrm{d}t} = \mu e^{-\mu t} \tag{4.26}$$

As in Equations (4.14) and (4.15), the mean value h and the variance σ_h^2 of the service time with exponential distribution are:

$$h = \int_0^{\infty} t f(t)\mathrm{d}t = \frac{1}{\mu} \tag{4.27}$$

$$\sigma_h^2 = \frac{1}{\mu^2} \tag{4.28}$$

In traffic theory, to simplify analytical equations, mean service time $h = 1$ (one average service time) and $\mu = 1$, is frequently adopted as the measuring unit of the service time in a given system.

4.3.2.4 Probability of Service of i Calls in Time Interval t

With the assumption that the service stream has an exponential distribution, completion moments of the service do not depend on the moments of arrivals of new calls (memorylessness of the exponential distribution). Suppose that, in a given system, there are k servers occupied at the moment t. The probability of an event where i servers in time Δt finish the service can be determined on the basis of the Bernoulli distribution for i successful random attempts, with the total number of attempts equal to k [3]:

$$P_i(k, \Delta t) = \binom{k}{i} p^i (1-p)^{k-i} \tag{4.29}$$

where p is the probability of service of one call in time Δt. Probability $P_i(k, \Delta t)$ is equal to the cumulative distribution function, so, on the basis of Equation (4.25) we get:

$$P_i(k, \Delta t) = \binom{k}{i} \left(1 - e^{-\mu t}\right)^i e^{-\mu t(k-i)} \tag{4.30}$$

4.3.2.5 Service Stream Parameter

Assume that in Equation (4.30), $i = 0$ and $t = \Delta t$. Then we obtain the probability of an event, where in time Δt there will be no release of any of k occupied servers:

$$P_0(k, \Delta t) = e^{-k\mu\Delta t} \tag{4.31}$$

The probability of an opposite event, namely finishing the service of at least one call, is:

$$\pi_1(\Delta t) = 1 - P_0(k, \Delta t) = 1 - e^{-k\mu\Delta t} \tag{4.32}$$

In accordance with the definition of the stream parameter (Equation (4.17)), we can write:

$$N(t) = \lim_{\Delta t \to 0} \frac{\pi_1(\Delta t)}{\Delta t} \tag{4.33}$$

The probability $\pi_1(\Delta t)$ can be rewritten by distributing the function $e^{-k\mu\Delta t}$ to the series:

$$\pi_1(\Delta t) = 1 - e^{-k\mu\Delta t} = 1 - \sum_{j=0}^{\infty} (-1)^j (k\mu\Delta t)^j \frac{1}{j!} = k\mu\Delta t + \theta(\Delta t) \tag{4.34}$$

where $\theta(\Delta t)$ is an infinitely small value in comparison to Δt (Equation (4.3)). After substituting Equation (4.34) into Equation (4.33), we obtain:

$$N(t) = \lim_{\Delta t \to 0} \left[k\mu + \frac{\theta(\Delta t)}{\Delta t} \right] = k\mu \tag{4.35}$$

4.3.2.6 Commentary

Equation (4.35) shows that the service stream parameter $N(t)$ depends exclusively on the service intensity μ and on the number of currently serviced calls. By determining the probability $\pi_2(\Delta t)$, the probability of the arrival of at least two calls in an infinite small time interval with the length Δt, it is easy to show that $\pi_2(\Delta t) = \theta(\Delta t)$. This means that the parameter μ can be interpreted as the intensity of the traffic source in the occupancy state. It can be assumed, therefore, that the service stream is characterized by properties similar to those of the Poisson stream with the intensity equal to $k\mu$ (for k occupied servers).

4.4 Markov Processes

4.4.1 Stochastic Processes

Consider a physical system that, along with time, changes its state, or transforms from one state to another randomly, in a no pre defined manner. The system is said to be involved in a stochastic process. A "physical system" can be understood as any device, or a group of devices, a company, or a living organism, and so forth. The bulk of processes occurring in real systems has, to a greater or lesser degree, random character.

The so-called Markov processes have a particular place in traffic theory. We say that a process occurring in a given system is a Markov process if, at any moment of time t_0, the probabilistic properties of the process in the future depend on its state at a given moment t_0 and do not depend on when and how the system has been placed in the state. In traffic theory, the so-called continuous Markov processes with discrete space of states are the most important. A process is discrete if it is characterized by a countable set of states. A process is continuous if moments of a transition from one state into another are not known and can occur at any moment. An example of such processes is the service process of calls in a group of links. The group is offered calls that can occur at any moment of time. Their service process is also continuous as the connection can be terminated at any moment of time.

Traffic theory proves that if all event streams changing the state of a system are Poisson streams, then the service process in this system is a Markov process [3, 8]. This is intuitively self-evident because Poisson stream is memoryless.

4.4.2 Markov Process as a Call Service Process in the Full-Availability Group

Let us consider a call service process in a group composed of V links (the so-called full-availability group) that is offered a call stream with the intensity λ, while the service time is exponential and its mean value is equal to $1/\mu$. Let us define the state of the process through the number of currently serviced links:

- state "0" – all links are free;
- state "i" – i links are occupied ($V - i$ links are free);
- state "V" – all links are occupied.

4.4.2.1 State Transition Diagram

Geometric diagrams called state transition diagrams are often used to analyze Markov processes with discretely spaced events. Appropriate states are designated by circles (or ellipses, rectangles, and so forth.) and possible transitions by, for instance, arrows. Figure 4.1 shows

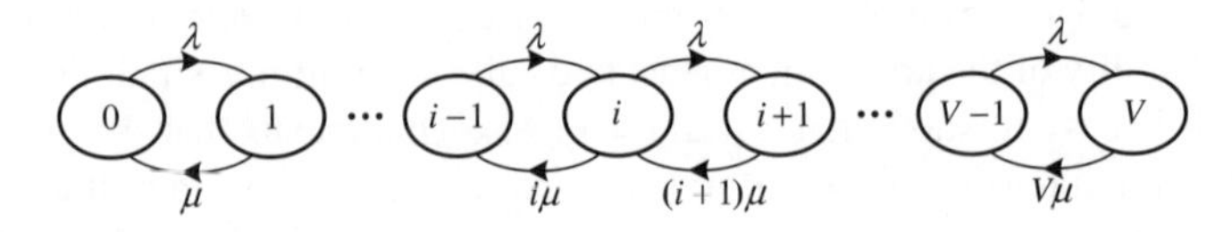

Figure 4.1 State transition diagram of the call service process in a group of links.

a state transition diagram of the call service process in a full-availability group composed of V links. The little arrows pointed "to the right," from the "younger" to the "older" states, indicate the changes in the group states at the moment of the arrival of new calls, whereas the little arrows pointed "to the left" indicate the changes in the group state at moments of completion of the service of calls. Individual arrows (transitions) indicate the corresponding values of the stream parameters transferring the system from one state to another.

4.4.2.2 Kolmogorov Equations

With a diagram of the Markov process, we are in position to construct a probabilistic model of a given system. Assume that the system (group) under consideration has $V + 1$ possible states. The probability of an event that, at the moment t, the system is in the state i, will be called the state i probability. This parameter will be designated by the symbol $p_i(t)$. At any moment of time, the sum of all state probabilities is equal to unity:

$$\sum_{i=0}^{V} p_i(t) = 1 \tag{4.36}$$

On the basis of the diagram of a given process it is possible to determine all state probabilities $p_i(t)$ within the time function t. To achieve this, a formation and a subsequent solution of the so-called Chapman-Kolmogorov equations [6], in short Kolmogorov equations, are to be applied. These are differential equations in which the unknown variables are the probabilities of a system being in the states determined by the discrete Markov process with continuous time.

4.4.2.3 Komogorov Equations in the Full-Availability Group

The process of obtaining such equations will be presented below with the example of a full-availability group composed of V links, whose state diagram is shown in Figure 4.1. Let us assume that the state probability $p_0(t)$ is known, and let us consider what the probability of an event is that after the time Δt the system will still be in the state "0," in other words the probability $p_0(t + \Delta t)$. There are two scenarios that will result in a situation that at the moment $t + \Delta t$ the system will be in state "0" (the corresponding streams are indicated in Figure 4.2 with uninterrupted line):

- The system was in state "0" at the moment t and, within the time interval Δt, did not change its state. The system could make a transition to state "1" at the moment of a new call arrival. The probability of an arrival of a call within the interval Δt is an elementary probability and, according to Equation (4.1), is $\lambda\Delta t$. Therefore, the probability of an event that within the interval Δt no new call arrives and the system will remain in state "0" is $1 - \lambda\Delta t$.

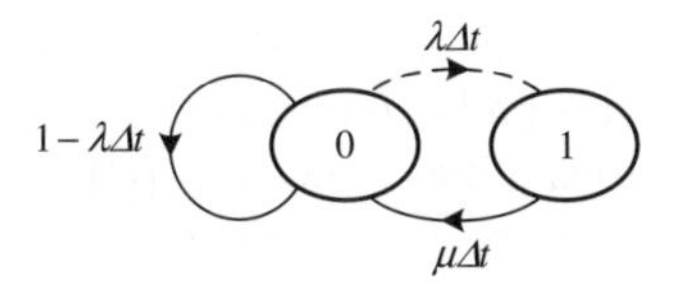

Figure 4.2 Determination of the probability $p_0(t)$.

- The system was in state "1" at the moment t and, within the time interval Δt, passed to state "0". The stream transferring the system from state "1" to state "0" is a service stream with intensity μ. It is also a Poisson stream, so the probability of the occurrence of a new event within the interval Δt is an elementary probability and is equal to $\mu\Delta t$ (Equation (4.22)).

The addition of the probabilities of the two presented scenarios will yield the probability $p_0(t+\Delta t)$:

$$p_0(t+\Delta t) = p_0(t)[1-\lambda\Delta t] + p_1(t)\mu\Delta t \tag{4.37}$$

Transferring $p_0(t)$ to the left side and dividing both sides of Equation (4.37) by $\Delta t \to 0$, will yield the following differential equation:

$$\frac{\mathrm{d}p_0(t)}{\mathrm{d}t} = -\lambda p_0(t) + \mu p_1(t) \tag{4.38}$$

Equation (4.38) is a Kolmogorov equation. An analogous procedure for states from "1" to "V" – with the condition (4.36) taken into account – leads to a system of Kolmogorov equations that describe the Markov process in a group with capacity V of the links:

$$\begin{cases} \frac{\mathrm{d}p_0(t)}{\mathrm{d}t} = -\lambda p_0(t) + \mu p_1(t) \\ \dots \\ \frac{\mathrm{d}p_i(t)}{\mathrm{d}t} = \lambda p_{i-1}(t) - (\lambda + i\mu)p_i(t) + (i+1)\mu p_{i+1}(t) \\ \dots \\ \frac{\mathrm{d}p_V(t)}{\mathrm{d}t} = \lambda p_{V-1}(t) - V\mu p_V(t) \\ \sum_{i=0}^{V} p_i(t) = 1 \end{cases} \tag{4.39}$$

4.4.2.4 Steady States and State Probabilities

To solve the system of Kolmogorov equations (4.39) it is necessary to make adequate and appropriate assumptions. These equations belong to the first order differential equations with constant coefficients and can be solved analytically. Analytical methods are convenient when the number of equations in a system does not exceed two or three. When the number of equations is much greater, it is far more convenient to use approximate numerical techniques to solve such systems of equations [7]. The complexity of the system of equations (4.39) results primarily from the fact that it describes the process in an unsteady state. In practice, it is usually sufficient to solve a problem in a steady state, when the parameters of the system are not dependent on the initial conditions any more. If for $t \to \infty$ there are limit values of state probabilities independent of initial conditions, then they are called steady-state probabilities.

It has been proven in the theory of stochastic processes that if the number of states of a system is finite and it is possible to move from each of the states, after a finite number of steps, to any other state, then the steady state probabilities do exist (it is a sufficient condition, though not necessary, for steady state probabilities to occur [4]). Let us assume that the condition is fulfilled and steady state probabilities exist. To distinguish steady state probabilities from state probabilities dependable on time, we designate the former with the symbol $[p_i]_V$. To calculate these probabilities, it is enough to notice that their derivatives in the system of differential equations (4.39) take up values equal to zero:

$$[p_i]_V = \lim_{t\to\infty} p_i(t) \Rightarrow \frac{\mathrm{d}p_i(t)}{\mathrm{d}t} = 0 \tag{4.40}$$

Thus, we obtain a far more simple system of linear algebraic equations:

$$\begin{cases} -\lambda[p_0]_V + \mu[p_1]_V = 0 \\ \dots \\ \lambda[p_{i-1}]_V - (\lambda + i\mu)[p_i]_V + (i+1)\mu[p_{i+1}]_V = 0 \\ \dots \\ \lambda[p_{V-1}]_V - V\mu[p_V]_V = 0 \\ \sum_{i=0}^{V}[p_i]_V = 1 \end{cases} \tag{4.41}$$

Since further considerations will exclusively concern solutions for steady states, for the sake of clarity and to simplify things, steady state probabilities will be called state probabilities. The system of linear algebraic equations that leads to a determination of these probabilities has been called, in traffic theory, the system of state equations or the system of statistical equilibrium equations.

4.4.2.5 Formation of State Equations

The state probability $[p_i]_V$ expresses the part of the total time of the process ($t \to \infty$) in which it remains in state i. If the parameter of the outgoing stream from the state is λ_i, then the total intensity of the outgoing stream is equal to $[p_i]_V\lambda_i$.

With such an interpretation, Equations (4.41) are analogous to Kirchhoff equations (the sum of the intensities of outgoing and ingoing currents from the node is identical) and express the effect of intensity equality of streams coming in and out from a given state. The formation diagram of each of the equations in Equations (4.41) is as follows:

the sum of the intensity of ingoing streams = the sum of the intensity of outgoing streams.

Figure 4.3 shows the formation of state equations of the Markovian process according to the above rule.

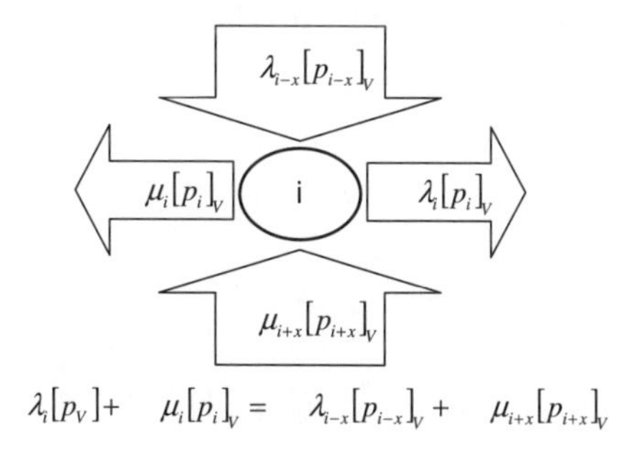

Figure 4.3 Formation of state equations in the Markov process.

4.4.2.6 Birth-and-Death Processes

The system of equations (4.41) that determines the Markov process [4, 8], is called the birth-and-death process. Its particular property is a mutual relation of all states in one chain in which each state is related to any neighboring state with one "left-side" and one "right-hand" arrow (Figure 4.1). A process can have a finite or infinite number of states (the number of states can be infinite though has to be countable, i.e. such that each state can unequivocally be ascribed a serial number [6]). The term "birth-and-death process" has its roots in the past and stems from the application of this type of process in biology to model population states.

The system of state equations (4.41) can, in elementary way, be reduced to the following form (through subsequent substitutions):

$$\begin{cases} \lambda[p_0]_V = \mu[p_1]_V \\ \cdots, \\ \lambda[p_{i-1}]_V = i\mu[p_i]_V \\ \cdots \\ \lambda[p_{V-1}]_V = V\mu[p_V]_V \\ \sum_{i=0}^{V}[p_i]_V = 1 \end{cases} \tag{4.42}$$

In traffic theory, the equations included in the system (4.42) are called local equilibrium equations and are characteristic of the birth-and-death process. Their formation (Figure 4.4) is elementary following the rule below:

stream intensity in the direction from state i *to state* $i+1$

$=$ *stream intensity in the direction from state* $i+1$ *to state* i.

This means that the streams between neighboring states are in balance (equilibrium).

4.4.2.7 Solution of the Birth-and-Death Process for the Full-Availability Group

Local equilibrium equations make it possible to determine the probability of a given state on the basis of the state directly preceding it. Let us solve the system of equations (4.42), in such a way as to obtain subsequent state probabilities in relation to the value $[p_0]_V$. On the basis of

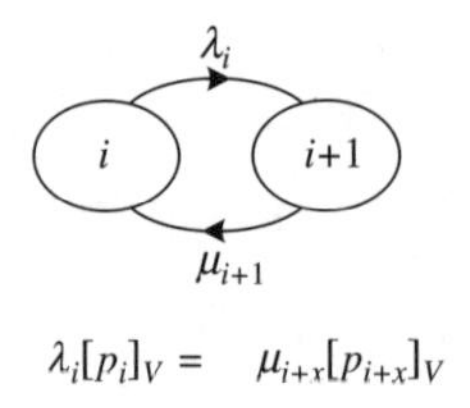

Figure 4.4 Formation of state equations in the birth-and-death process.

the first equation of local equilibrium we obtain:

$$[p_1]_V = \frac{\lambda}{\mu}[p_0]_V \tag{4.43}$$

whereas from the second local equilibrium equation (4.43) we have:

$$[p_2]_V = \frac{\lambda}{2\mu}[p_1]_V = \frac{\lambda\lambda}{\mu 2\mu}[p_0]_V = \frac{\lambda^2}{2\mu^2}[p_0]_V \tag{4.44}$$

In a series of similar k steps, we obtain the expression describing the probability $[p_k]_V$:

$$[p_k]_V = [p_0]_V \prod_{i=1}^{k} \frac{\lambda}{i\mu_i} = C_k[p_0]_V \tag{4.45}$$

where:

$$C_k = \prod_{i=1}^{k} \frac{\lambda}{i\mu_i} \tag{4.46}$$

Notice Equations (4.45) and (4.46). The numerator has the product of all the call intensities associated with "right-hand" arrows from state "0" to state "k" (Figure 4.1), whereas the denominator has the product of all intensities of the service streams related to "left-hand" arrows from state "k" to state "0." The process is said to be characterized by the so-called product form solution of state equations. After taking into consideration that $\sum_{i=0}^{V}[p_i]_V = 1$, we eventually obtain:

$$[p_k]_V = \frac{C_k}{\sum_{i\geq 0} C_i} \tag{4.47}$$

The state probabilities expressed by Equation (4.47) determine the average time of the system in a given occupancy state of a group.

4.5 The Concept of Traffic

4.5.1 Introductory Information

The task of a network is to transport information between its users. Telecommunication traffic (teletraffic) creates a flow of information, which causes network resources, links and channels, to be occupied. One can thus define the concept of traffic as a certain random process $c(t)$, whose trajectory is described by the number of links simultaneously serviced in the system at the moment t.

There are three types of traffic identified in telecommunication systems:

- offered traffic – the concept that defines a hypothetical traffic (this is not a measurable quantity) that is created by calls appearing at the input of a system;

- carried traffic – determines this part of traffic that will be serviced, in other words transferred by the telecommunication systems; for instance, all calls that will arrive at free links in a group will create traffic carried by the group and this type of traffic is measurable;
- lost traffic – determines the part of traffic that will not be serviced by the telecommunication systems, for example, calls that arrive at occupied links in a group will be rejected and will create lost traffic.

4.5.2 Traffic and Traffic Intensity

For a quantitative description of telecommunication traffic a parameter is used called the intensity of telecommunication traffic. Its unit is one erlang (1 erl).

One erlang corresponds to the time of a duration of one connection that equals one hour when the observation period is one hour.

Momentary traffic intensity – in line with the recommendation [9] – is equal to the number of simultaneously occupied resources of a system, for instance links in a group at a given moment of time t, which will be denoted by the symbol $c(t)$. The average traffic intensity in observation time t_{obs} can be then written in the following form:

$$Y = \frac{1}{t_{\mathrm{obs}}} \int_0^{t_{\mathrm{obs}}} c(t)\mathrm{d}t \tag{4.48}$$

In Equation (4.48) the expression below represents traffic value Q (this parameter is frequently defined as traffic load):

$$Q = \int_0^{t_{\mathrm{obs}}} c(t)\mathrm{d}t \tag{4.49}$$

The basic traffic unit is 1 SM – speech-minute. The unit 1 erlh – erlang-hour is often used, when

$$1 \text{ erlh} = 60 \text{ SM}$$

It is easy to prove [10] that the traffic load – determined by the integer (4.49) – is equal to the sum of occupancy distributions of all connections $N(t_{\mathrm{obs}})$ carried out during the observation time t_{obs} :

$$Q = \sum_{i=1}^{N(t_{\mathrm{obs}})} t_i \tag{4.50}$$

The value t_i determines the duration time of a connection with the serial number i, where

$$0 < i \leqslant N(t_{\mathrm{obs}})$$

Equation (4.50) explains why the value of offered traffic is not a measurable value. Offered traffic is composed of carried traffic and lost traffic, which is "produced" by calls that are not serviced. The service times of those calls are unknown and cannot be included in the sum (4.50).

4.5.3 Definitions of the Average Intensity of Carried Traffic

There are four equivalent definitions of the average intensity of carried traffic [11, 12], which will be given and commented on below. For better understanding of those individual definitions,

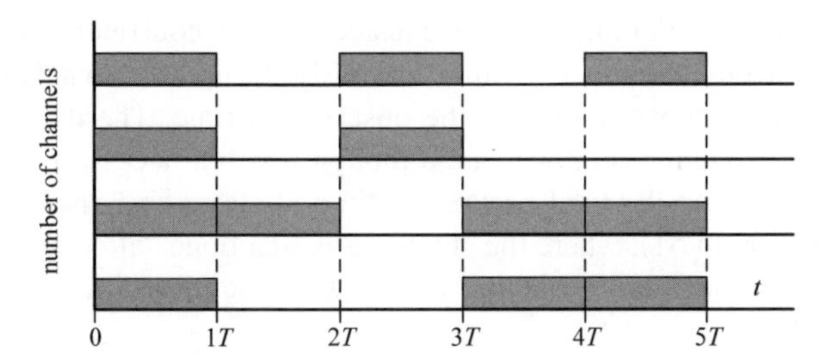

Figure 4.5 Example results of hypothetical observation of a group with the capacity V.

we will use Figure 4.5, which shows examples of results of a hypothetical observation of a group composed of four channels. To simplify, it is assumed that the service time of all connections can be expressed as a multiple of a certain time-interval T. This assumption makes it possible to adopt a simple interpretation of particular definitions without taking into consideration the influence of unfinished connections at the borders of the observation interval. In other cases, a sufficient duration time of observation is assumed, in other words one in which the influence of unfinished connections within the observation time can be omitted.

Let us adopt the following designations:

c – average number of connections within a time unit;
h – average service time of a connection;
$t_{i,j} = T$ – length of i-th occupancy period in channel j;
V – total number of observed channels;
N – number of observed occupancy periods;
N_j – number of carried out connections in channel j within the observation time t_{obs};
t_{obs} – observation time of the group.

Traffic intensity can be defined in the following way:

Definition 1 *Traffic intensity is equal to the average number of simultaneously occupied channels.*

After summing up all occupied channels in the subsequent observation periods we get the total number of occupied channels during the observation time. The average number of simultaneously occupied channels is obtained when the total number of occupied channels is divided by the total observation time. Thus, our considerations can be reduced to the following mathematical form:

$$Y(Definition 1.) = \frac{\sum_{i=1}^{N} \sum_{j=1}^{V} t_{i,j}}{t_{obs.}} = \frac{\sum_{i=1}^{N} \sum_{j=1}^{V} T}{NT} \tag{4.51}$$

Definition 2 *Traffic intensity is the ratio of the sum of channel occupancy time during the observation time to the observation time.*

It is easy to prove that the definition under consideration is equivalent to the previous definition because the average number of simultaneously occupied channels can be expressed as a ratio of the total occupancy time to the observation time. The difference lies only in that that in the case of Definition 2 we can, a though it is not necessary, sum up "along the channels" (lines), summing the total occupancy time of successive channels, as opposed to Definition 1 – Equation (4.51), where the summing is first done "after successive occupancy periods." Therefore, in the considered case, we can write the following:

$$Y(Definition\ 2) = \frac{\sum_{j=1}^{V}\sum_{i=1}^{N} t_{i,j}}{t_{\text{obs.}}} = \frac{\sum_{j=1}^{V}\sum_{i=1}^{N} T}{NT} \tag{4.52}$$

Definition 3 *Traffic intensity is equal to the product of the average number of connections that have been set up in a given time interval and the average duration time of a connection.*

The average number of connections that are set up within a time interval is equal to the ratio of the number of all connections set up in a group within the observation time to the observation time, i.e. the number of time intervals. Thus:

$$c = \frac{\sum_{j=1}^{V} N_j}{t_{\text{obs.}}} = \frac{\sum_{j=1}^{V} N_j}{NT} \tag{4.53}$$

The average service time of a connection is equal to the quotient of the total sum of the occupancy time of channels in a group within the observation time to the sum of the number of connections set up in the group:

$$h = \frac{\sum_{j=1}^{V}\sum_{i=1}^{N} t_{i,j}}{\sum_{j=1}^{V} N_j} = \frac{\sum_{j=1}^{V}\sum_{i=1}^{N} T}{\sum_{j=1}^{V} N_j} \tag{4.54}$$

Taking into account the results obtained on the basis of Equations (4.53) and (4.54) we obtain the value of the average traffic intensity calculated on the basis of Definition 3:

$$Y(Definition\ 3.) = ch = \frac{\sum_{j=1}^{V}\sum_{i=1}^{N} T}{NT} \tag{4.55}$$

Notice that by substituting Equations (4.53) and (4.54) into Equation (4.55), we get Equation (4.52), equivalent, in turn, to Equation (4.51).

Definition 4 *Traffic intensity is equal to the average number of connections within the period equal to the average duration time of a connection.*

The average number of connections in time unit c can be determined on the basis of Equation (4.53). The average duration time of a connection, calculated on the basis of Equation (4.54), is in fact indicated by the number of time units comprising the average duration time of a connection. Thus, the traffic intensity, according to Definition 4 is calculated in the same way as in the case of Definition 3.

With the comments to the definitions given above we have proved that they are identical and lead to identical results. In engineering practice, the most commonly used definition is Definition 4.

4.5.4 *Definition of the Average Intensity of Offered Traffic*

Let us notice that the values of the traffic intensity, determined on the basis of the definition presented in Section 4.5.3, are related to carried traffic – hence, we talk about the intensity of carried traffic Y because we consider carried connections, i.e. those that have been serviced. If, in Definition 4, we substitute the concept "connection" with the concept "call," then we obtain the definition of the intensity of the offered traffic:

Definition 5 *The intensity of the offered traffic is equal to the average number of calls in a period equal to the average duration time of the connection.*

Definition 5 allows the determination of the intensity of offered traffic through the intensities of the call stream and the service stream. If the average duration time of a connection is $1/\mu$, and the intensity of the call stream is λ, then the value of the intensity of offered traffic A can be determined with the following formula:

$$A = \lambda/\mu \tag{4.56}$$

4.6 Quality of Service in Telecommunication Systems

The quality of service in telecommunication networks and systems is described by a group of GoS (grade of service) parameters and a group of QoS (quality of service):

- GoS parameters – in line with recommendation E.600 ITU-T – determine qualitatively the grade of service of telecommunication traffic by network resources;
- QoS parameters – in line with recommendation E.800 ITU-T – determine the level of satisfaction of the user with the service offered.

This arrangement of the quality parameters stems from different ways of viewing and evaluating the operation of a network. The QoS parameters represent the point of view of the user of a network, while the GoS parameters that of the "network" (operator).

In designing and modeling network systems, the GoS parameters play the principal role, while a success or a failure in any introduction of a new service to a network is inseparably connected with the evaluation of the service by subscribers, which is, in turn, related to the QoS parameters. In many cases it is difficult to determine clearly the relations between "subjective" QoS parameters and statistically objective GoS parameters, so the so-called service level agreement (SLA) approach [10] has been worked out. This is, in fact, a kind of agreement between the user and the operator of a network. The agreement clearly defines what, in reality, particular parameters mean – for instance which values they can represent. Definitions and appropriate values of the parameters should be constructed in such a way that they be unequivocally understood both by the operator and the user of a network. The agreement should additionally detail relevant procedures following situations resulting from any failure to keep to the earlier stipulated conditions of the agreement.

4.6.1 *Basic GoS Parameters in Loss Systems*

In traffic theory the quality of service on the call stream's level, characterizes a possibility of setting up connections, or the time needed for a connection to be set up, depending on the way a system responds to the lack of free resources. If all resources of a system are occupied and a new call arrives, then, in the case of the loss system, this call will be lost (rejected).

For the quantitative evaluation of the quality of service in loss systems, the GoS parameters most frequently used are the loss coefficient (often called the call congestion) and the blocking coefficient (often called the time congestion), where:

- loss coefficient $B(t_1, t_2)$ within the time interval (t_1, t_2) determines the ratio of the number of lost calls $N_{\text{lost}}(t_1, t_2)$ to the number of all offered calls $N_{\text{of}}(t_1, t_2)$ in the system:

$$B(t_1, t_2) = \frac{N_{\text{lost}}(t_1, t_2)}{N_{\text{of}}(t_1, t_2)} \tag{4.57}$$

- blocking coefficient $E(t_1, t_2)$ determines the ratio of the time $T_{\text{bl}}(t_1, t_2)$, during which all resources of a system are occupied to the total time of observation $T(t_1, t_2)$:

$$E(t_1, t_2) = \frac{T_{\text{bl}}(t_1, t_2)}{T(t_1, t_2)} \tag{4.58}$$

In Equations (4.57) and (4.58), such parameters as the number of offered and lost calls or the occupancy time of all resources of a system, during each measurement (observation time) will take on different values. Therefore, they are random variables. If we substitute the values of the parameters with their average values, we obtain general characteristics of the system defined as the loss probability B and the blocking probability E:

$$B = \lim_{t_2 \to \infty} B(t_1, t_2) \tag{4.59}$$

$$E = \lim_{t_2 \to \infty} E(t_1, t_2) \tag{4.60}$$

4.6.2 Traffic Load-Carrying Capacity and Traffic Load of Communication Systems

To make a comparison of different telecommunication systems possible and feasible, the concept of traffic load-carrying capacity is used in practical engineering. The concept defines the traffic intensity offered to a system that the system can carry with assigned loss probability or blocking probability. The traffic load-carrying capacity depends on a series of factors including, first of all, the properties of the call stream, distribution of service time, structure and capacity of a system.

The traffic load of a system is defined as the intensity of traffic carried by the system.

References

[1] Akimuru, H. and Kawashima, K. (1999) *Teletraffic: Theory and Application*, Springer.

[2] Brockmeyer, E. (1948) A survey of A.K. Erlang's mathematical works. *Danish Academy of Technical Sciences*, **2**, 101–26.

[3] Khinchin, A. (1963) *Raboty po matematicheskoj teorij massovogo obsluzhivanija*. Izdatel'stvo Fiziko-Matematicheskoi Literatury.

[4] Syski, R. (1986) Introduction to congestion theory in telephone systems. *Studies in Telecommunication*, North Holland.
[5] Brockmeyer, E., Halstrom, H.L., and Jensen, A. (1960) The life and works of A.K. Erlang. *Acta Polytechnika Scandinavia*, **6**(287), 23–100.
[6] Feller, W. (1961) *An Introduction to Probability Theory and its Applications*. John Wiley & Sons, Inc.
[7] Shneps, M.A. (1979) *Sistemy raspredelenija informacii*. Metody raschyota. Radio i svjaz'.
[8] Markow, A. (1905) Issledovanie obščego slučaâ ispytanij svâzanych v cep'. *Zapiski Imperatorskoj Akademii Nauk*, **25**(3),1–33.
[9] ITU-T. Traffic Intensity Unit. Recommendation B.18, 1993.
[10] V.B. Iversen (ed.) (2005) *Teletraffic Engineering Handbook*. ITU-D, Study Group 2, Geneva, June 2001 (draft, published online by International Telecommunication Union). https://www.itu.int/ITU-D/study_groups/SGP_19982002/SG2/StudyQuestions/Question_16/RapporteursGroupDocs/teletraffic.pdf (accessed 19 July 2010).
[11] Bear, D. (1988) *Principles of Telecommunication Traffic Engineering*, Peter Peregrinus Ltd., 1988.
[12] Elldin, A. and Lind, G. (1967) Traffic measurements and traffic supervision, in *Elementary Telephone Traffic Theory*, (eds Elldin, A. and Lind, G.), L.M. Ericsson AB.

5

Basic Elements of Traffic Engineering used in Mobile Networks

5.1 Introduction

In this chapter we shall discuss basic models of systems in single-service networks. This means that each call always "demands" the same volume of network resources to set up a connection. An example of a "single-rate network" is, for instance, the "classic" public network, in which each call requires a channel with a capacity of 64 kbps. The models presented apply to planning and dimensioning cellular (mobile) networks of older generations (GSM), which require the same resources for each connection. In particular cases, the models presented can also be used to estimate traffic properties of cells in UMTS and LTE networks.

In mobile networks, the group is understood to be the resources available in one cell of the network. For example, there are 25 channels of 25 kHz each available in a cell of a GSM network. Thus, we can consider the cell as a full-availability group with the capacity of 25 channels.

In Sections 5.2 and 5.3, Erlang and Engset models of the full-availability group will be discussed. These models will be presented within a context of dimensioning hard and soft capacity of cellular networks.

5.2 Erlang Model

5.2.1 Assumptions of the Model

Let us consider a full-availability group with the following assumptions:

- capacity of the system is V servers (links, channels), each of which is available for any call if it is not occupied;

Modeling and Dimensioning of Mobile Networks: From GSM to LTE
Maciej Stasiak, Mariusz Głąbowski, Arkadiusz Wiśniewski and Piotr Zwierzykowski

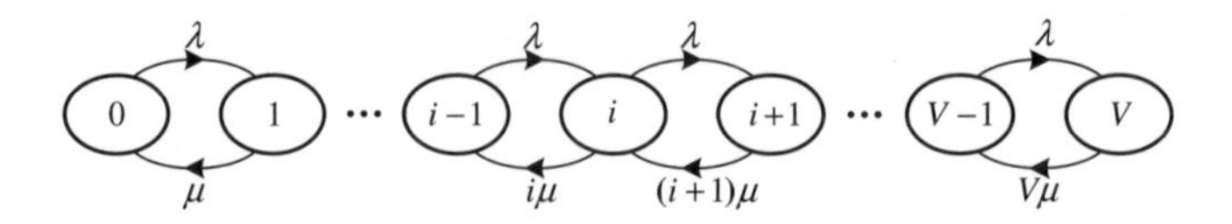

Figure 5.1 State transition diagram for Erlang model.

- the calls come from an unlimited number of traffic sources and create a Poisson stream with an intensity of λ;
- service time is described by an exponential distribution, the mean value of the service time is $1/\mu$;
- a rejected call due to the lack of free servers is lost.

The type of traffic in the model considered here, namely the traffic in which we assume that the service times are exponentially distributed and the arrival process is a Poisson process is called PCT1 (Pure Chance Traffic Type One) traffic. This type of traffic is known also as Erlang traffic.

5.2.2 Diagram of the Service Process

Section 4.4.2 discusses a Markov process that occurs in a full-availability group with capacity V. Figure 5.1 (and Figure 4.1) show the birth and the death processes that take place in a full-availability group satisfying the assumptions of the Erlang model (Section 5.2.1). Each state of the process is identified by the number of occupied servers (channels):

- state "0" – all channels are free;
- state "i" – i channels are occupied ($V - i$ channels are free);
- state "V" – all channels are occupied.

5.2.3 State Equations

In the process of birth-and-death state equation can be derived on the basis of the local equilibrium equations (Section 4.4.2). On the basis of the diagram from Figure 5.1 we obtain the following system of equations:

$$\begin{cases} \lambda[p_0]_V = \mu[p_1]_V \\ \cdots, \\ \lambda[p_{i-1}]_V = i\mu[p_i]_V \\ \cdots, \\ \lambda[p_{V-1}]_V = V\mu[p_V]_V \\ \\ \sum_{i=0}^{V}[p_i]_V = 1 \end{cases} \tag{5.1}$$

5.2.4 Occupancy Probability of Arbitrarily Chosen i Channels in the Group – Erlang Distribution

The solution of the equations in (5.1) can be rewritten in the following form (the method for solving the system of state equations for the birth-and-death process is presented in Section 4.4.2):

$$[p_k]_V = \frac{\left(\frac{\lambda}{\mu}\right)^k}{k!} \Bigg/ \sum_{i=0}^{V} \frac{\left(\frac{\lambda}{\mu}\right)^i}{i!} \tag{5.2}$$

Equation (5.2) expresses the state probability k and is, in accordance with the definition of states adopted in the model, the occupancy probability of any k channels (i.e. k servers) in the group with the capacity of V channels. Taking into consideration the definition of the offered traffic intensity expressed by Equation (4.56), Equation (5.2) can be rewritten in the following way:

$$[p_k]_V = \frac{A^k}{k!} \Bigg/ \sum_{i=0}^{V} \frac{A^i}{i!} \tag{5.3}$$

where $A = \lambda/\mu$ is the intensity of offered traffic.

In traffic theory, the probability distribution expressed by Equation (5.3) is called the occupancy distribution of the full-availability group or, in short, the Erlang distribution [1, 2].

5.2.5 Blocking Probability – Erlang Formula

The occupancy probability of all links in a group, in accordance with the definitions (4.58) and (4.60), is the blocking probability of the full-availability group:

$$E = E_V(A) = \frac{A^V}{V!} \Bigg/ \sum_{i=0}^{V} \frac{A^i}{i!} \tag{5.4}$$

Equation (5.4) is widely known as the Erlang formula or the Erlang B-formula. Blocking probability in the full-availability group with the capacity V and the offered traffic A is denoted in the literature of the subject – as in Equation (5.4) – by the symbol $E_V(A)$.

5.2.6 Loss Probability

The number of calls offered to the full-availability group within a time unit is determined by the intensity of the call stream λ. With a determined blocking probability $E_V(A)$, the average number of lost calls within a time unit is $\lambda E_V(A)$. Comparing the number of lost calls to those offered we obtain:

$$B = \frac{\lambda E_V(A)}{\lambda} = E_V(A) = E \tag{5.5}$$

Equation (5.5) determines that – in the case of the Erlang model for the full-availability group – the blocking probability is equal to the loss probability of calls.

5.2.7 Occupancy Probability of x Precisely Determined Channels in a Group – Palm-Jacobaeus Formula

The Palm-Jacobaeus formula [3, 4] determines the occupancy probability of particular channels in a full-availability group. Assume that, in a group with the capacity V, x channels are considered to be precisely determined. The following notation is then introduced:

- $P(x|i)$ – conditional occupancy probability of x determined channels, under condition that there are i arbitrary chosen occupied channels in the group;
- $H(x)$ – unconditional occupancy probability of x determined channels in the group.

On the basis of the total probability theorem, the probability $H(x)$ can be written in the following form:

$$H(x) = \sum_{i=x}^{V} P(x|i) \left[p_i\right]_V \tag{5.6}$$

where $\left[p_i\right]_V$ is the occupancy probability of any i channels in the full-availability group determined by the Erlang distribution (5.3).

Let us consider a conditional occupancy probability $P(x|i)$ of x determined channels in the group. The probability is a quotient of the number of $(i - x)$-combinations of occupied channels from a $(V - x)$-element set of occupied channels (because x channels from among i occupied channels have already been determined, i.e. chosen *a priori*) to the number of possible i-combinations of occupied channels from a V-element set of all channels of the group:

$$P(x|i) = \binom{V-x}{i-x} \Big/ \binom{V}{i} = \frac{(V-x)!}{V!} \frac{i!}{(i-x)!} \tag{5.7}$$

On the basis of Equations (5.3), (5.6) and (5.7) it is possible to determine the unconditional probability of the occupancy of x precisely determined channels in the group. After simple transformations we obtain:

$$H(x) = \frac{E_V(A)}{E_{V-x}(A)} \tag{5.8}$$

In traffic theory, Equation (5.8) is called the Palm-Jacobaeus formula. The formula was first introduced by Palm in 1938, although it was published as late as 1943 [4]. This important formula was earlier used for the approximation of blocking probability in grading groups (with limited-availability). At present, it is often used to approximate occupancy distributions of different telecommunication systems, for example, outgoing groups of switching networks.

5.2.8 Erlang Tables

The task of dimensioning groups in a telecommunication network consists of determining the appropriate capacity of a group so as not to allow losses in a group to exceed the assigned level E with the assigned offered traffic. This is one of the fundamental problems that designers of network systems have to face and resolve. To simplify the solution of this problem, the Erlang formula is presented in special tables called the Erlang tables. The tables for scientific

Table 5.1 Erlang tables

Capacity V	Blocking probability (E)			
	$E = 0.02$	$E = 0.01$	$E = 0.005$	$E = 0.001$
1	$A \approx 40.02$	$A \approx 0.01$	$A \approx 0.005$	$A \approx 0.001$
2	$A \approx 0.22$	$A \approx 0.15$	$A \approx 0.105$	$A \approx 0.046$
3	$A \approx 0.60$	$A \approx 0.45$	$A \approx 0.35$	$A \approx 0.19$
4	$A \approx 1.10$	$A \approx 0.90$	$A \approx 0.70$	$A \approx 0.44$
5	$A \approx 1.70$	$A \approx 1.40$	$A \approx 1.10$	$A \approx 0.80$
6	$A \approx 2.30$	$A \approx 1.90$	$A \approx 1.60$	$A \approx 1.10$
7	$A \approx 2.90$	$A \approx 2.50$	$A \approx 2.20$	$A \approx 1.60$
8	$A \approx 3.60$	$A \approx 3.10$	$A \approx 2.70$	$A \approx 2.10$
9	$A \approx 4.30$	$A \approx 3.80$	$A \approx 3.30$	$A \approx 2.60$
10	$A \approx 5.10$	$A \approx 4.50$	$A \approx 4.00$	$A \approx 3.10$

Dependence between E, V and the offered traffic (A).

scheduling are to be found in practically all textbooks on traffic theory. Table 5.1 shows an extract from the Erlang tables that are used in Section 5.2.9 to clarify the representation of the so-called *group conservation principle*.

The use of Erlang tables is very simple. While dimensioning groups, we first choose an appropriate column corresponding to the assigned level of blocking. Then, we find the same row in which the value of the offered traffic is equal or slightly greater than the assigned value. The value of the capacity of the group matching the level is the minimum number of channels satisfying the assumption on the adopted blocking probability.

5.2.9 Group Conservation Principle

While analyzing Erlang tables it is noticeable that, for an assigned value of blocking probability, the number of channels becomes higher when the traffic load-carrying capacity falling into one channel $a = A/V$ increases. This observation forms the base for the so-called group conservation principle, which will be illustrated here with an example. Let us consider two groups with the load-carrying capacity of 0.34 erl and the capacity of $V = 5$ channels. With the assigned blocking level 0.02, the load-carrying capacity corresponds to the first column and the fifth row in Table 5.1 ($0.34 \times 5 = 1.7$ erl.). If we combine the two groups into one, then, with the same individual load-carrying capacity, the total traffic offered to the group is equal to $0.34 \times 10 = 3.4$ erl. It reveals that, for transferring traffic with the adopted blocking probability, eight channels will suffice.

Let us sum up the conclusions resulting from the above example. Assume that two groups with the capacities V_1, V_2 service traffic with the values equal to A_1, A_2. If we now consider blocking probabilities of individual groups, then we obtain the following dependence:

$$E_{V_1+V_2}(A_1 + A_2) < \max(E_{V_1}(A_1), E_{V_2}(A_2)) \tag{5.9}$$

The dependence expressed by Equation (5.9) shows the principle of group conservation. The principle states that:

Blocking probability in a group created as the result of a combination of a certain number of groups is always lower than the maximum blocking probability in component groups functioning independently.

In practice, where component groups are of similar capacity, the inequality (5.9) is reduced to the following form:

$$E_{V_1+V_2}(A_1 + A_2) < (E_{V_1}(A_1), E_{V_2}(A_2)) \tag{5.10}$$

This means that blocking probability in a group created in a combination of a number of groups of similar capacity is always lower than blocking probability in component groups functioning independently.

A very important conclusion for designers of telecommunication networks follows from the group principle, namely that it is favorable, wherever it is possible, to combine small groups and form large groups, as this will reduce the probability of blocking or, in other words, effect in an increase in the traffic loadability of network systems.

5.2.10 Recursive Properties of the Erlang Formula

A direct description of the blocking probability in the full-availability group with the help of the Erlang formula (5.4) is complex due to the existence of the factorial (taking on very large values for large capacities of groups). However, the Erlang formula can be easily converted into a recursive form, which makes it easily programmable:

$$E_V(A) = \frac{AE_{V-1}(A)}{V + AE_{V-1}(A)} \tag{5.11}$$

where the initial value $E_0(A) = 1$, because $A^0 = 1$ and $0! = 1$.

5.2.11 Traffic Carried by the Full-Availability Groups

Let us determine the average number of concurrently occupied channels for the full-availability group in the Erlang model. On the basis of the Erlang distribution, we get:

$$Y_V = \sum_{k=0}^{V} k\,[P_k]_V = \frac{\sum_{k=0}^{V} k\frac{A^k}{k!}}{\sum_{k=0}^{V}\frac{A^k}{k!}} = \frac{A\sum_{k=0}^{V-1}\frac{A^k}{k!}}{\sum_{k=0}^{V}\frac{A^k}{k!}} = A[1 - E_V(A)] \tag{5.12}$$

In accordance with Definition 1 (Section 4.5.3), the average number of concurrently occupied channels determines the traffic intensity. In this particular case, it is traffic serviced by the full-availability group. The mean value of lost traffic R is the difference between offered and carried traffic. In keeping with Equation (5.12), we get:

$$R = A - Y_V = AE_V(A) \tag{5.13}$$

5.2.12 Traffic Carried by One Channel of the Full-Availability Group

In the case of the random hunting strategy, in which each channel of a group can be occupied with identical probability, traffic carried by one channel will be determined by the following equation:

$$y = \frac{Y_V}{V} = \frac{A[1 - E_V(A)]}{V} \tag{5.14}$$

Let us determine traffic carried by one channel in the group for the sequential hunting strategy.[1] Let us consider a group in which n channels are occupied. Traffic carried by n channels can be determined in exactly the same way as traffic carried by a group with the capacity of n channels. Hence, on the basis of Equation (5.12), we have:

$$Y_n = A[1 - E_n(A)] \tag{5.15}$$

Traffic carried by $n - 1$ channels will be determined in a similar way:

$$Y_{n-1} = A[1 - E_{n-1}(A)] \tag{5.16}$$

Now we can evaluate traffic carried by the channel n in the full-availability group with the sequential hunting strategy by subtracting the values of traffic determined by Equations (5.15) and (5.16):

$$y_n = Y_n - Y_{n-1} = A[E_{n-1}(A) - E_n(A)] \tag{5.17}$$

It should be emphasized that traffic serviced by the channel with the serial number n is independent of the capacity of the group. Consequently, traffic carried by "older" channels than the n channel does not have any influence on traffic serviced by channel n.

5.3 Engset Model

The Erlang formula (5.4) was derived with the assumption that the call stream intensity did not depend on the state of the group. If the number of traffic sources N is such that a decrease in the number by the value V (maximum number of occupied links in a group) has an effect on the intensity of the call stream, then the use of the Erlang model can eventually lead to inaccurate evaluation of blocking probabilities in a system. In practice, it is believed that for $N < 15V$, the use of Erlang formula should be restricted [5].

5.3.1 Assumptions of the Model

The influence of a finite number of traffic sources on the blocking probability is taken into account in the Engset model [6]. This model makes the following assumptions:

- capacity of the system is equal to V channels and each of them is available for any arbitrarily chosen call if only it is not occupied;
- the calls come from a finite number of traffic sources N ($N > V$);
- the intensity of calls in state i is directly proportional to the number of free $N - i$ traffic sources and is equal to $(N - i)\gamma$, where γ is the call intensity of one free source;
- service (holding) time is expressed by the exponential distribution; the mean value of the service time is $1/\mu$;
- a rejected call due to the lack of free channels is lost.

[1]Note that in the sequential hunting strategy channels are occupied subsequently in accordance with the channel indexing adopted earlier.

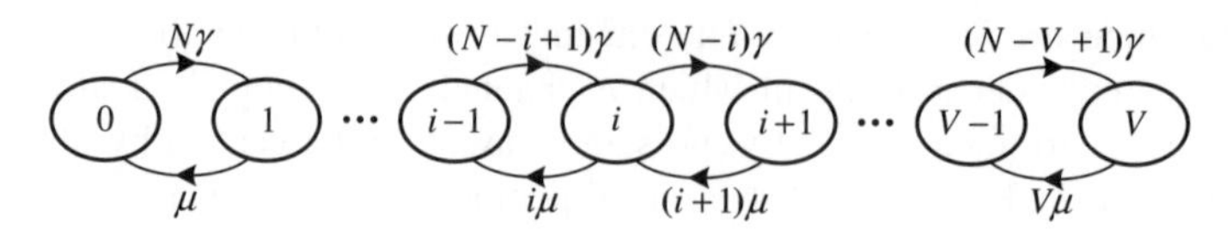

Figure 5.2 State transition diagram for Engset model.

The type of traffic in the model considered, in other words traffic in which we assume that the service times are exponentially distributed and the arrival process is formed by the limited number of sources, is called pure chance traffic type two (PCT2) traffic. This type of traffic is also known as Engset traffic.

5.3.2 Diagram of the Service Process

It can be proved (as in Section 4.3.2) that the service stream parameter in state i of the Engset model is equal to the service intensity of calls in a given state and is $i\mu$. The diagram of the birth-and-death process occurring in the full-availability group that satisfies the assumptions for the Engset model (Section 5.3.1) therefore takes on the form presented in Figure 5.2. Each state of the process is identified by the number of occupied channels:

- state "0" – all channels are free, all sources are free;
- state "i" – i channels are occupied, $(V - i)$ channels are free, i sources are serviced, $(N - i)$ sources are free;
- state "V" – all channels are occupied, V sources are serviced, $(N - V)$ sources are free.

5.3.3 State Equations

The system of local equilibrium equations of the birth-and-death process, whose diagram is presented in Figure 5.2, can be written in the following form:

$$\begin{cases} N\gamma \left[p_0\right]_V = \mu \left[p_1\right]_V \\ \cdots, \\ (N - i + 1)\gamma \left[p_{i-1}\right]_V = i\mu \left[p_i\right]_V \\ \cdots, \\ (N - V + 1)\gamma \left[p_{V-1}\right]_V = V\mu \left[p_V\right]_V \\ \\ \sum_{i=0}^{V} \left[p_i\right]_V = 1 \end{cases} \tag{5.18}$$

5.3.4 Occupancy Probability of Arbitrarily Chosen i Channels in the Group – Engset Distribution

The solution of the system of equations (5.18) can be expressed in the following form:

$$\left[p_i\right]_V = \binom{N}{i}(\gamma/\mu)^i \Big/ \sum_{j=0}^{V}\binom{N}{j}(\gamma/\mu)^j \tag{5.19}$$

By introducing a notion of traffic α, offered by one free source:

$$\alpha = \frac{\gamma}{\mu} \tag{5.20}$$

Formula (5.19) can be rewritten in the following form:

$$[p_i]_V = \binom{N}{i}\alpha^i \Big/ \sum_{j=0}^{V}\binom{N}{j}\alpha^j \tag{5.21}$$

Equation (5.21) gives the probability of state i and is, in accordance with the definition of states adopted for the model, the occupancy probability of arbitrarily chosen i channels in the group with the capacity of V channels which is offered a call stream generated by a finite number of traffic sources. The distribution probability is expressed by Equation (5.21) and, in short, is called the Engset distribution [6].

5.3.5 *Blocking Probability*

The occupancy state of all V links of the group being occupied determines the blocking probability in the Engset model:

$$[p_V]_V = E(\alpha, V, N) = \binom{N}{V}\alpha^V \Big/ \sum_{j=0}^{V}\binom{N}{j}\alpha^j \tag{5.22}$$

Equation (5.22) expresses the blocking probability, which is often denoted with the symbol $E(\alpha, V, N)$, which means: this is the blocking probability in a full-availability group with the capacity of V channels, which is offered traffic from N traffic sources ($N > V$), while the intensity of the traffic offered by one free source is α erl.

5.3.6 *Loss Probability*

The loss probability of calls in the Engset model is, according to the definition, equal to the ratio of the lost calls stream (offered to the group during the blocking state) to the offered calls stream. Notice that in state i the number of serviced sources is i and the remaining ones, in other words free traffic sources, generate a call stream with the intensity $(N - i)\gamma$. In state V this stream is lost. Thus, the loss probability can be determined by the following equation:

$$B(\gamma, V, N) = \frac{(N - V)\gamma\,[p_V]_V}{\sum_{j=0}^{V}(N - j)\gamma\,[p_j]_V} \tag{5.23}$$

After the substitution of appropriate values of the probabilities determined with Equations (5.21) into (5.23), we obtain:

$$B(\alpha, V, N) = \binom{N}{V}\alpha^V(N - V)\gamma \Big/ \sum_{j=0}^{V}\binom{N}{j}\alpha^j(N - j)\gamma$$

Taking into consideration the equality of the combinatorial expressions:

$$\binom{N}{V}(N-V) = N\binom{N-1}{V}$$

the loss probability can be written in the following form:

$$B(\alpha, V, N) = \binom{N-1}{V}\alpha^V \Bigg/ \sum_{j=0}^{V}\binom{N-1}{j}\alpha^j \tag{5.24}$$

Equation (5.24) expresses the loss probability, often denoted with the symbol $B(\alpha, V, N)$, which means: this is the loss probability in the full-availability group with the capacity of V channels, which is offered traffic from N traffic sources ($N > V$), while the intensity of traffic offered by one free source is α erl.

The result (5.24) can be interpreted in the following way: the loss probability in the group that is offered traffic from N sources is equal to the blocking probability in the group which is offered traffic from $N - 1$ sources:

$$B(\alpha, V, N) = E(\alpha, V, N-1) \tag{5.25}$$

Equations (5.22) and (5.24) are called the Engset formulas. From Equation (5.25) it is possible to show that in the Engset model – as opposed to the Erlang model – the blocking probability is always higher than the loss probability. In the Erlang model, those two probabilities take on the same values (Section 5.2).

5.3.7 Alternative Notation of the Engset Formula

Frequently, in the Engset formula (5.24), instead of the parameter α determining traffic offered by one free traffic source, the following expression is used $a/(1-a)$:

$$B = B(a, V, N) = \binom{N-1}{V}\left(\frac{a}{1-a}\right)^V \Bigg/ \sum_{j=0}^{V}\binom{N-1}{j}\left(\frac{a}{1-a}\right)^j \tag{5.26}$$

In order to interpret the parameter a, let us compare α with the expression $a/(1-a)$. If, at the same time, we include the definition (5.20), then we obtain:

$$a = \frac{\alpha}{1+\alpha} = \frac{(1/\mu)}{(1/\gamma)+(1/\mu)} \tag{5.27}$$

Equation (5.27) expresses the ratio of the average service time of a source to the sum of the average occupancy time and the mean time between the moment of releasing the source and the moment of commencing the occupancy associated with a generation of a next call by the source. Thus, the parameter a can be interpreted as the average traffic offered by one source. Note that the parameter α denotes traffic offered by one free source.

5.3.8 Relationship between the Erlang and Engset Distributions

While analyzing Equation (5.21) it can be observed that it is a generalization of the Erlang formula (5.2). Let us assume that the number of traffic sources N tends to infinity, while the parameter γ decreases in such a way that the product $N\gamma$ remains constant:

$$\lim_{N\to\infty} N\gamma = \lambda \tag{5.28}$$

With the above assumptions, for any arbitrarily chosen i, we have:

$$\lim_{N\to\infty} \binom{N}{i}\left(\frac{\gamma}{\mu}\right)^i = \lim_{N\to\infty} \frac{N(N-1)\ldots(N-i+1)}{i!}\left(\frac{\gamma}{\mu}\right)^i = \frac{1}{i!}\left(\frac{\lambda}{\mu}\right)^i \tag{5.29}$$

It follows that for $N \to \infty$, the Engset distribution becomes the Erlang distribution.

5.3.9 *Occupancy Probability of x Precisely Determined Channels in a Group*

Given the assumption that the occupancy distribution of a group is the Engset distribution, we can estimate, as in the case of the Erlang model (Section 5.2.7), the occupancy probability of precisely determined channels in the full-availability group. Let us assume that in a group with the capacity V, x channels have been determined. The occupancy probability of x determined channels in the full-availability group that is offered a stream from a finite number of sources will be determined by Equation (5.6), in which the conditional occupancy probability of precisely determined channels is independent of the occupancy distribution of the group and is expressed by Equation (5.7). The occupancy probability of i of any arbitrarily chosen links will be determined on the basis of the Engset distribution (5.22). After introducing all appropriate substitutions and elementary transformations, we obtain:

$$H(x) = \frac{E(\alpha, V, N)}{E(\alpha, V, N-x)} \tag{5.30}$$

Equation (5.30) is the equivalent of the Palm-Jacobaeus formula for the full-availability group, which is offered a call stream generated by a finite number of traffic sources. The equation was published for the first time in [7]. At present, it is often used to approximate occupancy distributions of outgoing groups in switching networks that are offered a call stream from a finite number of traffic sources, for example, in [8–10].

5.3.10 *Traffic Carried by the Full-Availability Group*

The intensity of traffic carried will be determined in keeping with Definition 1 (Section 4.5.3), which states that the traffic intensity is equal to the average number of simultaneously occupied channels. Hence, we have:

$$Y_V = \sum_{i=1}^{V} i\left[p_i\right]_V = \frac{\alpha N[1 - B(\alpha, V, N)]}{1 + \alpha N[1 - B(\alpha, V, N)]} \tag{5.31}$$

5.3.11 *Recursive Properties of the Engset Formula*

A direct determination of blocking probability in the group with the help of the Engset formula (5.21) is complex due to the occurrence of the factorial (taking on very large values for large capacities of groups). However, the Engset formula can be easily presented in a recursive form, which makes it easily programmable:

$$E(\alpha, V, N) = \frac{\alpha(N - V + 1)E(\alpha, V, N)}{V + \alpha(N - V + 1)E(\alpha, V, N)} \tag{5.32}$$

where the initial value $E(\alpha, 0, N) = 1$, since $\alpha^0 = 1$ and $\binom{N}{0} = 1$ (cf. Equation (5.22)) .

5.3.12 *Commentary to the Average Traffic Intensities*

The intensity of traffic carried in the Engset model can be determined on the basis of the definition, according to which the traffic intensity is equal to the average number of simultaneously occupied channels in the group with the capacity of V channels:

$$Y_V = \sum_{i=1}^{V} i\,[P_i]_V = \frac{\alpha N[1 - B(\alpha, V, N)]}{1 + \alpha[1 - B(\alpha, V, N)]} \tag{5.33}$$

where $[P_i]_V$ is the Engset distribution (Section 5.3.5, Equation (5.19)).

The intensity of traffic carried in the Engset model can also be written in another, more convenient, form, namely:

$$\begin{aligned} Y_V &= \sum_{i=1}^{V} i\,[P_i]_V = \sum_{i=1}^{V} \frac{\gamma}{\mu}(N - i + 1)\left[P_{i-1}\right]_V = \sum_{i=1}^{V-1} \alpha(N - i)\,[P_i]_V = \\ &= \left\{\sum_{i=1}^{V} \alpha(N - i)\,[P_i]_V\right\} - \alpha(N - V)\,[P_V]_V = \\ &= \alpha(N - Y_V) - \alpha(N - V)E(\alpha, V, N) \end{aligned} \tag{5.34}$$

By solving Equation (5.34) in relation to Y_V, we obtain the following expression – equivalent to Equation (5.33) – for the intensity of carried traffic:

$$Y_V = \frac{\alpha}{1 + \alpha}\,[N - (N - V)E(\alpha, V, N)] \tag{5.35}$$

The average intensity of offered traffic is defined as the average number of occupied channels in a group with the capacity of N channels:

$$A = Y_N = \sum_{i=1}^{N} i\,[P_i]_N = N\frac{\alpha}{1 + \alpha} = Na \tag{5.36}$$

where a is the intensity of traffic offered by *one source*, while α is the intensity of traffic offered by *one free source*:

$$a = \frac{\alpha}{1 + \alpha} \tag{5.37}$$

Knowing the average values of the intensity of offered traffic (Equation (5.36)) and carried traffic (Equation (5.35)), we are in position to determine the intensity of lost traffic:

$$R = A - Y_V = \frac{\alpha}{1 + \alpha}(N - V)E(\alpha, V, N) \tag{5.38}$$

With some issues in traffic theory, the probability of traffic losses is used. This probability is defined as the ratio of lost traffic to offered traffic. In the Engset model, the probability of traffic losses – on the basis of Equations (5.36) and (5.38) – is determined by the equation:

$$C = C(\alpha, V, N) = \frac{N - V}{N} E(\alpha, V, N) \tag{5.39}$$

Let us determine, now, a call stream parameter averaged after the states for a finite number of sources. The stream parameter – in this particular case – expresses the average number of calls falling into a time unit, in other words the average call intensity:

$$\Lambda = \sum_{i=1}^{V} \gamma(N - i)\,[P_i]_V = \gamma(N - Y) \tag{5.40}$$

The average call intensity, resulting from the estimated value of the intensity of offered traffic on the basis of Equation (5.36), is:

$$\Lambda_A = A\mu = N \frac{\gamma}{1 + \alpha} \tag{5.41}$$

Let us now interpret the difference $\Delta = \Lambda - \Lambda_A$. From Equations (5.40) and (5.41) we obtain:

$$\Delta = \Lambda - \Lambda_A = \frac{\gamma}{1 + \alpha}\,[N\alpha - Y(1 + \alpha)] \tag{5.42}$$

Substituting Equation (5.35) for the parameter Y in Equation (5.42), we eventually obtain:

$$\Delta = \gamma \frac{\alpha N}{1 + \alpha} \left[\frac{N - V}{N} E(\alpha, V, N) \right] = \gamma A C \tag{5.43}$$

The difference (5.43) is greater than zero and its occurrence results from the following reasoning. If a given traffic source cannot be serviced due to blocking in a group, then it immediately becomes free and can, along with other free sources, generate new calls. This means that the call stream parameter will increase by the value $\Delta = \gamma AC$. Note that the product AC determines the intensity of lost traffic, in other words the average number of sources that will become free as the result of blocking, while γ is the call intensity of one free source. If we were to adopt the hypothesis that each blocked source during the mean service time $1/\mu$ will not be serviced, then the increase in call intensities by the value Δ will not occur either, and the equality $\Lambda = \Lambda_A$ will follow. Summing up, we can state that Λ models the average call stream intensity, with the assumption that each call lost due to blocking immediately releases the source that generated the particular call. The parameter Λ_A models, in turn, the average call stream intensity, under the assumption that each lost call due to blocking will block the source that generated the call, for the time of its hypothetical service.

The above example shows how due care is required with any interpretation of the averaged traffic characteristics in the Engset model.

5.4 Comments

The analytical models presented in this chapter can be used for modeling and dimensioning radio interfaces in single-service networks that are characterized by so-called hard capacity (for example, a GSM network servicing speech traffic). Moreover, the Engset model makes it possible to take into consideration, in traffic calculations, a finite number of subscribers serviced by a given cell or a set of cells. Such a possibility is particularly important for cells servicing areas with a relatively low density of population, or microcells and indoors installations. The application of the selected models for modeling of the radio interface is presented in Chapter 11.

References

[1] Brockmeyer, E., Halstrom, H.L., and Jensen, A. (1960) The life and works of A. K. Erlang. *Acta Polytechnika Scandinavia*, **6** (287), 138–55.

[2] Erlang, A.K. (1917) Solution of some problems in the theory of probabilities of significance in automatic telephone exchanges. *Elektrotechnikeren*, **13**, 5.

[3] Jacobaeus, C. (1950) A study on congestion in link-systems. *Ericsson Technics*, **48**, 1–68.

[4] Palm, C. (1943) Nagra foljdsatser urde Erlang'ska formlerna. *Tekniska meddelanden frän Kungliga Telgrafstyrelsen*, **1–3**, 6–8.

[5] Shneps, M.A. (1979) *Sistemy raspredelenija informacii*. Metody raschyota. Radio i svjaz', Moscow.

[6] Engset, T. (1918) Die Wahrscheinlichkeitsrechnung zur Bestimmung der Wählerzahl in automatischen Fernsprechämtern. *Elektrotechnische Zeitschrift (ETZ)*, **31**: 304–8.

[7] Elldin, A. (1955) Automatic Telephone Exchanges with Crossbar Switches (Switch Calculations). *Technical Report B 11265*, LM Ericsson, Stockholm.

[8] Bazlen, D., Kampe, G., and Lotze, A. (1973) *On the Influence of Hunting Mode and Link Wiring on the Loss of Link Systems*. Proceedings of 7th International Teletraffic Congress, Stockholm, 1973 International Teletraffic Congress, Stockholm.

[9] Lotze, A. (1963) Bericht uber Verkehrtheoretische Untersuchungen CIRB. *Technical report, Inst. für Nachrichten-Vermittlung und Datenverarbeitung der Technischen Hochschule*, University of Stuttgart.

[10] Lotze, A. (1967) *History and Development of Grading Theory*. Proceedings of 5th International Teletraffic Congress, New York. International Teletraffic Congress, New York.

6

Modeling of Systems with Single-Rate Overflow Traffic

6.1 Introduction

This chapter presents the basics of the analysis of telecommunication systems employing the strategy of redirecting traffic via alternative routes – systems with traffic overflow. Modeling telecommunication networks with alternative routes is a complex matter and requires two basic problems to be solved, namely: determination of characteristics of traffic that overflows from direct groups, and determination of the number of links in alternative groups, where the loss coefficient will not exceed the assigned value. The chapter presents the Riordan method for a determination of the parameters of overflow traffic and then, the equivalent random traffic method and the Hayward method for dimensioning systems that are offered overflow traffic.

6.2 Basic Information on Overflow Systems

6.2.1 Simplified Classification of Groups in Telecommunication Networks

In the hierarchical telecommunication network, two types of groups are used:

- primary groups;
- overflow groups (alternative groups).

In primary groups, traffic is carried with congestion greater than the assumed loss probability. Traffic that is not carried in the group overflows to an alternative group. In these groups the congestion is not higher that the assumed loss probability. Traffic that is not carried in the alternative group is lost.

6.2.2 Alternative Paths

The basic problem in the analysis of network with alternative traffic routing is the proper determination of the number of links in alternative groups. This problem is closely associated with the fact that the overflow traffic stream is not described by the Erlang distribution. Let us

Modeling and Dimensioning of Mobile Networks: From GSM to LTE
Maciej Stasiak, Mariusz Głąbowski, Arkadiusz Wiśniewski and Piotr Zwierzykowski

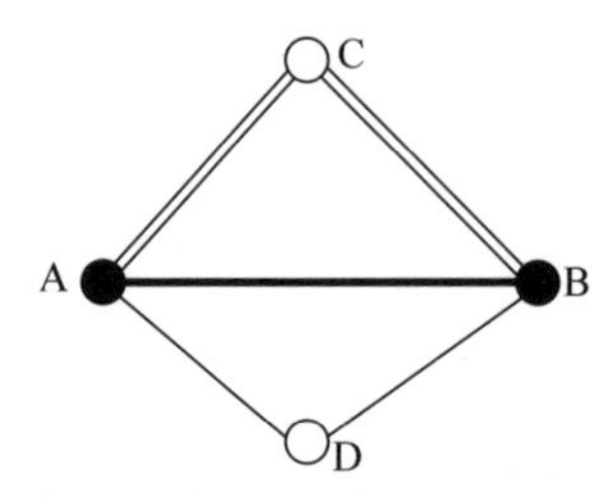

Figure 6.1 Alternative paths for the AB group.

consider possibilities of setting up a connection in a telecommunication network consisting of four switching nodes: A, B, C, D (Figure 6.1). Setting up connections between nodes A and B can be carried out by the links of the direct group AB. If all links of the direct group happen to be occupied, then the ACB alternative path can be used (first alternative path – double line in Figure 6.1) by the occupation of one link in the AC and in the CB group. If in at least one of the considered groups (AC, CB) all links are occupied, the second alternative path, ADB, can be used, and when the latter is also fully occupied the third alternative path ADCB is chosen. The sequence of choosing the alternative paths is written in routing tables of network nodes.

6.2.3 Overflow Traffic

Most traffic offered to the subscribers of switching node B from the subscribers of switching node A, in other words traffic A_{AB}, will be carried by the links of the group AB (Figure 6.1). Some of this traffic, which cannot be serviced by the links of the group AB, will be offered to groups AC and CB, which form the first alternative path. It is just this part of traffic A_{AB} that is called overflow (alternative) traffic and is denoted by the symbol R_{AB}. If offered traffic is of the PCT1 type, then traffic that overflows from this group will be of a totally different nature.

Overflow traffic can appear only in the occupancy time of all links of the group AB. This means that the overflow stream is more "concentrated" in certain time intervals; in other words it is characterized by "peakedness," as compared with PCT1 traffic. The call stream that overflows therefore cannot be approximated by the Poisson distribution. If identical values of offered traffic and congestion are assumed, then a greater number of links is required for servicing overflow traffic than for servicing PCT1 traffic. Figure 6.2 shows a typical traffic diagram during a peak hour for a full-availability primary group. The figure shows respectively distributions of offered, carried, and overflow traffic.

The following parameters can be used for the statistical evaluation of the overflow traffic stream: the mean value R of overflow traffic and the variance σ^2 of overflow traffic. With the help of those two parameters it is possible to determine "unevenness" of the overflow stream by introducing the concept of the peakedness coefficient Z, which is equal to the ratio of the variance σ^2 to the mean value R of overflow traffic:

$$Z = \sigma^2/R \tag{6.1}$$

The "unevenness" of the overflow stream can also be evaluated by using the coefficient D, which is equal to the difference between the variance and the mean value of overflow traffic:

$$D = \sigma^2 - R \tag{6.2}$$

The values of the parameters D and Z are presented in Table 6.1.

Table 6.1 Types of traffic in the telecommunication network

Parameter	Offered traffic	Carried traffic	Overflow traffic
Z	$Z = 1$	$Z < 1$	$Z > 1$
D	$D = 0$	$D < 0$	$D > 1$

This table presents the values of the parameters Z and D for different types of telecommunication traffic.

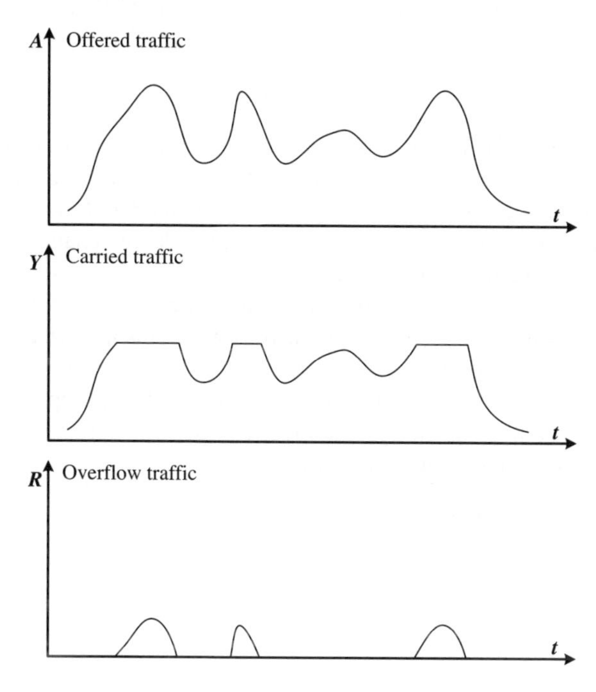

Figure 6.2 Types of traffic in the overflow system.

The service process in the overflow system can thus be characterized by four parameters: A, V, R, σ^2 (σ^2 can be replaced by Z or D). The stream offered to the group is here determined by one parameter A – the mean value of offered traffic, whereas the overflow traffic stream by two: the mean value of overflow traffic R and its variance σ^2.

6.3 Models of Alternative Groups

In this section we will discuss the problems involved in modeling groups that are offered traffic which overflows from primary groups – in other words overflow traffic.

6.3.1 Analytic Model of the System with Overflow Traffic

In Section 6.2.3 we show that traffic that overflows from the primary group, which is offered PCT1 traffic, can be characterized with the help of the following two parameters: the mean

limited primary group $\xrightarrow{A}$ 1 … V $\xrightarrow{R, \sigma^2}$ unlimited alternative group 1 … ∞

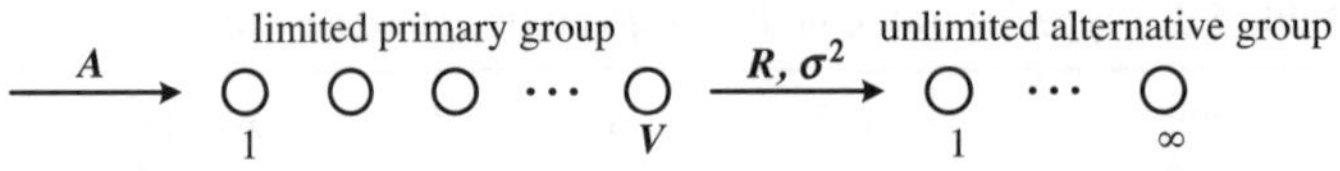

Figure 6.3 Model of the system with overflow traffic.

value of overflow traffic R and its variance σ^2 (or the coefficients Z or D). In order to evaluate analytically these parameters we will consider the following model: a full-availability group with the capacity of V channels (the primary group) is offered PCT1 traffic with the mean intensity A:

$$A = \lambda/\mu \tag{6.3}$$

Further we assume that traffic that is not carried because of the occupancy of all the channels of the considered group overflows to a full-availability group with an unlimited number of channels (an unlimited overflow group) – Figure 6.3. The model of the overflow system presented here is used to determine the average number of occupied channels R in the overflow group – mean value of overflow traffic, and its variance σ^2.

The model of the system with overflow traffic, presented in Figure 6.3 and composed of two full-availability groups, is determined by the two-dimensional service process: $\{i(t),\ j(t)\}$, where $i(t)$ is the number of occupied channels in the primary group at the point of time t, whereas $j(t)$ is the number of occupied channels in the overflow group at the point of time t. The steady-state probabilities of the system under consideration are denoted by the symbols $\left[p_{i,j}\right]_{V,\infty}$ and are defined in the following way:

$$\left[p_{i,j}\right]_{V,\infty} = \lim_{t\to\infty} P\left\{i(t) = i,\ j(t) = j\right\} \tag{6.4}$$

where $(0 \leqslant i \leqslant V)$ and $(0 \leqslant j \leqslant \infty)$.

6.3.2 State Equations

The probabilities $\left[p_{i,j}\right]_{V,\infty}$ can be determined on the basis of a system of state equations that, for the considered process, takes the following form:

$$\begin{gathered}\ldots,\\ -(\lambda + j\mu)\left[p_{0,j}\right]_{V,\infty} + \mu\left[p_{1,j}\right]_{V,\infty} + (j+1)\mu\left[p_{0,j+1}\right]_{V,\infty} = 0\end{gathered} \tag{6.5}$$

$$\begin{gathered}\ldots,\\ -(\lambda + i\mu + j\mu)\left[p_{i,j}\right]_{V,\infty} + \lambda\left[p_{i-1,j}\right]_{V,\infty}\\ +(i+1)\mu\left[p_{i+1,j}\right]_{V,\infty} + (j+1)\mu\left[p_{i,j+1}\right]_{V,\infty} = 0\end{gathered} \tag{6.6}$$

$$\begin{gathered}\ldots,\\ -(\lambda + V\mu + j\mu)\left[p_{V,j}\right]_{V,\infty} + \lambda\left[p_{V-1,j}\right]_{V,\infty}\\ +\lambda\left[p_{V,j-1}\right]_{V,\infty} + (j+1)\mu\left[p_{V,j+1}\right]_{V,\infty} = 0\end{gathered} \tag{6.7}$$

$$\begin{gathered}\ldots,\\ \sum_{j=0}^{\infty}\sum_{i=0}^{V}\left[p_{i,j}\right]_{V,\infty} = 1\end{gathered} \tag{6.8}$$

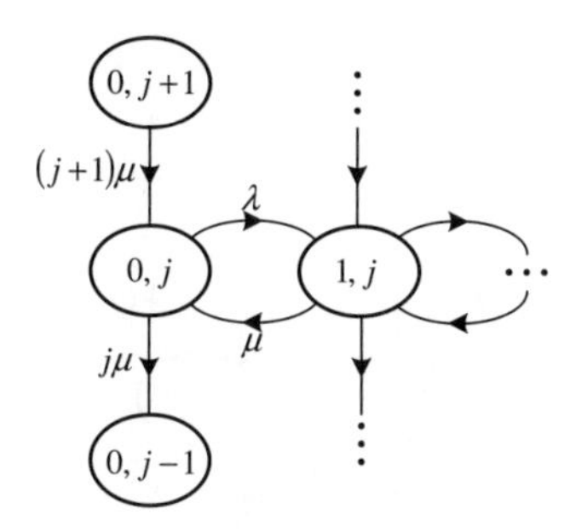

Figure 6.4 A fragment of the diagram of the Markov process in the system with overflow traffic (illustration of Equation (6.5)).

The system of Equations (6.5)–(6.8) is introduced according to a standard procedure based on the analysis of the diagram of the process under consideration, following the rule that for each state the sum of the intensity of outgoing streams is equal to the sum of the intensity of incoming streams. Appropriate fragments of the diagram corresponding to Equations (6.5)–(6.7), are presented in Figures 6.4–6.6.

Figure 6.4 shows available transitions of the system in the neighborhood of state $(0, j)$, in other words when the primary group does not service any calls while j calls are serviced in the overflow group. It is noticeable that in this state new calls are directed exclusively to the primary group (calls can be redirected to the overflow group only in the case of the occupancy of all channels of the primary group), whereas in the overflow group only disconnections can appear. This diagram corresponds to Equation (6.5).

Figure 6.5 illustrates available transitions of the system in neighborhood of state (i, j) for $i < V$, in other words when the primary group services i calls, while the overflow group services j calls. In this state, new calls are directed exclusively to the primary group that

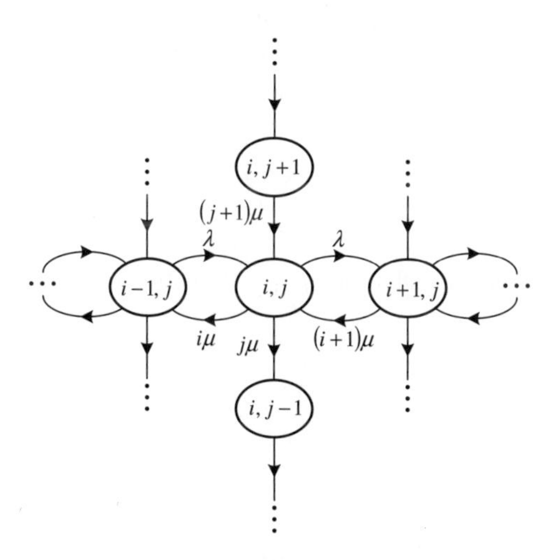

Figure 6.5 A fragment of the diagram of the Markov process in the system with overflow traffic (illustration of Equation (6.6)).

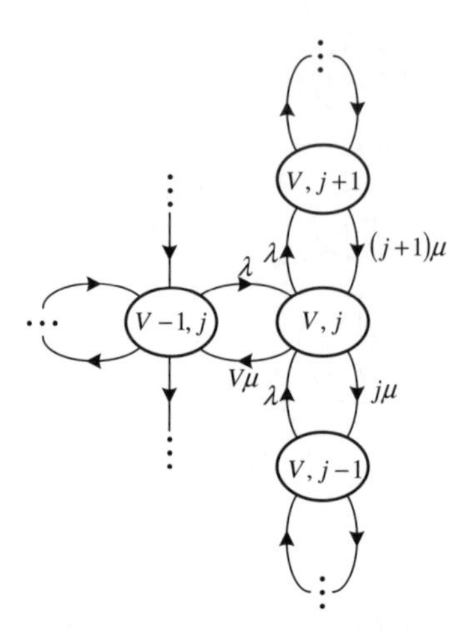

Figure 6.6 A fragment of the diagram of the Markov process in the system with overflow traffic (illustration of Equation (6.7)).

still has free channels. In the overflow group there are only disconnections. Figure 6.5 also illustrates Equation (6.6).

Figure 6.6 shows, in turn, the available transitions in the neighborhood of state (V, j), in other words when the primary group is totally occupied, whereas the overflow group services j calls. In this state all new calls will be directed to an overflow group. The diagram presented in Figure 6.6 corresponds to Equation (6.7).

6.3.3 Determination of Overflow Traffic Parameters – Riordan Formulas

Having obtained the solutions to Equations (6.5)–(6.8) we can determine all essential characteristics of the system with traffic overflow. However, the determination of the parameters R and σ^2, related to the overflow group with unlimited capacity, can be simplified. Note that to determine parameters R and σ^2 it is not necessary to know all probabilities $\left[p_{i,j}\right]_{V,\infty}$, but it is sufficient to know only those probabilities $\left[p_j\right]_\infty$ that relate to the overflow group, namely:

$$\left[p_j\right]_\infty = \sum_{i=0}^{V} \left[p_{i,j}\right]_{V,\infty} \tag{6.9}$$

On the basis of the occupancy distribution $\left[p_j\right]_\infty$ (Equation (6.9)), it is possible to determine the parameters R and σ^2:

$$R = \sum_{j=0}^{\infty} j \left[p_j\right]_\infty \tag{6.10}$$

$$\sigma^2 = \sum_{j=0}^{\infty} j^2 \left[p_j\right]_\infty - R^2 \tag{6.11}$$

Let us introduce the following characteristic function $\Psi(s)$:

$$\Psi(s) = \sum_{j=0}^{\infty} [p_j]_{\infty} s^j \tag{6.12}$$

Considering Equation (6.12), Equation (6.8) for distribution $[p_j]_{\infty}$ can be expressed in the following form:

$$\Psi(1) = \sum_{j=0}^{\infty} [p_j]_{\infty} = 1 \tag{6.13}$$

The series (6.12) is thus convergent for $s = 1$ and can be differentiated. For the differential of the first and second order of the function $\Psi(s)$, we obtain the following expressions:

$$\frac{\mathrm{d}\Psi(s)}{\mathrm{d}s} = \sum_{j=0}^{\infty} j \, [p_j]_{\infty} s^{j-1} \tag{6.14}$$

$$\frac{\mathrm{d}\Psi^2(s)}{\mathrm{d}s^2} = \sum_{j=0}^{\infty} (j^2 - j) \, [p_j]_{\infty} s^{j-2} \tag{6.15}$$

Comparing Equations (6.14) and (6.15) with the definitions (6.10) and (6.11), we can write:

$$R = \sum_{j=0}^{\infty} j \, [p_j]_{\infty} = \frac{\mathrm{d}\Psi(1)}{\mathrm{d}s} \tag{6.16}$$

$$\sigma^2 = \sum_{j=0}^{\infty} j^2 \, [p_j]_{\infty} - R^2 = \frac{\mathrm{d}\Psi^2(s)}{\mathrm{d}s^2} + R - R^2 \tag{6.17}$$

It was proved [1–3] that, in this case, the second-order differential for $s = 1$ takes the following form:

$$\frac{\mathrm{d}\Psi^2(s)}{\mathrm{d}s^2} = \frac{RA}{V + 1 - A + R} \tag{6.18}$$

Substituting Equation (6.18) into Equation (6.17) we obtain the final result derived by Riordan [3]:

$$R = AE_V(A) \tag{6.19}$$

$$\sigma^2 = R\left(\frac{A}{V + 1 - A + R} + 1 - R\right) \tag{6.20}$$

6.3.4 Comment on Riordan Formulas

In hierarchical telecommunication networks, alternative groups usually carry traffic that overflows from several primary groups. To determine the parameters of the total overflow traffic offered to the alternative path let us assume that the PCT1 traffic streams carried by primary groups are statistically mutually independent. Consequently, traffic streams that overflow from these groups are also independent and the parameters of the total traffic offered to the alternative groups can be determined by the following equations:

$$R = \sum_{i=0}^{k} R_i \tag{6.21}$$

$$\sigma^2 = \sum_{i=0}^{k} \sigma_i^2 \tag{6.22}$$

where R_i = mean value of overflow traffic from i-th group, σ_i^2 = variance of overflow traffic from i-th group.

6.4 Equivalent Groups

From Equations (6.19) and (6.20), on the basis of the parameters A and V we can determine unequivocally the parameters of overflow traffic R and σ^2. At the same time, the equations can be used to solve an "inverse problem" – to determine the parameters A and V of traffic offered to the primary group on the basis of the parameters R and σ^2 of the overflow traffic. The possibility of determination of the parameters of offered traffic on the basis of overflow traffic underlies the idea for the equivalent random traffic (ERT) method, proposed by Wilkinson [3] and Bretschneider [2].

The ERT method is based on determining the fictitious equivalent full-availability group with the equivalent capacity of V^* channels carrying equivalent PCT1 traffic with a mean value A^*. The determined mean value A^* of equivalent PCT1 traffic offered to the fictitious group (with the capacity of V^*) should cause an overflow of traffic with a given mean value R and variance σ^2. According to such an approach, the system with many primary groups initially described by the pairs of parameters A_i and V_i can be simplified to a system with a single fictitious group described by one pair of parameters only (A^*, V^*). The parameters of this equivalent group can be used to determine the number N of required channels for service calls with assumed blocking probability E. Therefore, overflow traffic requires $V_{\text{alt}} = (N - V^*)$ channels in the alternative group. The following sections present the ERT method in the form of the most frequently used calculational algorithm.

6.4.1 Formulation of the Problem of Dimensioning Alternative Groups

Let us formulate the problem in the following way: it is necessary to determine the number of channels V_{alt} in the alternative full-availability group (Figure 6.7) if this group is offered k overflow streams. These streams overflow from k full-availability groups, determined by the pairs (A_i, V_i), where ($1 \leqslant i \leqslant k$). The blocking probability in this system should not exceed a given value E.

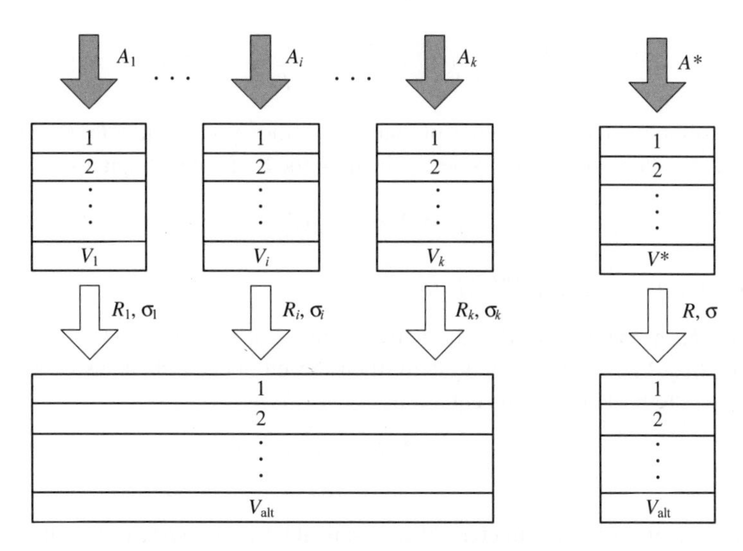

Figure 6.7 Overflow diagram in the ERT method.

6.4.2 Equivalent Random Traffic (ERT) Method

The solution to the problem stated in the previous section can be obtained on the basis of the ERT method, which is most frequently presented in the form of the following algorithm:

- The mean value and the variance of each of k traffic streams that overflow to the alternative group is determined on the basis of Equations (6.19)–(6.20):

$$R_i = A_i E_{V_i}(A_i) \tag{6.23}$$

$$\sigma_i^2 = R_i \left(\frac{A_i}{V_i + 1 - A_i + R_i} + 1 - R_i \right) \tag{6.24}$$

- Assuming statistical independence of overflow streams, the parameters of the total stream that overflows to the considered alternative group are determined:

$$R = \sum_{i=1}^{k} R_i \tag{6.25}$$

$$\sigma^2 = \sum_{i=1}^{k} \sigma_i^2 \tag{6.26}$$

- On the basis of the values obtained for R and σ^2, the parameters A^* i V^* of the equivalent group are then determined. These parameters can be determined by providing solution to the Riordan system of equations:

$$R = A^* E_{V^*}(A^*) \tag{6.27}$$

$$\sigma^2 = R\left(\frac{A^*}{V^* + 1 - A^* + R} + 1 - R\right) \tag{6.28}$$

- With the assumed value of blocking probability in the system equal to E, the number of channels of the alternative group is determined on the basis of the equation:

$$E = E_{(V^* + V_{\text{alt}})}(A^*) \tag{6.29}$$

6.4.3 Comments on the ERT Method

- If the blocking probability in the alternative group E_{alt} is given, instead of the blocking probability E of the whole of the system, then the number of channels of the alternative group V_{alt} is determined from the equation:

$$E_{\text{alt}} = \frac{A^* E_{(V^* + V_{\text{alt}})}(A^*)}{R} \tag{6.30}$$

- The determination of the parameters of the equivalent group (A^*, V^*) requires the application of complex, iterative algorithms [4]. Consequently, to simplify the calculations, an approximate solution of the system of equations (6.27) and (6.28), proposed by G. Rapp [5], can be used:

$$A^* = \sigma^2 + 3\frac{\sigma^2}{R}\left(\frac{\sigma^2}{R} - 1\right) \tag{6.31}$$

$$V^* = A^* \frac{(R^2 + \sigma^2)}{R^2 + \sigma^2 - R} - R - 1 \tag{6.32}$$

- In most cases groups of telecommunication networks carry both their own traffic (PCT1) and overflow traffic. The number of channels in such groups is then determined according to the ERT method presented above, treating them as alternative groups. In this case, the variance and the mean values of the PCT1 traffic are equal.
  Let us assume that a given alternative group is offered its own traffic PCT1 with the mean value of A_2 and the overflow traffic from a full-availability group (A_1, V_1) with the parameters R_1 and σ_1^2. It results from the ERT method that the parameters for the total traffic that overflows to the alternative group will be equal:

$$R = R_1 + A_2 \tag{6.33}$$

$$\sigma^2 = \sigma_1^2 + A_2 \tag{6.34}$$

  Further calculations will be carried out in keeping with the algorithm presented in Section 6.4.2.

6.4.4 Overflow Group Decomposition Scheme

Let us consider a full-availability group with the capacity of V channels, which is offered overflow traffic with the mean value R and variance σ^2. The peakedness of offered traffic is

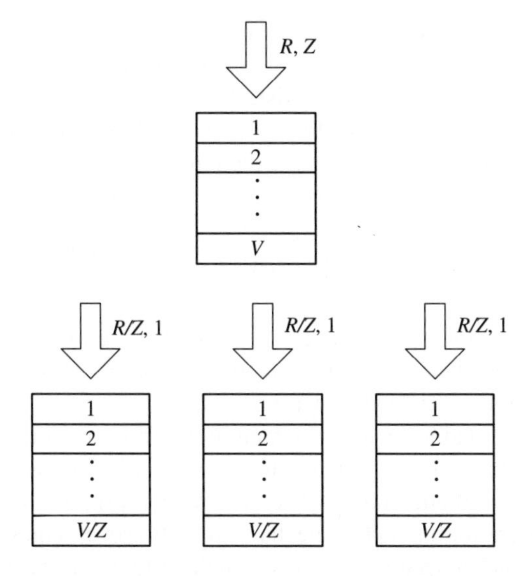

Figure 6.8 Decomposition of the (V, R, Z) system into Z subsystems $(V/Z, R/Z, 1)$.

then (Equation (3.1)):

$$Z = \sigma^2/R \tag{6.35}$$

Let us transform the overflow group in the following way (Figure 6.8):

- The group is divided into Z identical full-availability groups (subsystems), each one with the capacity:

$$V_e = V/Z \tag{6.36}$$

- Each group is then offered traffic with the mean value:

$$R_e = R/Z \tag{6.37}$$

- Variance of the traffic R_e is then determined by the formula:

$$\sigma_e^2 = \left(\frac{1}{Z}\right)^2 \sigma^2 \tag{6.38}$$

Equation (6.38) results from a property of the variance, the variance of the random variable multiplied by a constant value is equal to a squared constant value multiplied by the variance of the random variable. Now we can determine the peakedness coefficient of traffic offered to any individual subsystem. Taking into account Equations (6.35), (6.37) and (6.38), we obtain:

$$Z_e = \frac{\sigma_e^2}{R_e} = \frac{\sigma^2}{RZ} = 1 \tag{6.39}$$

The peakedness coefficient equal to one means that traffic R_e is PCT1 traffic. Thus, we have decomposed the full-availability group – described by the parameters (R, V, Z) – into Z subsystems (full-availability groups) – described by the parameters $(R/Z, V/Z, 1)$ – that are offered PCT1 traffic. All groups are identical, so the blocking probability in all groups will also be identical. In Hayward work [6] it is assumed that the blocking probability in the group $(R/Z, V/Z, 1)$ will be the same as in the overflow group (R, V, Z). Therefore, we can write:

$$E(R, V, Z) \approx E(R/Z, V/Z, 1) \approx E_{\frac{V}{Z}}\left(\frac{R}{Z}\right) \tag{6.40}$$

Equation (6.40) is a modification of the Erlang formula (Section 5.2.5) that takes into consideration the non-Poisson character of the call stream offered to the group. In traffic theory, this formula is called the Fredericks–Hayward formula.

The interpretation of Equation (6.40) presented in the above reasoning assumes mutual independence of traffic offered to subsystems. In fact, a distribution of the traffic stream into several identical streams without an application of an appropriate call-allocation mechanism is not possible. The operation of such a mechanism is, however, equivalent to the introduction of mutual correlation between the streams, which, in turn, can be interpreted as a lack of independence of the traffic streams offered to the subsystems. This phenomenon makes Equation (6.40) to be an approximate formula. It should be stressed, though, that it is characterized by high accuracy [6–8].

Note that in the present section we consider the division of the traffic stream and not that of the call stream, since it is the traffic stream that takes into consideration not only the characteristics of the call stream but also the characteristics of the service stream. For instance, in the stream offered to the subsystem, the service time will have an influence on new calls offered to this subsystem: a "long" service time of a call in a given subsystem will result in a lower number of new calls offered to this subsystem within the successive time intervals. This means that the mechanisms of call allocations to appropriate subsystems require – in order to retain the same parameters of the traffic offered to subsystems – the service time of individual calls to be taken into consideration. Indeed, this phenomenon is responsible for the mutual correlation of traffic streams offered to subsystems.

6.4.5 Fredericks–Hayward Method

Equation (6.40) forms the basis of the Fredericks–Hayward method [6], which can be described in the form of the following algorithm:

- The mean value and the variance of each of k traffic streams that overflows to an alternative group can be determined based on Equations (6.8) and (6.20):

$$R_i = A_i E_{V_i}(A_i) \tag{6.41}$$

$$\sigma_i^2 = R_i\left(\frac{A_i}{V_i + 1 - A_i + R_i} + 1 - R_i\right) \tag{6.42}$$

- Assuming statistical independence of streams that overflow, the parameters R and σ^2 of the total overflow traffic to the alternative group and the peakedness of the overflow traffic are determined:

$$R = \sum_{i=1}^{k} R_i \tag{6.43}$$

$$\sigma^2 = \sum_{i=1}^{k} \sigma_i^2 \tag{6.44}$$

$$Z = \sigma^2 / R \tag{6.45}$$

- With the assigned blocking probability in the system, equal to E, the number of channels of the alternative group is determined on the basis of the Fredericks–Hayward formula:

$$E = E_{\frac{V_{\text{alt}}}{Z}}\left(\frac{R}{Z}\right) \tag{6.46}$$

The Fredericks–Hayward method is far simpler than the ERT method because it requires only calculations based on the Erlang formula. The formula is used in two steps of the algorithm – with the determination of mean values of traffic that overflows to the alternative group (Equation (6.41)) and, in the form of Fredericks–Hayward formula, with the determination of the capacity of the alternative group (Equation (6.46)).

Both methods, Fredericks–Hayward and ERT, are characterized by high accuracy. The results obtained using Fredericks–Hayward method are very close to the results obtained using the ERT method.

6.5 Modeling of Overflow Traffic in Systems with Finite Number of Traffic Sources

This chapter is devoted to presenting an analytical method for determining the mean value and the variance in systems with traffic overflowing from primary groups servicing PCT2 traffic streams [9]. The basis of this method is the application of the ERT method to convert the traffic stream generated by a PCT2 traffic stream to the equivalent PCT1 traffic stream [9, 10].

In order to present the idea of the method let us consider a group with the capacity of V channels servicing N traffic sources. According to the consideration presented in Section 5.3, the arrival rate in the state of n occupied channels can be expressed by the following formula:

$$\lambda(n) = (N - n)\gamma \tag{6.47}$$

where n is a number of calls being serviced in state n and γ is the mean arrival rate generated by an idle source. In the model considered here we assume additionally that the service time for calls of particular classes has an exponential distribution. Thus, the intensity of traffic α offered by an idle source is equal to:

$$\alpha = \frac{\gamma}{\mu} \tag{6.48}$$

where $1/\mu$ is the mean service time of calls.

Based on the results presented in [9] and [10] we can determine the mean value R_{PCT2}, the variance σ^2_{PCT2} and the coefficient D_{PCT2} of the number of occupied channels in the group being considered:

$$R_{\mathrm{PCT2}} = \frac{N\alpha}{1+\alpha} \tag{6.49}$$

$$\sigma^2_{\mathrm{PCT2}} = \frac{N\alpha}{(1+\alpha)^2} \tag{6.50}$$

$$D_{\mathrm{PCT2}} = \sigma^2_{\mathrm{PCT2}} - R_{\mathrm{PCT2}} = -N\frac{\alpha^2}{(1+\alpha)^2} \tag{6.51}$$

The traffic described by Equations (6.49), (6.50) and (6.51) can be treated as an equivalent PCT1 stream with intensity A^* offered to the equivalent group with the capacity equal to V^* channels. The values of the fictitious parameters A^* and V^* can be obtained on the basis of Riordan formulas, according to the ERT method:

$$R_{\mathrm{PCT2}} = A^* E_{V^*}\left(A^*\right) \tag{6.52}$$

$$D_{\mathrm{PCT2}} = R_{\mathrm{PCT2}}\left[\frac{A^*}{V^*+1-A^*+R_{\mathrm{PCT2}}} - R_{\mathrm{PCT2}}\right] \tag{6.53}$$

Equations (6.52) and (6.53) are solved with the help of the following recurrent Erlang formula for negative values of link capacity [9, 11]:

$$E_{V-1}(A) = \frac{VE_V(A)}{A(1-E_V(A))} \tag{6.54}$$

where the initial solution, for $V=-1$, takes the following form:

$$E_{-1}(A) = [-Ei(-A)Ae^A]^{-1} \tag{6.55}$$

and function $Ei(A)$ is defined as follows:

$$Ei(x) = -\int_x^{\infty}(At+A)^{-1}e^{At+A}\mathrm{d}(At+A) \tag{6.56}$$

With the parameter of the fictitious PCT1 traffic A^* (equivalent to the original PCT2 traffic) we can calculate, on the basis of Equations (6.19) and (6.20), the parameters of the traffic overflowing from the primary group servicing originally PCT2 traffic streams, in other words the variance σ^2 and the mean value R. Subsequently, on the basis of the parameters of traffic overflowing from primary groups, we can dimension the overflow group with the help of the Hayward method or the ERT method.

6.6 Comments

The methods presented in this chapter can be applied to determine the intensity of overflow single-service traffic between cells (sectors) or different layers within one cellular network.

Such analyses are particularly useful for operators that have at their disposal frequencies in the bandwidth of, for example, 900 and 1800 MHz, or operators that separate cells into the micro and the macro layer. A determination of overflow traffic for such cases makes it possible to determine the appropriate capacity of cells servicing overflow traffic that will secure the quality requirements to be fulfilled by the cellular network – the blocking or loss probability. The model can be also used for optimization of the structure of an existing network through an indication of cells for, for example, traffic handover mechanisms with the appropriate capacity for the intensity of overflow traffic. An example of the application of the model for dimensioning of a single-service network is proposed in Chapter 13.

References

[1] Akimuru, H. and Kawashima, K. (1993) *Teletraffic: Theory and Application*, Springer.

[2] Bretschneider, G. (1956) Die Berechnung von Leitungsgruppen für berfließenden Verkehr in Fernsprechwählanlagen. *Nachrichtentechnische Zeitung (NTZ)*, **9**, 533–40.

[3] Wilkinson, R.I. (1956) Theories of toll traffic engineering in the USA. *Bell System Technical Journal*, **40**, 421–514.

[4] Ott, K.W. and de los Rios, F.J. (1968) Computation of urban trunking networks with alternate routing by computer. *Electrical Communication*, **43** (2), 157–62.

[5] Rapp, Y. (1964) *Planning of Junction Network in a Multi-exchange Area*. Proceedings of 4th International Teletraffic Congress, London.

[6] A. Fredericks. Congestion in blocking systems – a simple approximation technique. *Bell System Technical Journal*, **59** (6), 805–27.

[7] Holtzmann, J.M. (1973) *The Accuracy of the Equivalent Random Method with Renewal Inputs*. Proceedings of 7th International Teletraffic Congress, Stockholm. International Teletraffic Congress, Stockholm.

[8] Iversen, V.B. (ed.) (2001) *Teletraffic Engineering Handbook*. ITU-D, Study Group 2, Geneva, June 2001 (draft, published online by International Telecommunication Union), https://www.itu.int/ITU-D/study_groups/SGP_19982002/SG2/StudyQuestions/Question_16/RapporteursGroupDocs/teletraffic.pdf (accessed 19 July 2010).

[9] Bretschneider, G. (1973) *Extension of the Equivalent Random Method to Smooth Traffics*. Proceedings of 7th International Teletraffic Congress, Stockholm. International Teletraffic Congress, Stockholm.

[10] Šneps, M.A. (1979) *Sistemy raspredeleniâ informacii*. Metody rasčëta. Radio i Swâz', Moskva.

[11] Syski, R. (1986) *Introduction to Congestion Theory in Telephone Systems*. North Holland.

7

Models of Links Carrying Multi-Service Traffic

7.1 Introduction

The calculation procedure for a determination of the traffic characteristics of networks servicing multi-service traffic streams, such as integrated services digital networks (ISDN), is a complex issue [1–5]. Multi-service traffic is a mixture of different classes of multi-rate traffic, each of which demands a certain number of channels to service its own call. In stochastic analysis of network systems that are offered multi-service traffic streams it is thus necessary to take into consideration the class of a call and the number of channels required to service the call. This, in consequence, leads to a complex description of service processes in network systems to which such call streams are offered.

The situation is even more complicated in the case of a description of broadband multi-service network systems [1, 5–7] that service not only constant bit rate (CBR) traffic sources, but also variable bit rate (VBR) traffic sources. These sources require a bigger number of parameters to describe traffic streams. Therefore, in order to simplify the description of broadband networks – in which virtual circuit switching is used either partly (as in IP networks) or entirely (as in asynchronous transport mode (ATM) networks) – a notion of the so-called equivalent bandwith [7, 8] has been introduced for those traffic classes that are characterized by variable bit rate. Equivalent bandwidths are certain constant bit rates assigned to VBR sources that are a multiplicity of a certain basic bit rate (basic bandwidth unit – BBU) [7, 9]. This approach makes it possible to evaluate traffic characteristics of broadband integrated services digital networks (B-ISDN) by methods that have been worked out for narrow band networks (ISDN) with circuit switching [7].

The UMTS system – from the point of view of the bit rates of carried services – can be also regarded as a multi-rate network [10–12]. In the further considerations presented in this chapter we will use the universal notation of BBUs to express the system resources demanded by particular calls.[1]

[1]Detailed definition of BBU for the UMTS system will be given in Section 11.1.3.

Modeling and Dimensioning of Mobile Networks: From GSM to LTE
Maciej Stasiak, Mariusz Głąbowski, Arkadiusz Wiśniewski and Piotr Zwierzykowski

7.2 Multi-Dimensional Erlang Distribution

This chapter will present the model of a system to which several classes of PCT1 traffic streams are offered. It is also assumed that a call of each class demands one BBU. This means that this is not multi-rate traffic. However, considering the wide usage of this model, we will present its most important properties.

7.2.1 Assumptions

Let us consider a model of the full-availability group that is offered M independent Poisson call streams (from an infinite number of traffic sources) with the intensities $\lambda_1, \ldots, \lambda_2, \ldots, \lambda_M$. Figure 7.1 shows a model of this group with the capacity of V BBUs. We assume that a call of each class demands one BBU. Service times of particular classes have an exponential character with mean values respectively equal to $1/\mu_1, 1/\mu_2, \ldots, 1/\mu_M$. Thus, the mean traffic intensity offered to the group by a class i stream is equal to:

$$A_i = \frac{\lambda_i}{\mu_i} \tag{7.1}$$

7.2.2 Process Diagram at the Microstate Level

Figure 7.2 shows a fragment of the diagram of the multi-dimensional Markov process in system being considered. The figure shows several neighboring microstates. A microstate is defined by the ordered set of M integers $\{x_1, x_2, \ldots, x_M\}$, whereas each of them determines the number of calls of a given class serviced by the group:

- state $\{0, \ldots, 0, \ldots, 0\}$ – all BBUs are free;
- state $\{x_1, \ldots, x_i, \ldots, x_M\}$ – x_1 calls of class 1 are serviced, ..., x_i calls of class i are serviced, ..., x_M calls of class M are serviced.

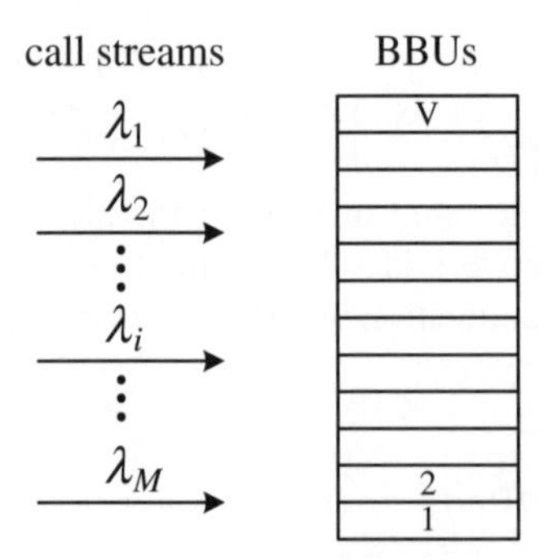

Figure 7.1 The full-availability group with several call streams.

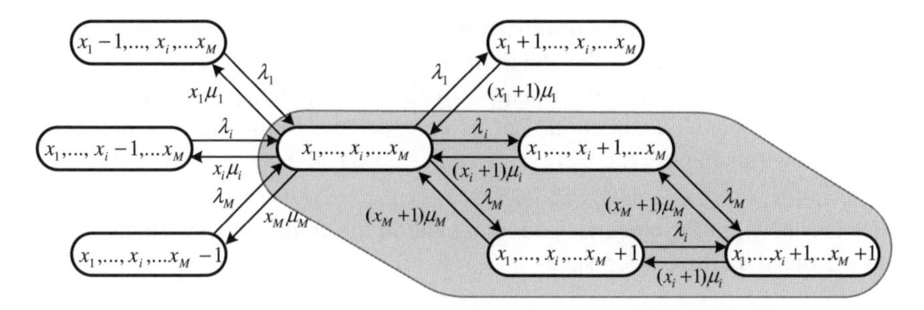

Figure 7.2 A fragment of the diagram of the multi-dimensional Markov process in the full-availability group.

The total number of busy BBUs satisfies the condition:

$$\sum_{i=1}^{M} x_i \leqslant V \tag{7.2}$$

The microstate probability will later be denoted by the symbol $\left[p(x_1,\dots,x_i,\dots,x_M)\right]_V$. For clarity, the number of busy BBUs of particular classes: $x_1,\dots,x_i,\dots,x_M$ is written within parenthesis.

7.2.3 *Reversibility of the Multi-Dimensional Erlang Process*

An interesting property of the multi-dimensional Erlang distribution is its reversibility [13, 14]. According to [15], a stationary Markov process is reversible then and only then, when there exists such a state probability distribution $p(X)$, whose elements satisfy the equilibrium conditions for each pair of states X and Y:

$$p(X)\sigma(X,Y) = p(Y)\sigma(Y,X) \tag{7.3}$$

where X and Y are any states that belong to the state space of the process, whereas $\sigma(X,Y)$ denotes the intensity of the transition from state X to state Y, and $\sigma(Y,X)$ denotes the intensity of the transition from state Y to state X.

Equilibrium equations (7.3) allow us to verify reversibility of the stationary Markov process on the basis of the state probability distribution of relevant states and associated transitions. As a result of investigations carried out by Kolmogorov [16] it was proved that the stationary Markov process is reversible only when the intensities of transitions satisfy the following condition:

$$\sigma(X_1,X_2)\cdots\sigma(X_{n-1},X_n)\sigma(X_n,X_1) = \sigma(X_1,X_n)\cdots\sigma(X_3,X_2)\sigma(X_2,X_1) \tag{7.4}$$

for any finite sequence of states $X_1, X_2, \dots, X_n$ that belong to state space of the process.

In line with the criterion proposed by Kolmogorov, in the case of the Markov process that occurs in the full-availability system servicing M Poisson streams, a necessary and sufficient condition for reversibility of the process is equilibrium of the call and service streams between any two randomly selected microstates for any cycle that connects the microstates.

Thus, the reversibility property results from the fact that when considering the adjacent states related to one another in the diagram of the Markov process, the stream intensity product in the clockwise direction is equal to the corresponding stream intensity product in the counterclockwise direction. Let us consider the following four probabilities of microstates shown in Figure 7.2: $[p(x_1, \ldots, x_i, \ldots, x_M)]_V$, $[p(x_1, \ldots, x_i + 1, \ldots, x_M)]_V$, $[p(x_1, \ldots, x_i + 1, \ldots, x_M + 1)]_V$ and $[p(x_1, \ldots, x_i, \ldots, x_M + 1)]_V$ and the streams (stream intensities products), in the clockwise direction and in the counterclockwise direction, in the quadrangle formed by these microstates:

- Clockwise direction:

$$\begin{aligned}&\left\{\lambda_i \left[p(x_1, \ldots, x_i, \ldots, x_M)\right]_V\right\} \times \left\{\lambda_M \left[p(x_1, \ldots, x_i + 1, \ldots, x_M)\right]_V\right\} \\ &\times \left\{(x_i + 1)\mu_i \left[p(x_1, \ldots, x_i + 1, \ldots, x_M + 1)\right]_V\right\} \\ &\times \left\{(x_M + 1)\mu_M \left[p(x_1, \ldots, x_i, \ldots, x_M + 1)\right]_V\right\}\end{aligned} \tag{7.5}$$

- Counterclockwise direction:

$$\begin{aligned}&\left\{\lambda_M \left[p(x_1, \ldots, x_i, \ldots, x_M)\right]_V\right\} \times \left\{\lambda_i \left[p(x_1, \ldots, x_i, \ldots, x_M + 1)\right]_V\right\} \\ &\times \left\{(x_M + 1)\mu_M \left[p(x_1, \ldots, x_i + 1, \ldots, x_M + 1)\right]_V\right\} \\ &\times \left\{(x_i + 1)\mu_i \left[p(x_1, \ldots, x_i + 1, \ldots, x_M + 1)\right]_V\right\}\end{aligned} \tag{7.6}$$

Comparing Equations (7.5) and (7.6) and dividing them by microstate probabilities we obtain a proof that the clockwise stream and the counterclockwise stream are identical and are equal to:

$$\lambda_i \times \lambda_M \times (x_i + 1)\mu_i \times (x_M + 1)\mu_M = \lambda_M \times \lambda_i \times (x_M + 1)\mu_M \times (x_i + 1)\mu_i \tag{7.7}$$

Equation (7.7) expresses the reversibility criterion of Markov process, known as the Kolmogorov criterion [15]: *A necessary and sufficient condition for the reversibility of the process is the equilibrium of streams (stream parameter products) in the clockwise direction and in the counterclockwise direction in the quadrangle formed by four neighboring states.*

The reversibility property entails local equilibrium equations between any two microstates of the process. Hence, if there is a possibility of reaching a microstate $(x_1, \ldots, x_i + 1, \ldots, x_M)$ from microstate $(x_1, \ldots, x_i, \ldots, x_M)$, then there is also a possibility to reach microstate $(x_1, \ldots, x_i, \ldots, x_M)$ from microstate $(x_1, \ldots, x_i + 1, \ldots x_M)$, with the values of corresponding streams being the same. In particular, it is possible to determine local equilibrium equations between any two neighboring microstates of the process – just as in the case of the one-dimensional birth-and-death process (Section 4.4.2):

$$\lambda_i \left[p(x_1, \ldots, x_i, \ldots, x_M)\right]_V = (x_i + 1)\mu_i \left[p(x_1, \ldots, x_i + 1, \ldots, x_M)\right]_V \tag{7.8}$$

$$\ldots,$$

$$\lambda_M \left[p(x_1, \ldots, x_i, \ldots, x_M)\right]_V = (x_M + 1)\mu_M \left[p(x_1, \ldots, x_i, \ldots, x_M + 1)\right]_V \tag{7.9}$$

7.2.4 Multi-Dimensional Erlang Distribution at Microstate Level

Let us express the probability of microstate $\left[p(x_1, \ldots, x_i, \ldots, x_M)\right]_V$ with the probability $\left[p(0, \ldots, 0, \ldots, 0)\right]_V$ by choosing the path: $(0, \ldots, 0, \ldots, 0)$, $(1, \ldots, 0 \ldots, 0)$, ..., $(x_1, \ldots, 0, \ldots, 0), (x_1, \ldots, 1, \ldots, 0), \ldots, (x_1, \ldots, x_i, \ldots, x_M)$. Using local equilibrium equations and through successive substitutions we obtain:

$$\left[p(x_1, \ldots, x_i, \ldots, x_M)\right]_V = \frac{\left(\frac{\lambda_1}{\mu_1}\right)^{x_1}}{x_1!} \cdots \frac{\left(\frac{\lambda_i}{\mu_i}\right)^{x_i}}{x_i!} \cdots \frac{\left(\frac{\lambda_M}{\mu_M}\right)^{x_M}}{x_M!} \left[p(0, \ldots, 0, \ldots, 0)\right]_V$$

$$= \frac{A_1^{x_1}}{x_1!} \cdots \frac{A_i^{x_i}}{x_i!} \cdots \frac{A_M^{x_M}}{x_M!} \left[p(0, \ldots, 0, \ldots, 0)\right]_V \tag{7.10}$$

As all call streams offered to the group are independent, whereas the process is reversible, we can write the probability of each microstate in the product form:

$$\left[p(x_1, \ldots, x_i, \ldots, x_M)\right]_V = G_V \prod_{j=1}^{M} \left[p(x_j)\right]_V = G_V \prod_{j=1}^{M} \frac{A_j^{x_j}}{x_j!} \tag{7.11}$$

where G_V is the normalization constant:

$$G_V = \frac{1}{\sum\limits_{z_1=0}^{V} \cdots \sum\limits_{z_j=0}^{l_j} \cdots \sum\limits_{z_M=0}^{l_M} \prod\limits_{i=1}^{M} \frac{A_i^{z_i}}{z_i!}} \tag{7.12}$$

The limits for the sums in Equation (7.12) are determined by the following formula:

$$l_j = V - \sum_{k=1}^{j-1} z_k \tag{7.13}$$

The distribution (7.11) is the so-called multi-dimensional Erlang distribution [17] for the full-availability group servicing several traffic streams from an infinite number of sources.

7.2.5 Macrostate Probability

It is convenient to consider the multi-dimensional Erlang process at the level of the so-called macrostates (Section 7.2.4). Each macrostate determines the number of busy BBUs in a considered group independently on the number of serviced calls of particular classes. The macrostate probability $[P_n]_V$ is then the occupancy probability of n BBUs of the group and can be expressed with appropriate microstate probabilities:

$$[P_n]_V = \sum_{\Omega(n)} [p(x_1, x_2, \ldots, x_M)]_V \tag{7.14}$$

where $\Omega(n)$ is the set of all microstates $\{x_1, x_2, \ldots, x_M\}$ that satisfies the condition:

$$n = \sum_{i=1}^{M} x_i \tag{7.15}$$

On the basis of Newton's formula defining the power of the sum of the finite number of elements, we can write:

$$\left(\sum_{i=1}^{M} A_i\right)^n = n! \sum_{\Omega(n)} \prod_{i=1}^{M} \frac{A_i^{x_i}}{x_i!} \tag{7.16}$$

Taking into consideration (7.16) and (7.12), the macrostate probability can be written in the following form:

$$[P_n]_V = \frac{\left(\sum_i^M A_i\right)^n}{n!} \Bigg/ \sum_{k=0}^{V} \frac{\left(\sum_i^M A_i\right)^k}{k!} \tag{7.17}$$

7.2.6 Interpretation of Macrostate Distribution

Equation (7.17) expresses the Erlang distribution for offered traffic $A = \sum_{i=1}^{M} A_i$. This distribution can be treated as a full-availability group model [18] which is offered one call stream with an intensity equal to:

$$\lambda = \sum_{i=1}^{M} \lambda_i \tag{7.18}$$

and the hyperexponential service time distribution (the weighted sum of exponential distributions), with its mean value is equal to:

$$\frac{1}{\mu} = \sum_{i=1}^{M} \frac{\lambda_i}{\lambda}\left(\frac{1}{\mu_i}\right) = \frac{1}{\lambda}\sum_{i=1}^{M}\left(\frac{\lambda_i}{\mu_i}\right) = \frac{1}{\lambda}\sum_{i=1}^{M} A_i \tag{7.19}$$

7.2.7 Blocking and Loss Probability

With the case of the multi-dimensional Erlang distribution, blocking states for each traffic class are identical. Therefore, the blocking (loss) probability for each call class is exactly the same and is determined by the so-called multi-dimensional Erlang formula [17]:

$$E = B = E_V\left(\sum_{i=1}^{M} A_i\right) \tag{7.20}$$

7.2.8 Recursive Notation of the Multi-Dimensional Erlang Distribution

The multi-dimensional Erlang distribution has a number of interesting properties. One of the most important properties is the possibility of presenting the macrostate probability (7.17) with a simple recursive dependence [18]:

$$n\,[P_n]_V = \sum_{i=1}^{M} A_i \left[P_{n-1}\right]_V \tag{7.21}$$

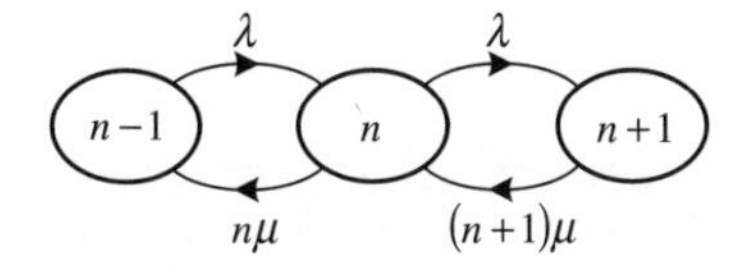

Figure 7.3 Fragment of the diagram of the birth-and-death process in the full-availability group.

The dependence (7.21) makes it possible to determine quickly the occupancy distribution of the multi-dimensional Erlang distribution, for example, on the basis of the following algorithm [18]:

$$\begin{cases} q(0) = 1 \\ q(n) = \frac{1}{n} \sum\limits_{i=1}^{M} A_i q(n-1) \\ Q(V) = \sum\limits_{i=0}^{V} q(i) \\ [P_n]_V = \frac{q(n)}{Q(V)} \end{cases} \tag{7.22}$$

Let us note that for $M = 1$, Equation (7.21) comes down to the local equilibrium equation of the Markov process in the full-availability group (Equation (4.42)).

7.2.9 Interpretation of the Recursive Notation of the Multi-Dimensional Erlang Distribution

Let us consider again the diagram of the one-dimensional birth-and-death process in the full-availability group servicing one call stream. A fragment of the diagram is shown in Figure 7.3.

The arrival rate λ and service rate μ are expressed in Figure 7.3 in relation to the same time unit. Let us rescale the birth-and-death process in the full-availability group in such a way that we adopt the mean service time $1/\mu$ as a time unit. With this approach, the value of the call stream intensity will be equal to the intensity of the offered traffic: $A = \lambda(1/\mu)$ (cf. Equation (4.56)), whereas the value of the service stream intensity will be equal to the number of serviced calls in a given state: $n = n\mu(1/\mu)$. A fragment of the diagram of such rescaled process is shown in Figure 7.4. Note that with the process being defined in this way we can talk about the traffic stream, which is expressed in erlangs, instead of the call stream, which is expressed in the number of calls per time unit.

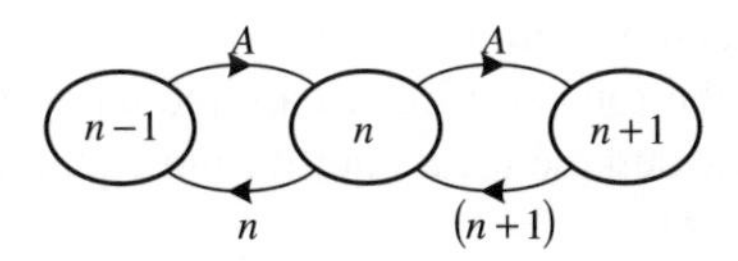

Figure 7.4 Fragment of a diagram of the rescaled birth-and-death process in the full-availability group.

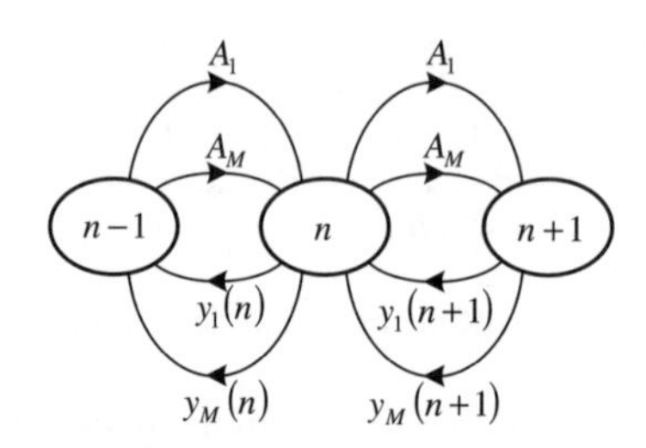

Figure 7.5 Fragment of the diagram interpreting the recursive notation of the multi-dimensional Erlang distribution.

Figure 7.5 shows a graphical interpretation of Equation (7.21). It is a one-dimensional Markov process in the full-availability group with M traffic streams. The number of calls of class i serviced in macrostate n of the process is denoted by $y_i(n)$. From the considerations presented above it is evident that the diagram obtained can be treated as the process in which each balance associated with a given i call stream is rescaled by $1/\mu_i$.

7.2.10 Service Streams at the Macrostate Level

Each macrostate of a Markov chain in the full-availability group – with the relevant fragment of the diagram of the Markov process shown in Figure 7.5 – is satisfied by the following state equation:

$$[P_n]_V \left[\sum_{i=1}^{M} A_i + \sum_{i=1}^{M} y_i(n) \right] = \sum_{i=1}^{M} A_i \left[P_{n-1}\right]_V + \sum_{i=1}^{M} y_i(n+1) \left[P_{n+1}\right]_V \tag{7.23}$$

where $y_i(n)$ is the number of calls of class i serviced in the group in state n. In each state n there are n calls serviced, so the sum of all service streams from state n towards lower states is equal to n:

$$n = \sum_{i=1}^{M} y_i(n) \tag{7.24}$$

Taking into consideration Equations (7.21) and (7.24), Equation (7.23) can be rewritten in the following form:

$$\sum_{i=1}^{M} A_i \left[P_n\right]_V = \sum_{i=1}^{M} y_i(n+1) \left[P_{n+1}\right]_V \tag{7.25}$$

Equation (7.25) is a statistical equilibrium between the total stream outgoing from macrostate n towards state $(n+1)$, and the total service stream incoming to state n from state $(n+1)$. The equation is fulfilled only when the local equilibrium equations for each of the streams of class i are fulfilled ($1 \leqslant n \leqslant M$):

$$A_i \left[P_n\right]_V = y_i(n+1) \left[P_{n+1}\right]_V \tag{7.26}$$

On the basis of Equation (7.26), the number of calls of class i serviced in state $(n+1)$ can eventually be written in the following way:

$$y_i(n+1) = \begin{cases} A_i[P_n]_V/[P_{n+1}]_V & \text{for } n+1 \leq V \\ 0 & \text{for } n+1 > V \end{cases} \tag{7.27}$$

7.3 Full-Availability Group with Multi-Rate Traffic

7.3.1 Assumptions

Let us consider a full-availability group that is offered M independent call streams with the intensity: $\lambda_1, \lambda_2, \ldots, \lambda_M$. Unlike in the multi-dimensional Erlang distribution, we assume that a call of each of the classes requires a different number of BBUs. The demanded number of BBUs for calls of particular classes will be denoted by the symbols: $t_1, t_2, \ldots, t_M$. The service times of calls of all classes have exponential distributions with the intensities $\mu_1, \mu_2, \ldots, \mu_M$. Thus, the average intensity of the traffic offered by a stream of class i is expressed by Equation (7.1). Figure 7.6 shows a model of the group with the capacity of V BBUs that is offered M call streams. With the assumption that $t_1 = 1$, $t_2 = 2$, Figure 7.6 shows such a state of the group, in which it services one call of the first class and two calls of the second class.

7.3.2 Diagram of Markov Process at the Microstate Level

The diagram of a Markov process at the microstate level will be identical to the diagram shown in Figure 7.2. The figure shows several neighboring microstates. A microstate is defined by an ordered set that includes M integers $\{x_1, x_2, \ldots, x_M\}$, where each of them determines the number of calls of a given class serviced by the group:

- state $\{0, \ldots, 0, \ldots, 0\}$ – the group does not service calls;
- state $\{x_1, \ldots, x_i, \ldots, x_M\}$ – x_1 calls of class 1 are serviced,. . ., x_i calls of class i are serviced . . ., and x_M calls of class M are serviced in the group.

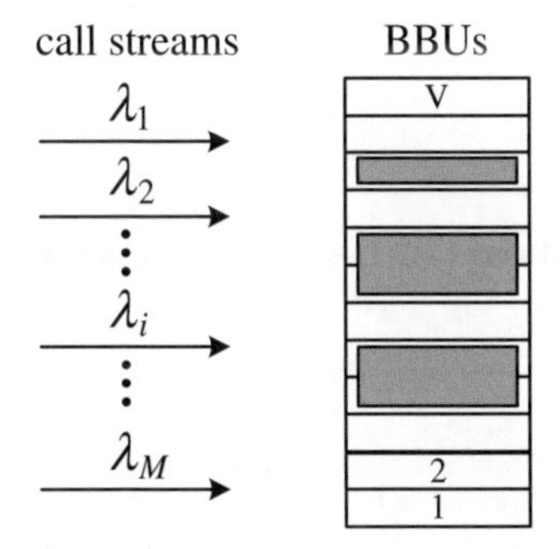

Figure 7.6 Full-availability group with several call streams with different demands.

The total number of busy BBUs in the group satisfies the condition:

$$\sum_{i=1}^{M} x_i t_i \leqslant V \tag{7.28}$$

7.3.3 *Reversibility of the Process at the Microstate Level*

The Markov process in a group with multi-rate traffic is a reversible process. All call streams offered to the group are independent, so – following the same reasoning as in Section 7.2.3 – each microstate can be written in the product form:

$$\left[p(x_1, \ldots, x_i, \ldots, x_M)\right]_V = G_V \prod_{j=1}^{M} \left[p(x_j)\right]_V = G_V \prod_{j=1}^{M} \frac{A_j^{x_j}}{x_j!} \tag{7.29}$$

where G_V is the normalization constant:

$$G_V = \frac{1}{\sum_{z_1=0}^{V} \cdots \sum_{z_j=0}^{l_j} \cdots \sum_{z_M=0}^{l_M} \prod_{i=1}^{M} \frac{A_i^{z_i}}{z_i!}} \tag{7.30}$$

In this case, when the group services call streams with different demands, the limit l_j in the sum takes on the following values:

$$l_j = \left\lfloor V - \sum_{k=1}^{j-1} z_k t_k \right\rfloor \tag{7.31}$$

where $\lfloor * \rfloor$ denotes the integer part of the expression $*$.

7.3.4 *Macrostate Probability*

A macrostate is defined by the number of serviced BBUs in the system:

$$[P_n]_V = \sum_{\Omega(n)} [p(x_1, x_2, \ldots, x_M)]_V \tag{7.32}$$

where $\Omega(n)$ is the set of all such subsets $\{x_1, x_2, \ldots, x_M\}$, that satisfy the equation:

$$n = \sum_{i=1}^{M} x_i t_i \tag{7.33}$$

Therefore, on the basis of Equations (7.29)–(7.32), the macrostate probability can be expressed by the following equation [19]:

$$[P_n]_V = \sum_{\Omega(n)} [p(x_1, x_2, \ldots, x_M]_V) = \sum_{\Omega(n)} \frac{\prod_{j=1}^{M} \frac{A_j^{x_j}}{x_j!}}{\sum_{z_1=0}^{V} \cdots \sum_{z_j=0}^{l_j} \cdots \sum_{z_M=0}^{l_M} \prod_{i=1}^{M} \frac{A_i^{z_i}}{z_i!}} \tag{7.34}$$

Equation (7.34) determines the occupancy distribution at the microstate level in the full-availability group with multi-rate traffic. This distribution is often called the occupancy distribution in the full-availability group with multi-rate traffic. It should be stressed that due to the complex structure of the formula, its usefulness in engineering calculations is very limited.

7.3.5 Recursive Notation of the Occupancy Distribution of the Full-Availability Group with Multi-Rate Traffic

Let us assume that the microstate $\{x_1, \ldots, x_i, \ldots, x_M\}$ presented in Figure 7.2 satisfies Equation (7.29). Each of the microstates on the left side of the microstate $\{x_1, \ldots, x_i, \ldots, x_M\}$ is related to such a macrostate in which the number of busy BBUs is decreased by t_i BBUs, necessary to set up a connection of class i. Thus, microstate $\{x_1, \ldots, x_i - 1, \ldots, x_M\}$ is related to a macrostate in which the number of busy BBUs is $n - t_i$. Therefore, the following equality is satisfied:

$$\sum_{\substack{j=1 \\ j \neq i}}^{M} x_j t_j + (x_i - 1) t_i = \sum_{j=1}^{M} x_j t_j - t_i = (n - t_i) \tag{7.35}$$

The process, the fragment of which is shown in Figure 7.2, is a reversible process (Section 7.3.3). It is proved in Section 7.2.3 that the reversibility property entails local equilibrium equations between each of the two neighboring states [18, 20]. Let us consider then a local equilibrium equation of a Markov process taking part in the considered system. Such an equation for the class i stream can be written in the following form:

$$x_i \mu_i [p(x_1, \ldots, x_i, \ldots, x_M)]_V = \lambda_i [p(x_1, \ldots, x_i - 1, \ldots, x_M)]_V \tag{7.36}$$

Call streams offered to the group are independent, so we can sum up all M equations of the type (7.36) for microstate $\{x_1, x_2, \ldots, x_M\}$. Then we obtain:

$$[p(x_1, \ldots, x_i, \ldots, x_M)]_V \sum_{i=1}^{M} x_i t_i = \sum_{i=1}^{M} A_i t_i [p(x_1, \ldots, x_i - 1, \ldots, x_M)]_V \tag{7.37}$$

Taking into consideration the definition of macrostate (7.32) and (7.33), Equation (7.37) can be transformed into the following form:

$$n \sum_{\Omega(n)} [p(x_1, \ldots, x_i, \ldots, x_M)]_V = \sum_{i=1}^{M} A_i t_i \sum_{\Omega(n)} [p(x_1, \ldots, x_i - 1, \ldots, x_M)]_V \tag{7.38}$$

which can be rewritten in the following form:

$$n\,[P_n]_V = \sum_{i=1}^{M} A_i t_i \left[P_{n-t_i}\right]_V \tag{7.39}$$

where $[P_{n-t_i}]_V = 0$, if $n < t_i$, and value $[P_0]_V$ results from the normalization condition:

$$\sum_{n=0}^{V} [P_n]_V = 1 \tag{7.40}$$

The recursive dependence (7.39) enables us to determine the occupancy distribution in the full-availability group with multi-rate traffic on the basis of the following algorithm:

$$\begin{cases} q(0) = 1 \\ q(n) = \frac{1}{n} \sum\limits_{i=1}^{M} A_i q(n - t_i) t_i \\ Q(V) = \sum\limits_{i=0}^{V} q(i) \\ [P_n]_V = \frac{q(n)}{Q(V)} \end{cases} \tag{7.41}$$

The recursive formula (7.39) was proposed by Kaufman [20] and Roberts [21]. Hence, occupancy distribution in the full-availability group, determined on the basis of the recursion (7.39), is usually called the Kaufman–Roberts distribution.

7.3.6 Blocking Probability and Loss Probability

In the case of the Kaufman–Roberts distribution the blocking and loss probabilities are identical because all streams of offered traffic classes are, by assumption, of the PCT1 type. Blocking in the full-availability group occurs when the group does not have t_i free BBUs necessary to set up a connection of class i. Thus, the blocking (loss) probability for a class i call can be written in the following form:

$$E_i = B_i = \sum_{n=V-t_i+1}^{V} [P_n]_V \tag{7.42}$$

7.3.7 Recursive Properties of the Kaufman–Roberts Distribution

Figure 7.7 shows a graphical interpretation of the occupancy distribution (7.39) for a group servicing two call streams ($t_1 = 1, t_2 = 2$). The figure shows the diagram of a one-dimensional Markov process in the full-availability group with multi-rate traffic. It is thus justifiable to say that the Kaufman–Roberts distribution resulting from the macrostate level is related to a one-dimensional distribution, while the occupancy distribution at the microstate level is related to a multi-dimensional distribution (Section 7.3.3, distribution (7.29)–(7.30)).

The diagram of the one-dimensional Markov process in the full-availability group with M traffic streams shown in Figure 7.7 can be treated as a rescaled process (see Section 7.2.9). The symbols y_1 and y_2 in Figure 7.7 denote the values of service streams of particular classes. Note that these parameters are not required for a direct determination of the occupancy distribution

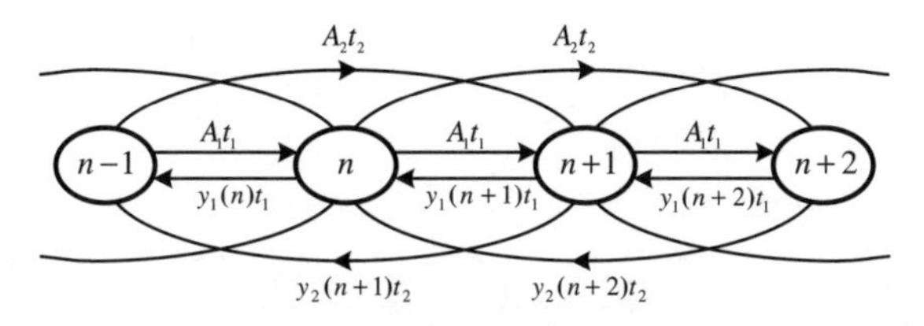

Figure 7.7 A fragment of the one-dimensional Markov chain in the full-availability group with two call streams ($t_1 = 1, t_2 = 2$).

in the full-availability group on the basis of Equation (7.39). The knowledge of service stream parameters is, however, indispensable for solving many other problems discussed in the following chapters. Formulas that enable the determination of service streams in the full-availability group with multi-rate traffic will therefore be derived in the next sections.

Erlang believed that the task of determining blocking probability in the full-availability group with losses and a single-rate traffic was the most important problem in traffic theory [22]. This applies just as well to multi-rate systems and, therefore, Equations (7.39) and (7.42) perform the same function as Erlang B-formula in systems with single-service traffic.

7.3.8 Delbrouck Formula

In the literature, for instance in [23] or [24], more general versions of the Kaufman–Roberts distribution have also been proposed. The iterative formula given in [23] is the most general one and allows us to determine the occupancy distribution in the full-availability group that is offered a mixture of traffic streams of Erlang, Engset and Pascal type (the so-called BBP type traffic [18] – binomial, Poisson and Pascal):

$$n\,[P_n]_V = \sum_{i=1}^{M} \left(\frac{A_i t_i}{Z_i}\right) \sum_{j=1}^{\lfloor n/t_i \rfloor} (-\alpha_i)^{j-1} \left[P_{n-jt_i}\right]_V \tag{7.43}$$

where α_i is traffic offered by one free source of class i, and Z_i is the peakedness coefficient of offered traffic determined by the ratio of variance to the mean value of the distribution related to a given call stream.

It should be emphasized that the recursive formulas (7.39) and (7.43) enable us to determine the occupancy distribution only in the full-availability group which is offered Poisson-type call streams (Equation (7.39)) or, at the maximum, a mixture of Poisson, Engset and Pascal streams (Equation (7.43)). These formulas cannot be though applied in the general case – for the *state-dependent systems*, in which call streams between neighboring states of the service process depend on the current state of this process. Such systems will be discussed in next chapters.

7.3.9 Service Streams in the Full-Availability Group with Multi-Rate Traffic

As was pointed out in Section 7.3.7, from Equation (7.39) it can be seen that the knowledge of service-stream parameters of particular classes in the calculations of the occupancy distribution in the full-availability group with multi-rate traffic is irrelevant. However, in many traffic issues, the value of the number of serviced calls of a given class in a given state of the group may turn

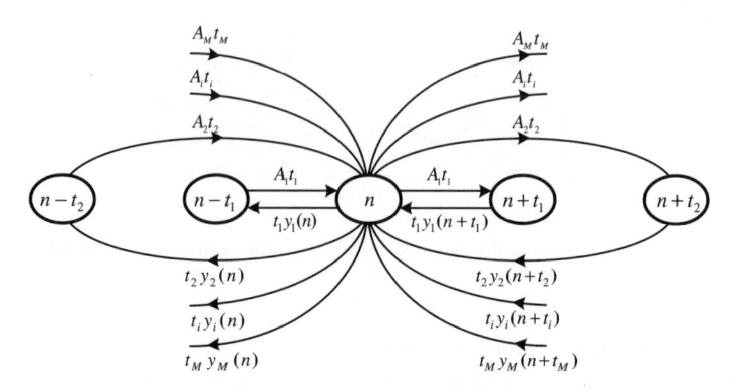

Figure 7.8 A fragment of the one-dimensional Markov chain in the full-availability group with multi-rate traffic.

out to be necessary. Let us consider then a fragment of the diagram of the one-dimensional Markov chain shown in Figure 7.8 and corresponding to the Kaufman–Roberts distribution (Equation (7.39)).

Each state of the Markov chain in the full-availability group (Figure 7.8) satisfies the following state equation:

$$[P_n]_V \left[\sum_{i=1}^{M} A_i t_i + \sum_{i=1}^{M} t_i y_i(n) \right] = \sum_{i=1}^{M} A_i t_i \left[P_{n-t_i} \right]_V + \sum_{i=1}^{M} t_i y_i(n+t_i) \left[P_{n+t_i} \right]_V \quad (7.44)$$

where $y_i(n)$ is the number of calls of a given class serviced in the group in state n. It follows from Figure 7.8 that the sum of all service streams from state n towards lower states is equal to n:

$$n = \sum_{i=1}^{M} t_i y_i(n) \quad (7.45)$$

After taking into account Equations (7.39) and (7.45), Equation (7.44) can be transformed into the following form:

$$\sum_{i=1}^{M} A_i t_i [P_n]_V = \sum_{i=1}^{M} t_i y_i(n+t_i) \left[P_{n+t_i} \right]_V \quad (7.46)$$

Equation (7.46) is a statistical equilibrium equation between the total call stream outgoing from state n and the total service stream incoming to state n. This equilibrium is fulfilled only when the local equilibrium equations for the streams of individual traffic classes are satisfied [20]:

$$A_i t_i [P_n]_V = t_i y_i(n+t_i) \left[P_{n+t_i} \right]_V \quad (7.47)$$

On the basis of Equation (7.47), the number of calls of class i in the group in state $(n + t_i)$ can eventually be written in the following way:

$$y_i(n+t_i) = \begin{cases} A_i \left[P_n\right]_V / \left[P_{n+t_i}\right]_V & \text{for } n+t_i \leq V \\ 0 & \text{for } n+t_i > V \end{cases} \tag{7.48}$$

7.3.10 Convolution Algorithm

The blocking probability for calls of particular classes in the full-availability group with multi-rate traffic can be determined – beside using Equations (7.39)–(7.42) – on the basis of the convolution algorithm [4, 18]. This algorithm for the full-availability group with multi-rate traffic can be presented in the following way:

1. **Determination of the occupancy distribution $[p]_V^{(i)}$ of each traffic class, with the assumption that only this class is offered to the group.**
   For example, for class i we obtain:

$$[p]_V^{(i)} = \left\{ [p_0]_V^{(i)}, [p_1]_V^{(i)}, \ldots, [p_V]_V^{(i)} \right\} \tag{7.49}$$

2. **Determination of the aggregated occupancy distribution $[Q]_V^{(-i)}$, obtained on the basis of the convolution operation carried out successively for all traffic classes except class i:**

$$[Q]_V^{(-i)} = [p]_V^{(1)} * [p]_V^{(2)} * \ldots * [p]_V^{(i-1)} * [p]_V^{(i+1)} * \ldots * [p]_V^{(M)} \tag{7.50}$$

   where the convolution operation between classes i and j is defined as follows:

$$[p]_V^{(i)} * [p]_V^{(j)} = \left\{ [p_0]_V^{(i)} [p_0]_V^{(j)}, \sum_{n=0}^{1} [p_n]_V^{(i)} [p_{1-n}]_V^{(j)}, \ldots, \sum_{n=0}^{V} [p_n]_V^{(i)} [p_{V-n}]_V^{(j)} \right\} \tag{7.51}$$

   It should be stressed at this point that if the distributions $[p]_V^{(i)}$ and $[p]_V^{(j)}$ are normalized, then the distribution resulting from the convolution operation of these distributions is not always normalized because the resulting distribution is truncated at state V. For this reason during the convolution operation of several distributions – after Equation (7.50) – it is recommended that the result of the convolution of each two successive distributions be normalized.

3. **Blocking probability calculation E_i and loss probability calculation B_i for the stream of class i.**
   The basis for such calculations is the convolution operation between the occupancy distribution $[Q]_V^{(-i)}$, obtained in step 2, and the distribution $[p]_V^{(i)}$:

$$[P]_V = [Q]_V^{(-i)} * [p]_V^{(i)} = \left\{ [q_{0,0}]_V^{(i)(-i)}, \sum_{n=0}^{1} [q_{n,1-n}]_V^{(i)(-i)}, \ldots, \sum_{n=0}^{V} [q_{n,V-n}]_V^{(i)(-i)} \right\} \tag{7.52}$$

where:

$$\sum_{n=0}^{k} \left[q_{n,k-n}\right]_V^{(i)(-i)} = \sum_{n=0}^{k} \left[p_n\right]_V^{(i)} \left[Q_{k-n}\right]_V^{(-i)} \tag{7.53}$$

The symbol $[q_{n,k-n}]_V^{(i)(-i)}$ denotes the occupancy probability of k BBUs in the convolution distribution with the assumption that n BBUs are occupied by calls of class i. Thus, the blocking probability – after the normalization process of the distribution $[P]_V$ – can be determined on the basis of Equation (7.42):

$$E_i = \sum_{n=V-t_i+1}^{V} [P_n]_V \tag{7.54}$$

If all traffic streams are Poisson streams, then the blocking probability is equal to the loss probability ($E_i = B_i$). If the traffic streams are Engset or Pascal streams, then the loss probability can be determined by the formula:

$$B_i = \frac{\sum\limits_{k=V-t_i+1}^{V} \sum\limits_{n=0}^{k} \lambda_i(n) \left[q_{n,k-n}\right]_V^{(i)(-i)}}{\sum\limits_{k=0}^{V} \sum\limits_{n=0}^{k} \lambda_i(n) \left[q_{n,k-n}\right]_V^{(i)(-i)}} \tag{7.55}$$

The parameter $\lambda_i(n)$ in Equation (7.55) is a call stream of class i in a state of the group in which n BBUs are occupied by calls of class i.

The convolution algorithm enables us to determine the characteristics of the full-availability group which is offered call streams with different demands of the Poisson, Engset and Pascal types (i.e. the BPP type of traffic).

7.4 State-Dependent Systems

Here we will discuss certain systems, important to traffic engineering, in which call streams between neighboring states of the service process depend on the current state of this process. Such systems are called *state-dependent systems*. There are a few types of groups that can be distinguished in multi-rate systems that are examples of state-dependent systems:

- A group of separated links (*the limited-availability group*). A group of this type is created as the result of joining several identical, separated links that are offered a common stream of multi-rate traffic. The system services a call only when it can be entirely serviced by free BBUs of one of the component links. Let us consider now, as an example, a group composed of two links. When a call of a given class demands, for instance, six BBUs to set up a connection and in one link that belongs to the group there are two free BBUs, and the other link has four free BBUs, then the connection will not be set up.

- Separated link with bandwidth reservation (*the full-availability group with reservation*). In groups of this type, part of resources is reserved for calls of particular traffic classes (or class). The objective of the introduction of the reservation mechanism is to ensure similar values of quality-of-service parameters for calls of different classes. The application of reservation mechanisms is particularly advisable in cases when call classes offered to the group differ significantly from one another in terms of the number of demanded BBUs. Two types of reservation are distinguished: static and dynamic. In the case of the static reservation, a certain part of the group's resources is reserved for particular traffic classes, whereas in the case of the dynamic reservation, the reservation mechanism is introduced only when the assumed occupancy state in the groups is exceeded.
- Separated link with threshold mechanism (*the full-availability group with threshold mechanism*). In this group traffic parameters can change depending on the load of the system. The threshold (or a number of thresholds) is introduced into a group. A threshold is a state, above which the call admission control function changes the conditions in service of calls, for example, allocates a lower number of BBUs. Two types of systems are usually considered: the single-threshold model (STM) and the multi-threshold model (MTM). In the STM system, one threshold for each call class is introduced, whereas in MTM systems a set of such thresholds for each call class is introduced.

7.4.1 Assumptions

Let us consider a system that is offered M independent call streams with the intensities: $\lambda_1, \lambda_2, \ldots, \lambda_M$. The demanded number of BBUs for the calls of particular classes will be denoted, as earlier, by the symbols: $t_1, t_2, \ldots, t_M$. The service times of calls of all classes have an exponential distribution with the parameters: $\mu_1, \mu_2, \ldots, \mu_M$. Hence, the mean traffic intensity offered by a stream of class i will be expressed by Equation (7.1).

We assume that call streams between given microstates depend on them (are dependent on these microstates). This dependence can be taken into account by the introduction of the so-called conditional transition coefficient $\sigma_i(x_1, \ldots, x_i, \ldots, x_M)$ between adjacent microstates, that determines which part of the incoming call stream λ_i will be transferred between the microstates $\{x_1, \ldots, x_i, \ldots, x_M\}$ and $\{x_1, \ldots, x_i + 1, \ldots, x_M\}$.

7.4.2 Diagram of the State-Dependent Process at the Microstate Level

Figure 7.9 shows a fragment of a Markov process in state-dependent systems. The figure presents several neighboring states. The microstate is defined – as in state-independent systems (Section 7.3.2) – by an ordered set that consists of M integers $\{x_1, x_2, \ldots, x_M\}$, where each determines the number of calls of a given class serviced by the group. The total number of busy BBUs in the group satisfies the condition:

$$\sum_{i=1}^{M} x_i t_i \leqslant V \tag{7.56}$$

Figure 7.9 also shows the indicated conditional transition coefficients between neighboring microstates of the diagram.

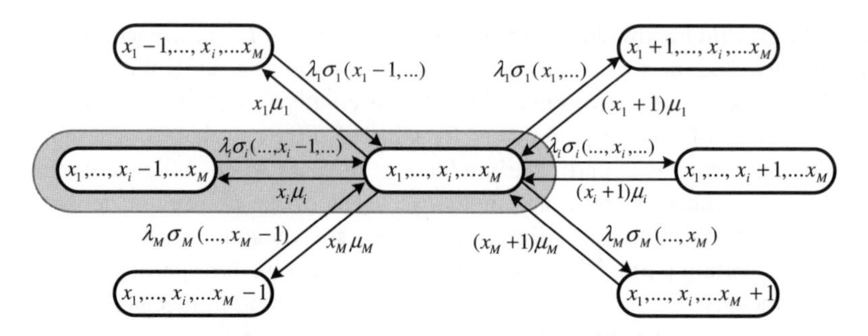

Figure 7.9 A fragment of the Markov process diagram in the state-dependent system .

7.4.3 *Reversibility of the State-Dependent Multi-Dimensional Process*

The multi-dimensional process in a state-dependent system is not always a reversible process. Let us consider in Figure 7.10 the four microstates: $(x_1, \ldots, x_i, \ldots, x_j, \ldots, x_M)$, $(x_1, \ldots, x_i + 1, \ldots, x_j \ldots, x_M)$, $(x_1, \ldots, x_i + 1, \ldots, x_j + 1, \ldots, x_M)$, $(x_1, \ldots, x_i, \ldots, x_j + 1, \ldots, x_M)$ and the streams (products of stream parameters) in the clockwise direction and in the counterclockwise direction in the quadrangle formed by these microstates.

Let us determine the stream values in the clockwise direction and in the reverse direction:

- The clockwise direction:

$$\begin{aligned} &\lambda_i\sigma_i(x_1, \ldots, x_i, \ldots, x_j, \ldots, x_M)(x_i + 1)\mu_i \\ &\quad\times\lambda_j\sigma_j(x_1, \ldots, x_i + 1, \ldots, x_j, \ldots, x_M)(x_j + 1)\mu_j \end{aligned} \tag{7.57}$$

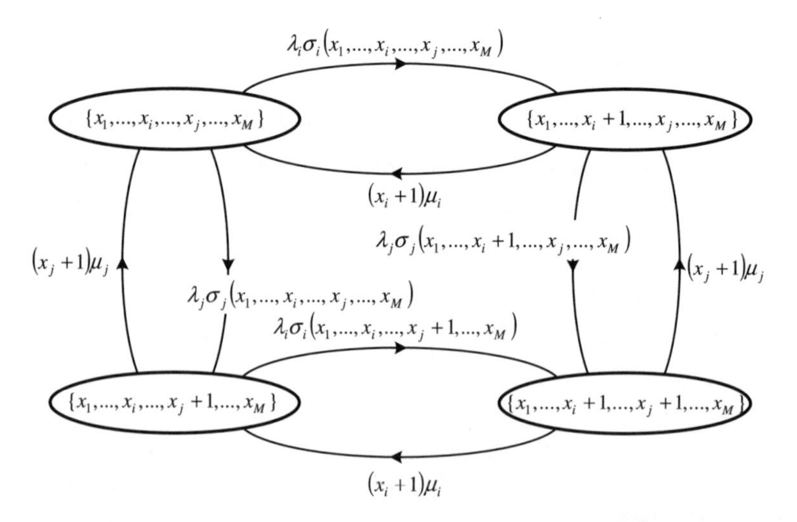

Figure 7.10 The microstate quadrangle in the state-dependent system.

- The counterclockwise direction:

$$\lambda_i\sigma_i(x_1,\ldots,x_i,\ldots,x_j+1,\ldots,x_M)(x_i+1)\mu_i$$
$$\times\lambda_j\sigma_j(x_1,\ldots,x_i,\ldots,x_j,\ldots,x_M)(x_j+1)\mu_j \quad (7.58)$$

It can be stated on the basis of Equations (7.57) and (7.58) that the Markov process in the state-dependent system is a reversible process if between the conditional transition coefficients – for each pair of streams – there exists the following dependence:

$$\sigma_i(x_1,\ldots,x_i,\ldots,x_j,\ldots,x_M)\sigma_j(x_1,\ldots,x_i+1,\ldots,x_j,\ldots,x_M)$$
$$=\sigma_i(x_1,\ldots,x_i,\ldots,x_j+1,\ldots,x_M)\sigma_j(x_1,\ldots,x_i,\ldots,x_j,\ldots,x_M) \quad (7.59)$$

7.4.4 Approximation of the State-Dependent Process by the Reversible Process

In most systems considered in teletraffic theory, Equation (7.59) is not satisfied, although the left side and the right side can – in some systems – differ only slightly. Thus, a question arises – in what conditions can the Markov process in state-dependent systems then be approximated by reversible processes?

The analysis of all transition probabilities $\sigma_i(x_1,\ldots,x_i,\ldots,x_M)$ related to a given macrostate n, in other words those that satisfy Equation (7.29), leads to the conclusion that the values of these parameters can be different depending on the occupancy state of the system $\{x_1,\ldots,x_i,\ldots,x_M\}$ imposed by calls of particular classes. Let us assume now that the differences between the parameters $\sigma_i(x_1,\ldots,x_i,\ldots,x_M)$ within one macrostate are omittable. Consequently, we can adopt the following assumption [25]:

Assumption 1

$$\sigma_i(x_1,\ldots,x_i,\ldots,x_M)=\sigma_i(n) \quad (7.60)$$

The adoption of the above assumption for all sets $\{x_1,\ldots,x_i,\ldots,x_M\}$ satisfying Equation (7.29) means that σ_i depends exclusively on the total number of busy BBUs and does not depend on the distribution of busy BBUs between particular classes of serviced calls. Taking into consideration Assumption 1, Equation (7.59) can be rewritten in the following form:

$$\frac{\sigma_i(n)}{\sigma_i(n+t_j)}=\frac{\sigma_j(n)}{\sigma_j(n+t_i)} \quad (7.61)$$

Equation (7.61) will be satisfied if we assume that:

$$\sigma_i(n)\approx\sigma_i(n+t_j) \quad (7.62)$$

$$\sigma_j(n)\approx\sigma_j(n+t_i) \quad (7.63)$$

The approximation, expressed with Equations (7.62) and (7.63) will be applicable if the conditional transition coefficient – for each class of calls – is the slowly varying function of n. We can therefore adopt the following assumption [2, 25]:

Assumption 2

$$\left|\frac{\sigma_i(n) - \sigma_i(n-1)}{\sigma_i(n-1)}\right| \ll 1 \tag{7.64}$$

The second assumption can also be treated as a complement to the first. It indicates the conditions in which the parameters $\sigma_i(x_1, \ldots, x_i, \ldots, x_M)$ do not depend on the distribution of busy BBUs between particular classes of calls. That is to say, if $\sigma_i(n)$ is a slowly changing function of n, then the dependence of the parameters $\sigma_i(x_1, \ldots, x_i, \ldots, x_M)$ within a given macrostate n can be omitted.

Summing up, we can state that if a state-dependent process satisfies Assumptions 1 and 2, then it can be approximated by a reversible process.

7.4.5 Generalized Kaufman–Roberts Distribution

The adoption of Assumptions 1 and 2 permits us to approximate the state-dependent process by the reversible process. Consequently, all streams can be treated independently and the process can be analyzed on the basis of local equilibrium equations. Let us apply, then, similar reasoning as in Section 7.3.5 for the state-dependent process (at the microstate level, Section 7.4.2) – in other words, let us consider the local equilibrium equation of the Markov process occurring in a given system. Such an equation, for a stream of class i (in Figure 7.9 the states included within the dotted line), can be written in the following form:

$$\begin{aligned} &x_i\mu_i[p(x_1, \ldots, x_i, \ldots, x_M)]_V \\ &\quad = \lambda_i\sigma_i(x_1, \ldots, x_i - 1, \ldots, x_M)[p(x_1, \ldots, x_i - 1, \ldots, x_M)]_V \end{aligned} \tag{7.65}$$

where $\sigma_i(x_1, \ldots, x_i - 1, \ldots, x_M)$ is the conditional transition probability between neighboring microstates of the process related to the stream of class i. Since call streams offered to the group are independent, we can sum up for microstate $\{x_1, x_2, \ldots, x_M\}$ all M equations of the type (7.65). Using the condition (7.60), the result of this operation will be the following equation:

$$[p(x_1, \ldots, x_i, \ldots, x_M)]_V \sum_{i=1}^{M} x_i t_i = \sum_{i=1}^{M} A_i t_i \sigma_i(n - t_i)[p(x_1, \ldots, x_i - 1, \ldots, x_M)]_V \tag{7.66}$$

After taking into consideration the definition of macrostate (7.32) and using the same procedure as in the case of the full-availability group (Section 7.3.5), Equation (7.66) can finally take the following form:

$$n\,[P_n]_V = \sum_{i=1}^{M} A_i t_i \sigma_i(n - t_i)\left[P_{n-t_i}\right]_V \tag{7.67}$$

where $[P_{n-t_i}]_V = 0$, if $n < t_i$, and value $[P_0]_V$ results from the condition (7.40).

Equation (7.67) determines the so-called generalized Kaufman–Roberts distribution. With the application of the distribution it is possible to determine iteratively the occupancy distribution $[P_n]_V$ in any state-dependent system that is offered a mixture of different multi-rate traffic. To achieve that, however, it is necessary to know the probabilities $\sigma_i(n)$. Assuming the knowledge of conditional transition probabilities, it is possible to determine the generalized Kaufman–Roberts distribution on the basis of the following algorithm:

$$\begin{cases} q(0) = 1 \\ q(n) = \frac{1}{n} \sum_{i=1}^{M} A_i t_i \sigma_i(n - t_i) q(n - t_i) \\ Q(V) = \sum_{i=0}^{V} q(i) \\ [P_n]_V = \frac{q(n)}{Q(V)} \end{cases} \tag{7.68}$$

7.4.6 *Blocking Probability*

The parameter $\sigma_i(n)$ is the conditional transition probability determining what part of a call stream will be transferred from state n to state $n+1$. This implies that $1 - \sigma_i(n)$ is the conditional blocking probability, in other words the blocking probability in state n. The total blocking probability for the call stream of class i can therefore be expressed by the following formula:

$$E_i = \sum_{n=0}^{V-t_i} [P_n]_V [1 - \sigma_i(n)] + \sum_{n=V-t_i+1}^{V} [P_n]_V \tag{7.69}$$

The first part of the sum (7.69) determines the blocking probability, with the assumption that the system has free BBUs to set up a connection of a given class, whereas this connection will not be set up due to certain properties of the system, for instance, its structure that make it dependent on states. The other part of the sum (7.69) determines blocking states following the absence of free BBUs in the system, when the system does not have t_i free BBUs necessary to set up a connection of class i.

7.4.7 *Interpretation of the Generalized Kaufman–Roberts Distribution*

Figure 7.11 shows the graphical interpretation of the distribution (7.67) for a system with two call streams. A call of the first class requires $t_1 = 1$ BBUs to set up a connection, whereas a call of the second class demands $t_2 = 2$ BBUs. The parameter $y_i(n)$ in Figure 7.11 defines the number of calls of class i in the system in state n. Following the reasoning in Section 7.3.9, we obtain:

$$y_i(n + t_i) = \begin{cases} A_i \sigma_i(n) [P_n]_V \,/\, \left[P_{n+t_i}\right]_V & \text{for } n + t_i \leq V \\ 0 & \text{for } n + t_i > V \end{cases} \tag{7.70}$$

Comparing Figure 7.9 with 7.11 one can notice that the transfer of the description of the system with multi-rate traffic from the microstate level to that of macrostate leads to an approximation of the multi-dimensional Markov process (microstates) – occurring in the system being

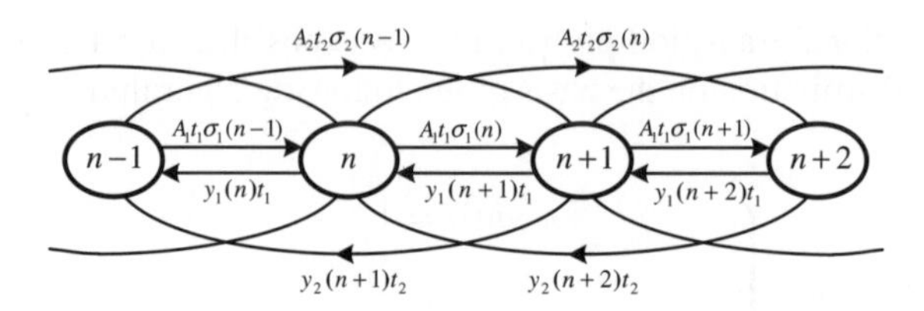

Figure 7.11 A fragment of the one-dimensional Markov chain in the state-dependent system with two call streams ($t_1 = 1, t_2 = 2$).

considered here – by one-dimensional Markov chains (macrostates). It should be emphasized that the accuracy of the approximation depends on the degree of fulfillment of Assumptions 1 and 2 by the parameter $\sigma_i(n)$. The occupancy distribution $[P_n]_V$ forms a basis for calculations of the blocking probability for call streams of particular classes in state-dependent systems. Further on in the chapter we will discuss the determination of the occupancy distributions and blocking probabilities in a number of selected state-dependent systems.

7.5 Systems with Finite and Infinite Number of Traffic Sources

In this section we will present an approximated recursive algorithm that allows us to determine the occupancy distribution for state-dependent and state-independent systems with multi-rate traffic that are offered several classes of Erlang and Engset traffic with different numbers of demanded BBUs to set up connections.

7.5.1 Assumptions

Let us consider a system with multi-service traffic with the capacity of V channels that is offered two types of traffic streams: M_1 Erlang streams (PCT1 streams) and M_2 Engset streams (PCT2 streams). The parameters of the streams will be denoted by two subscripts, i, and j, where i indicates the type of the stream (subscript "1" is attributed to PCT1 streams, subscript "2" to PCT2 streams), and the second subscript indicates the number of the given type of class.

The call intensity of Erlang stream of class i is $\lambda_{1,i}$, whereas the call intensity of Engset stream of class j is $\lambda_{2,j}(y_{2,j})$ and depends on the number of $y_{2,j}$ of currently serviced class j calls:

$$\lambda_2(y_{2,j}) = (N_{2,j} - y_{2,j})\gamma_{2,j} \tag{7.71}$$

where $\gamma_{2,j}$ is the call intensity from one free traffic source PCT2, while the number of PCT2 traffic sources of class j is equal to $N_{2,j}$.

The BBUs demanded for calls of particular classes will be denoted, as earlier, by symbols: $t_{1,i}$ and $t_{2,j}$. The service times of calls of all classes have exponential distributions with the parameters: $\mu_{1,i}$ and $\mu_{2,j}$. Therefore, the total traffic intensity of PCT1 of class i, offered to the system, is:

$$A_{1,i} = \frac{\lambda_{1,i}}{\mu_{1,i}} \tag{7.72}$$

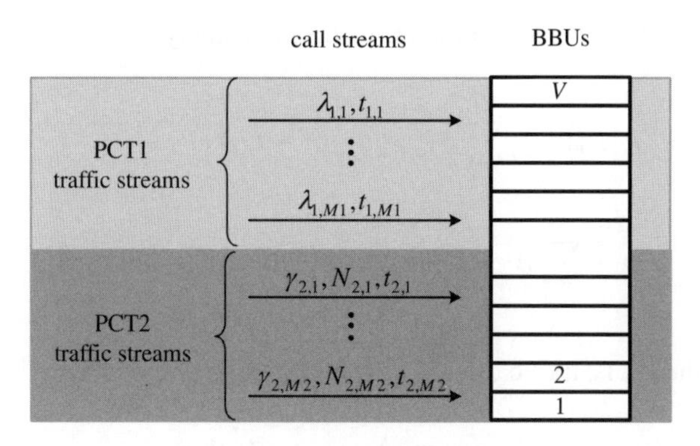

Figure 7.12 Multi-rate system with two types of different PCT1 and PCT2 traffic streams.

whereas the traffic intensity PCT2, offered by one free traffic source of class j, is equal to:

$$\alpha_{2,j} = \frac{\gamma_{2,j}}{\mu_{2,j}} \tag{7.73}$$

A model of the multi-service system under consideration here, which is offered two types of different traffic streams, Erlang and Engset streams, is shown in Figure 7.12.

In the case of the model under consideration here, just as in the case of the multi-dimensional Erlang distribution (Section 7.2), it is possible to determine the appropriate occupancy distribution at the microstate level and then, on the basis of this, determine the essential characteristics of the system. However, due to a very low applicability of such an approach, a description at the microstate level will be omitted in this chapter and we will provide the reader with an approximate solution [26, 27] – characterized by high accuracy – for the macrostate level.

It should be stressed that there is a well-known precise solution for the full-availability group with multi-rate PCT1 and PCT2 traffic expressed with Equation (7.43) in Section 7.3.8. Equation (7.43) cannot be used in the general case, in other words for state-dependent systems in which call streams between neighboring states of the service process depend on the current state of this process.

7.5.2 The Multi-Service Erlang-Engset Model

We now discuss the modeling of multi-service systems that are offered different traffic streams of the type PCT1 and PCT2. The basis for the modeling is formed by the distribution expressed by Equation (7.67), for the multi-dimensional distribution in the state-dependent system:

$$n\left[P_n\right]_V = \sum_{i=1}^{M} A_i t_i \sigma_i(n - t_i)\left[P_{n-t_i}\right]_V \tag{7.74}$$

It is assumed in the model [26] that the number of BBUs $y_{2,j}(n)$, occupied in every state n by a call of the Engset stream of class j is the same as the number of BBUs occupied by the equivalent Erlang stream generating offered traffic with the intensity $A_{2,j} = N_{2,j}\alpha_{2,j}$, equal in value to the traffic offered by all free sources of class j. If the number of BBUs occupied by Engset streams in subsequent states is known, then Equation (7.74) can be rewritten in a form that will also include the property of the PCT2 traffic, namely:

$$n\left[P_n\right]_V=\sum_{i=1}^{M_1} A_{1,i}t_{1,i}\sigma_{1,i}(n-t_{1,i})\left[P_{n-t_{1,i}}\right]_V$$
$$+\sum_{j=1}^{M_2} A_{2,j}(n-t_{2,j})t_{2,j}\sigma_{2,j}(n-t_{2,j})\left[P_{n-t_{2,j}}\right]_V \tag{7.75}$$

where $A_{2,j}(n)$ is the PCT2 traffic offered by the Engset stream of class j in macrostate n:

$$A_{2,j}(n)=[N_{2,j}-y_{2,j}(n)]\alpha_{2,j} \tag{7.76}$$

7.5.3 Calculation Algorithm

The concept presented in Section 7.5.2 defines the calculation algorithm that can be presented in four subsequent steps:

1. **Determination of the occupancy distribution of macrostates $[P_n]_V$ with the assumption that the offered call streams are Erlang streams.**
The distribution is determined on the basis of the recursive dependence (7.75), in which it is assumed that $A_{2,j}=N_{2,j}\alpha_{2,j}$:

$$n\left[P_n\right]_V=\sum_{i=1}^{M_1} A_{1,i}t_{1,i}\sigma_{1,i}(n-t_{1,i})\left[P_{n-t_{1,i}}\right]_V$$
$$+\sum_{j=1}^{M_2} N_{2,j}\alpha_{2,j}t_{2,j}\sigma_{2,j}(n-t_{2,j})\left[P_{n-t_{2,j}}\right]_V \tag{7.77}$$

2. **Determination of the number of busy channels $y_{2,j}(n)$ by calls of Engset stream in each of macrostates n ($0\leqslant n\leqslant V$).**
Having determined all macrostate probabilities $[P_n]_V$, the number of calls of the Engset stream of class j, i.e. $y_{2,j}(n)$, serviced in macrostate n, will be determined based on Equation (7.70), which in the case being considered here will be rewritten with the inclusion of traffic $A_{2,j}=N_{2,j}\alpha_{2,j}$ in the following way:

$$y_{2,j}(n)=\begin{cases} N_{2,j}\alpha_{2,j}\left[P_{n-t_{2,j}}\right]_V \Big/ \left[P_n\right]_V & \text{for } n\le V\\ 0 & \text{for } n>V\end{cases} \tag{7.78}$$

3. **Determination of the occupancy distribution of macrostates $[P_n]_V$ with the assumption that the offered call streams are Erlang and Engset streams.**
Knowing the parameters $y_{2,j}(n)$ we can once again determine the distribution of macrostates $[P_n]_V$, this time taking into consideration the state-dependent traffic PCT2. This distribution will be determined on the basis of the recursive dependence (7.75), in which it is adopted

that $A_{2,j}(n) = [N_{2,j} - y_{2,j}(n)]\alpha_{2,j}$:

$$n\,[P_n]_V = \sum_{i=1}^{M_1} A_{1,i}t_{1,i}\sigma_{1,i}(n-t_{1,i})\left[P_{n-t_{1,i}}\right]_V + \sum_{j=1}^{M_2}[N_{2,j} - y_{2,j}(n-t_{2,j})]\alpha_{2,j}t_{2,j}\sigma_{2,j}(n-t_{2,j})\left[P_{n-t_{2,j}}\right]_V \quad (7.79)$$

4. **Calculation of blocking and loss probabilities for Erlang and Engset call streams.** Erlang stream of class i:

$$E_{1,i} = B_{1,i} = \sum_{n=0}^{V-t_{1,i}}[P_n]_V\,[1-\sigma_{1,i}(n)] + \sum_{n=V-t_{1,i}+1}^{V}[P_n]_V \quad (7.80)$$

Engset stream of class j:

$$E_{2,j} = \sum_{n=0}^{V-t_{2,j}}[P_n]_V\,[1-\sigma_{2,j}(n)] + \sum_{n=V-t_{2,j}+1}^{V}[P_n]_V \quad (7.81)$$

In each macrostate n free PCT2 traffic sources of class j generate a call stream with the intensity: $[N_{2,j} - y_{2,j}(n)]\gamma_{2,j}$. The call stream lost in macrostate n is: $[N_{2,j} - y_{2,j}(n)]$ $[1-\sigma_{2,j}(n)]\gamma_{2,j}$ for $0 \le n \le V - t_{2,j}$ and is $[N_{2,j} - y_{2,j}(n)]\gamma_{2,j}$ for $n > V - t_{2,j}$. The loss probability can therefore be written in the following form:

$$B_{2,j} = \frac{\sum_{n=0}^{V-t_{2,j}}[P_n]_V\,[1-\sigma_{2,j}(n)][N_{2,j} - y_{2,j}(n)]\gamma_{2,j}}{\sum_{n=0}^{V}[P_n]_V\left[N_{2,j} - y_{2,j}(n)\right]\gamma_{2,j}} + \frac{\sum_{n=V-t_{2,j}+1}^{V}[P_n]_V\,[N_{2,j} - y_{2,j}(n)]\gamma_{2,j}}{\sum_{n=0}^{V}[P_n]_V\left[N_{2,j} - y_{2,j}(n)\right]\gamma_{2,j}} \quad (7.82)$$

The approximate recursive algorithm presented here is characterized by very high accuracy, which has been proven by simulation studies [26] and these justify its direct application in solving engineering problems. The advantage of the algorithm is its exclusive operation on macrostates, which much improves its computational power and efficiency.

7.5.4 Comments

The algorithm presented above enables us to model the radio interface of a single cell or other interfaces of a cellular network (for example, the Iub interface in the UMTS network) that service a mixture of different multi-service streams.

The model allows us to take into account the influence of the number of users belonging to different classes on the traffic characteristics of the system, such as blocking or loss probability. This differentiation in the number of users and services can be a result, for example, of the capability of mobile terminals (for instance, only some of users are equipped with terminals appropriate for videotelephony).

It is worth emphasizing that traffic generated by services that are used by a great number of users can be treated as a PCT1 stream, whereas the remaining services correspond to PCT2 streams. Examples of the application of the algorithm for the modeling and dimensioning of the radio interface with hard and soft capacity are presented in Chapter 11.

7.6 Limited-Availability Group

7.6.1 Basic Model of the Limited-Availability Group

The limited-availability group is a model of several identical separated links that are offered a mixture of different multi-rate traffic streams. These links are called subgroups. The system services a call only when this call can be entirely serviced by BBUs of one of the subgroups.

Limited-availability groups have been widely discussed, for example in: [3, 25, 28–31]. This wide interest in the models is closely related to the application of such models in studies and investigations into resource allocation in both narrow and broadband networks.

7.6.1.1 Assumptions

Figure 7.13 shows a diagram of the limited-availability group that is offered M independent call streams with the intensities: $\lambda_1, \lambda_2, \ldots, \lambda_M$. The BBUs demanded for calls of particular classes will be denoted, as earlier in the text, by the symbols: $t_1, t_2, \ldots, t_M$. Service times of calls of all classes have exponential distributions with the parameters: $\mu_1, \mu_2, \ldots, \mu_M$. Hence, the mean intensity of traffic offered by the class i stream is expressed by Equation (7.1). The group services a call only when it can be entirely serviced by BBUs of one of subgroups.

The limited-availability group is characterized by the following structural parameters:

k – number of links (subgroups) in the group;
f – link (subgroup) capacity, i.e. the number of BBUs in a subgroup;
V – total group capacity ($V = kf$).

7.6.1.2 Distribution of Free Basic Bandwidth Units

An approximated method for the occupancy distribution and the blocking probability calculation in the limited-availability group is proposed in [25]. The method applies the generalized Kaufman–Roberts distribution with appropriately defined conditional transition coefficients.

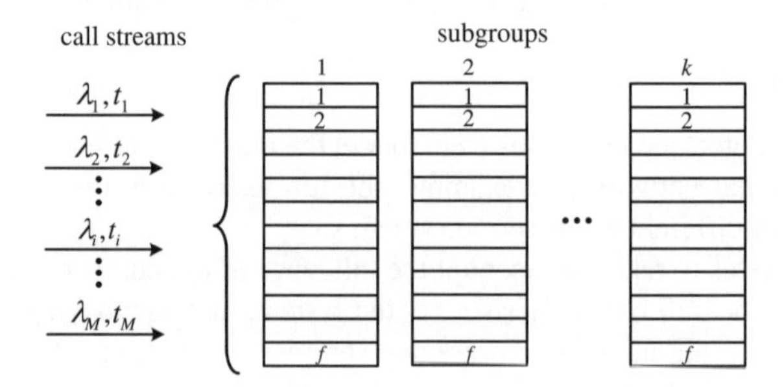

Figure 7.13 The limited-availability group.

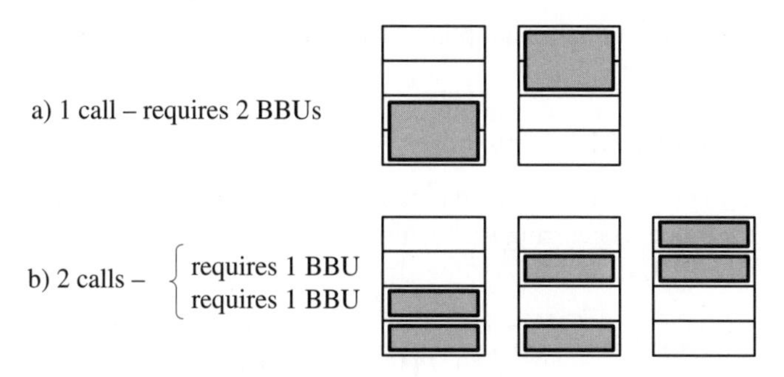

Figure 7.14 Call arrangements in the limited-availability group.

To determine these coefficients, a special combinatorial function is used that determines the number of arrangements of free BBUs in the group.

Let us consider a limited-availability group with the parameters (k, f, V). With the assumption that there are n busy and x free BBUs $(n + x = V)$ in the group, we will determine the combinatorial function that defines the number of arrangements of free BBUs. This number depends on the total number of busy BBUs and on the arrangements of busy BBUs between particular classes of calls. This phenomenon is presented in Figure 7.14. The figure shows an exemplary group composed of two subgroups, each having two BBUs. This group is offered two call streams with one demanding $t_1 = 1$ BBU, and the other demanding $t_2 = 2$ BBUs. Let us assume that the group is in a state where two BBUs are occupied $(n = 2)$. When occupied BBUs service a call of the second class, then the total number of possible arrangements is equal to 2 (Figure 7.14a). If BBUs are occupied by two calls of the first class, then the total number of arrangements is equal to 3 (Figure 7.14b).

In the method proposed in [25] it is adopted that the possible number of arrangements does not depend on the distribution of occupied BBUs between given classes of calls. Each allocation can be thus treated as a division of busy BBUs of one class $(t_1 = 1)$ between subgroups. The adoption of such an assumption has been verified with the help of simulation experiments. These simulations prove that if the relation $f \gg t_{max}$ between the capacity of one subgroup and the number of BBUs demanded by the call with maximum requirements is satisfied, then the number of real arrangements of occupied (free) BBUs in the group (in other words, the division of occupied BBUs between particular classes of calls taken into consideration) is similar to the number of allocations of occupied (free) BBUs in the group servicing only one, the youngest, class of calls $(t_1 = 1)$.

The number of possible allocations of x free BBUs in k links (subgroups), each having the capacity equal to f BBUs, with the additional assumption that there are at least g free BBUs in each subgroup, can be determined on the basis of the following combinatorial formula:

$$F(x, k, f, g) = \sum_{i=0}^{\left\lfloor \frac{x-kg}{f-g+1} \right\rfloor} (-1)^i \binom{k}{i} \binom{x - k(g-1) - 1 - i(f-g+1)}{k-1} \tag{7.83}$$

Equation (7.83) can be treated, in accordance with the classical combinatorial interpretation, as a number of possible allocations of x balls in k boxes, where each of which can accommodate f balls, with the additional assumption that g balls have been earlier placed in each of the boxes.

7.6.1.3 Conditional Transition Probability

Equation (7.83) allows us to determine the conditional transition probability between neighboring states of the process occurring in the limited-availability group in state of n busy BBUs (i.e. $x = V - n$ free BBUs):

$$\sigma_i(n) = \frac{F(V-n,k,f,0) - F(V-n,k,t_i-1,0)}{F(V-n,k,f,0)} = 1 - \frac{F(V-n,k,t_i-1,0)}{F(V-n,k,f,0)} \tag{7.84}$$

The parameter $F(V-n,k,f,0)$ in Equation (7.84) expresses the number of possible allocations of $V-n$ free BBUs in k subgroups, while the parameter $F(V-n,k,t_i-1,0)$ is the number of possible allocations in k subgroups, with the assumption that the capacity of each subgroup is limited to $t_i - 1$ BBUs. This means that $F(V-n,k,t_i-1,0)$ expresses the number of unfavorable allocations, i.e. those in which a class i call cannot be serviced. Thus, the numerator of Equation (7.84) defines the number of favorable allocations and $\sigma_i(n)$ is the probability of such an allocation of free BBUs in the occupancy state n in which a call of class i can be serviced. After the substitution of (7.83) into (7.84), we finally obtain:

$$\sigma_i(n) = \frac{-\sum_{j=0}^{\left\lfloor \frac{V-n}{t_i} \right\rfloor} (-1)^j \binom{k}{j} \left[\binom{V+k-n-jt_i-1}{k-1} - \binom{V+k-n-j(f+1)-1}{k-1} \right]}{\sum_{j=0}^{\left\lfloor \frac{V-n}{f+1} \right\rfloor} (-1)^j \binom{k}{j} \binom{V+k-n-j(f+1)-1}{k-1}} \tag{7.85}$$

7.6.1.4 The Occupancy Distribution in the Limited-Availability Group

Having determined the conditional transition probabilities $\sigma_i(n)$ it is possible to determine the occupancy distribution in the limited-availability group on the basis of Equation (7.67) and then, with the application of the dependence (7.69), determine the blocking probability for each class of calls. Further on in the text the occupancy distribution in the limited-availability group, described by Equations (7.67) and (7.85), will be labeled as the limited-availability group distribution.

The simulation studies [25], carried out for various structures of limited-availability groups and different streams of multi-service traffic, corroborated the high accuracy of the proposed model. The studies prove that with the increase of the parameter f (subgroup capacity), in comparison to the number of BBUs required by a call with maximum demands $t_{\max}$, the values $\sigma_i(n)$ become less and less receptive to the dependence on the number of busy BBUs and to the division of this number into different call classes. The probability $\sigma_i(n)$ becomes a slowly changing function of the parameter n. Therefore, for $f \gg t_{\max}$, the limited-availability group distribution will be a good approximation of the real occupancy distribution in the limited-availability group. In practical terms, for $f \geq 5t_{\max}$, the results of calculations following the proposed formulas can be regarded as accurate and precise. In [25] an exemplary system in which $f < 2t_{\max}$ is presented and it is proved that the relative error in such a group in extreme cases does not exceed 10%. The above confirms that the proposed formulas are also applicable in those groups for which the relation $f \geq 5t_{\max}$ is not satisfied.

7.6.2 Generalized Model of the Limited-Availability Group

Until now we have been considering a model of separated links with equal capacity. This chapter will discuss a model of the limited-availability group composed of links with different capacities.

7.6.2.1 Assumptions

Let us assume that the limited-availability group is composed of q types of links. Each type is unequivocally defined by the number k_q of links of a given type and the capacity f_q of each of the links of a given type (Figure 7.15). The total capacity of the limited-availability group with different capacities of links can thus be expressed by the formula below:

$$V = \sum_{s=1}^{q} k_s f_s \tag{7.86}$$

We assume, as in the previous model, that the group is offered M independent call streams with the intensities: $\lambda_1, \lambda_2, \ldots, \lambda_M$. The demanded BBUs for calls of particular classes will be denoted, as earlier, with the symbols $t_1, t_2, \ldots, t_M$. Service times of calls of all classes have exponential distributions with the parameters $\mu_1, \mu_2, \ldots, \mu_M$. The system under consideration here with multi-rate traffic can admit a call of a given class for service only when it can be entirely serviced by the resources of one of the links.

7.6.2.2 Occupancy Distribution

An approximated method of calculation of occupancy distribution and blocking probability in the limited-availability group with different capacities of subgroups is proposed in [28]. The

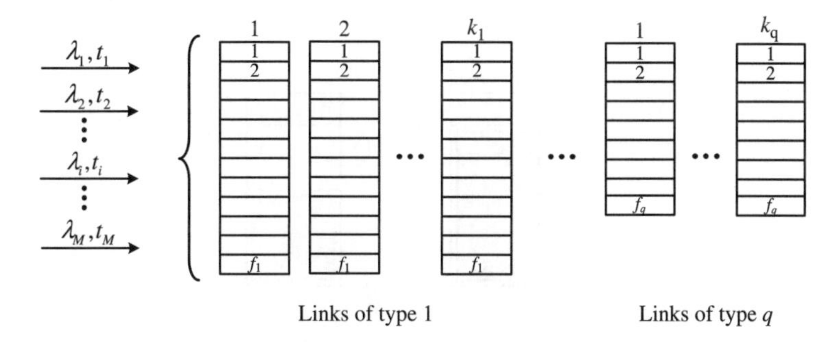

Figure 7.15 Generalized limited-availability group model.

method is based on the application of the distribution (7.67):

$$n\,[P_n]_V = \sum_{i=1}^{M} A_i t_i \sigma_i(n-t_i)\left[P_{n-t_i}\right]_V \tag{7.87}$$

where $[P_{n-t_i}]_V = 0$, if $n < t_i$, and the value $[P_0]_V$ results from the normalization condition (7.40).

The conditional transition probability $\sigma_i(n)$ for the stream of class i in the generalized limited-availability group model with the parameters: q, k_q, f_q, V, is determined – similarly as in the basic limited-availability group model (Section 7.6.1.3) – with the assumption that there are n busy BBUs in the considered group and that each allocation of free BBUs is treated as a division of busy BBUs of one class ($t = 1$) in the subgroups.

7.6.2.3 Allocations of Free BBUs in the Group with Two Types of Links

Let us consider first the limited-availability group composed of links of two types ($q = 2$). In the case being considered here, the total capacity of the group V can be presented as the sum of the capacities of the links of the first and the second type, in other words

$$V = V_1 + V_2, \quad \text{where: } V_1 = k_1 f_1,\ V_2 = k_2 f_2 \tag{7.88}$$

The problem of the determination of the number of all possible allocations of x free BBUs in the group, in this particular case, can be considered in two stages:

1. The first stage includes a determination of the number of all possible ways of allocating of x free BBUs in the links of two types, in other words all possibilities of a division of x BBUs into x_1 BBUs in the links of the first type and $x - x_1$ BBUs in the links of the second type.
2. In the second stage we determine the number of possible allocations of a given number of BBUs in the links of the same type, in other words the number of allocations of x_1 BBUs in the links of the first type (with the constraint in the capacity of the link to f_1 BBUs taken into consideration) and $x - x_1$ BBUs in the links of the second type (with the constraint in the capacity of the link to f_2 BBUs) (Figure 7.16).

The number of allocations determined in the second stage is determined on the basis of Equation (7.83). Eventually, the number of possible allocations of x free BBUs in the links of

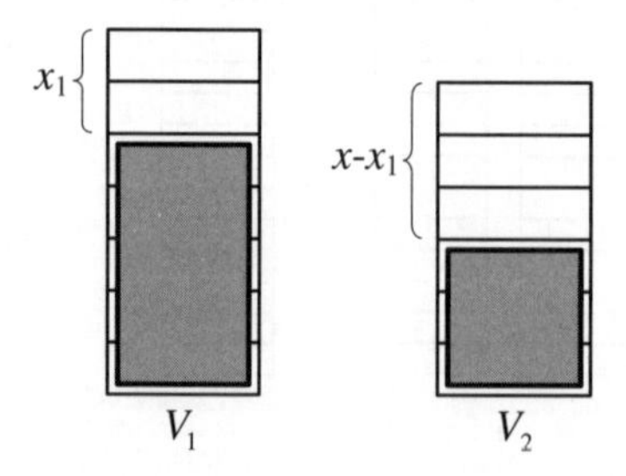

Figure 7.16 Possible allocations of x free BBUs in the links of two types.

the two types – with the assumption that each link of the first type has at least g_1 free BBUs and that there are at least g_2 free BBUs in each link of the second type – can be expressed by the formula:

$$F\{x, (k_1, k_2), (f_1, f_2), (g_1, g_2)\} = \sum_{x_1=0}^{x} F(x_1, k_1, f_1, g_1)F(x - x_1, k_2, f_2, g_2) \quad (7.89)$$

7.6.2.4 Allocations of Free BBUs in the Group with q Types of Links

Continuing our considerations from Section 7.6.2, we can determine – in exactly the same way as for the group composed of links of two types – the number of possible allocations of x free BBUs in groups composed of links of three types:

$$F\{x, (k_1, k_2, k_3), (f_1, f_2, f_3), (g_1, g_2, g_3)\}$$
$$= \sum_{x_1=0}^{x} \sum_{x_2=0}^{x-x_1} F(x_1, k_1, f_1, g_1)F(x_2, k_2, f_2, g_2)F(x - x_1 - x_2, k_3, f_3, g_3) \quad (7.90)$$

Let us generalize Equation (7.90) for groups composed of links of q types:

$$F\{(x, (k_1, k_2, \ldots, k_q), (f_1, f_2, \ldots, f_q), (g_1, 2_2, \ldots, g_q)\}$$
$$= \sum_{x_1=0}^{x} \sum_{x_2=0}^{x-x_1} \cdots \sum_{x_{q-1}=0}^{x-\sum_{r=1}^{q-2} x_r} F(x_1, k_1, f_1, g_1)F(x_2, k_2, f_2, g_2)$$
$$\ldots F(x_{q-1}, k_{q-1}, f_{q-1}, t_{q-1})F(x - \sum_{r=1}^{q-1} x_r, k_q, f_q, g_q) \quad (7.91)$$

Equation (7.91) can eventually be rewritten in the following form:

$$F\{(x, (k_1, k_2, \ldots, k_q), (f_1, f_2, \ldots, f_q), (g_1, g_2, \ldots, g_q)\}$$
$$= \sum_{x_1=0}^{x} \cdots \sum_{x_{q-1}=0}^{x-\sum_{r=1}^{q-2} x_r} \left\{ \prod_{z=1}^{q-1} F(x_z, k_z, f_z, g_z) \cdot F(x - \sum_{r=1}^{q-1} x_r, k_q, f_q, g_q) \right\} \quad (7.92)$$

7.6.2.5 Conditional Transition Probability

Equation (7.92) enables us to determine the conditional transition probability between neighboring states of the process occurring in the limited-availability group with q types of links in state of n busy BBUs (i.e. $x = V - n$ free BBUs). Following the same procedure as in Section 7.6.1, the conditional transition probability $\sigma_i(n)$ in the generalized limited-availability group model (composed of q types of links) can be determined on the basis of Equation (7.84) modified the following way:

$$\sigma_i(n) = 1 - \frac{F\{(V-n, (k_1 \ldots k_q), (t_i - 1, \ldots, t_i - 1), (0 \ldots 0)\}}{F\{(V-n, (k_1 \ldots k_q), (f_1 \ldots f_q), (0 \ldots 0)\}} \tag{7.93}$$

in which the value of the combinatorial function $F(\cdot)$ is determined on the basis of Equation (7.92).

After determining all probabilities $\sigma_i(n)$ (Equation (7.93)), it is possible, following the application of Equation (7.87), to determine the occupancy distribution $[P_n]_V$ in the limited-availability group with different types of links.

7.6.2.6 Blocking Probability and Loss Probability

The blocking state in the generalized limited-availability group model occurs when none of the links of a group has a sufficient number of free BBUs to service a call of class i. This means that each occupancy state n in a link of the type q, such as: $(f_q - t_i + 1 \leq n \leq f_q)$, is a blocking state. Any possible blocking states for a stream of class i in the limited-availability group, composed of links of q types, can be thus determined by the condition below:

$$V - \sum_{s=1}^{q} k_s(t_i - 1) \leq n \leq V \tag{7.94}$$

On the basis of the values determined for the conditional transition probabilities $\sigma_i(n)$ (Equation (7.93)) and the occupancy distribution $[P_n]_V$ (Equation (7.87)), the blocking probability for calls of class i can be determined by the following formula:

$$E_i = B_i = \sum_{n=V-\sum_{s=1}^{q} k_s(t_i-1)}^{V} [1 - \sigma_i(n)][P_n]_V \tag{7.95}$$

7.6.3 Comments

The models of the limited-availability group presented here can be generalized by taking into account the service of PCT1 and PCT2 traffic streams. If this is the case, then the occupancy distribution expressed by the dependence (7.87) will undergo changes and takes on the following distribution form (7.79):

$$\begin{aligned} n\,[P_n]_V &= \sum_{i=1}^{M_1} A_{1,i} t_{1,i} \sigma_{1,i}(n - t_{1,i}) \left[P_{n-t_{1,i}}\right]_V \\ &+ \sum_{j=1}^{M_2} [N_{2,j} - y_{2,j}(n - t_{2,j})] \alpha_{2,j} t_{2,j} \sigma_{2,j}(n - t_{2,j}) \left[P_{n-t_{2,j}}\right]_V \end{aligned} \tag{7.96}$$

The parameters $\sigma_{1,i}$ and $\sigma_{2,j}$ in Equation (7.96) can be determined on the basis of the dependence (7.85), for PCT1 traffic streams and PCT2 traffic streams. A change will also follow in the method for a determination of blocking and loss probability (7.95). For this purpose, the dependencies (7.80) and (7.81) will be used for a determination of the blocking probability, while Equations (7.80) and (7.82), with the dependence (7.85) taken into consideration, will be used to determine the loss probability.

The model of the limited-availability group presented in this chapter can be applied to modeling the phenomenon of transferring connections in the radio interface in cellular networks. Examples of practical applications of the discussed model are presented in Section 13.4.1. This model can be also used for the traffic analysis in the GSM system that uses quick data transmission with HSCSD channel switching (Section 1.5), in which there is a number of transivers available in one cell.

7.7 Full-Availability Group with Reservation

Systems with multi-service traffic should ensure high quality of service for calls of different traffic classes. In broadband networks with integrated services, the reservation mechanism is associated with the call admission control function [7]. The goal of this function is to ensure optimal access to network resources for calls of different traffic classes. The efficiency of the call admission control function depends, then, on the adopted strategy of network resources allocation for different traffic classes. One of possible strategies is the bandwidth reservation for particular traffic streams. The advantage of reservation mechanisms is the provision of similar values of quality of service parameters, for instance loss probability, for calls of different classes of offered traffic.

7.7.1 Bandwidth Reservation

Bandwidth reservation consists in a determination of the so-called reservation threshold Q for each of traffic classes. This parameter defines such a boundary occupancy state of a group for calls of a given class, in which servicing calls of this class will still be possible. All "older" states than the reservation threshold belong to the so-called reservation space R, in which calls of a given class will be blocked. The idea of bandwidth reservation in the full-availability group with multi-rate traffic is presented in Figure 7.17.

The problem of dynamic reservation has been widely discussed in literature of the subject, for example in: [9, 31–36]. The article [33] gives the most commonly used approximate and recursive method for the occupancy distribution calculation and for the calculation of the equalized blocking probability in the full-availability group with multi-rate traffic and reservation. The article [33] provides a formulation of the so-called blocking probability equalization rule that was successively generalized in [35].

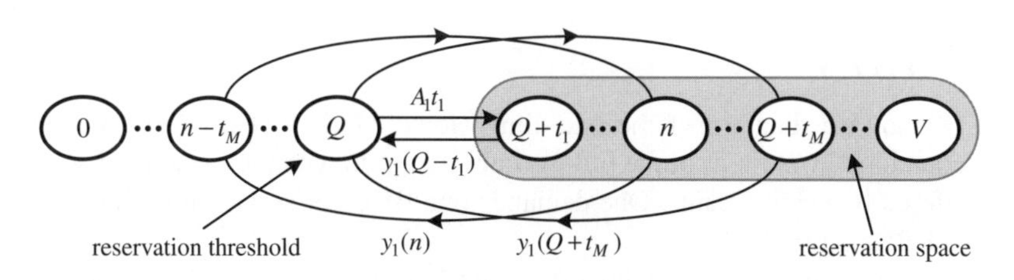

Figure 7.17 Reservation threshold in the full-availability group with multi-rate traffic.

7.7.2 Blocking Probability Equalization Rule

Let us consider a full-availability group model (Figure 7.17) with multi-rate traffic [33]. We introduce reservation thresholds Q_i in a full-availability group with the capacity V for each traffic class independently. The parameter Q_i determines such a boundary state of the system in which servicing a call of class i is still possible. All states higher than Q_i belong to reservation space R_i, in which class i calls will be blocked:

$$R_i = V - Q_i \tag{7.97}$$

According to the blocking probability equalization rule [33], the blocking probability of calls for all traffic classes in the group will be equalized if the reservation threshold for all traffic classes is the same and its value is equal to the difference between the group capacity and the maximum number of BBUs required by a call of the oldest class (demanding $t_M = t_{\max}$ BBUs):

$$Q = V - t_{\max} \tag{7.98}$$

Such a definition of the reservation threshold implies that, in state $V - t_{\max}$, the group can service a call of any class, while in state $V - t_{\max} + 1$ all calls are blocked.

7.7.3 Occupancy Distribution in the Group with Reservation

The analytical model for blocking probability calculation in the group with reservation proposed in [33] involves a determination of the occupancy distribution with the application of the generalized Kaufman–Roberts formula (7.67) in which the parameter $\sigma_i(n)$ is defined as below:

$$\sigma_i(n) = \begin{cases} 0 & \text{for } n > Q \\ 1 & \text{for } n \leq Q \end{cases} \tag{7.99}$$

The equalized blocking probability E for calls of all classes is determined by the following formula:

$$E = \sum_{n=Q+1}^{V} [P_n]_V \tag{7.100}$$

Equations (7.67) and (7.99) determine the occupancy distribution in the full-availability group with reservation hereafter referred to as the distribution of the full-availability group with reservation.

7.7.4 Comments

Figure 7.18 shows a diagram of the one-dimensional Markov chain corresponding to the recursive equation (7.67) for a simple full-availability group with the capacity of $V = 3$ BBUs which is offered two call streams. One demands one BBU for servicing its call ($t_1 = 1$), the other – two BBUs ($t_2 = 2$).

Figure 7.19 presents – in line with Equations (7.67) and (7.99) – a diagram of a one-dimensional Markov chain approximating the processes occurring in the group, with the

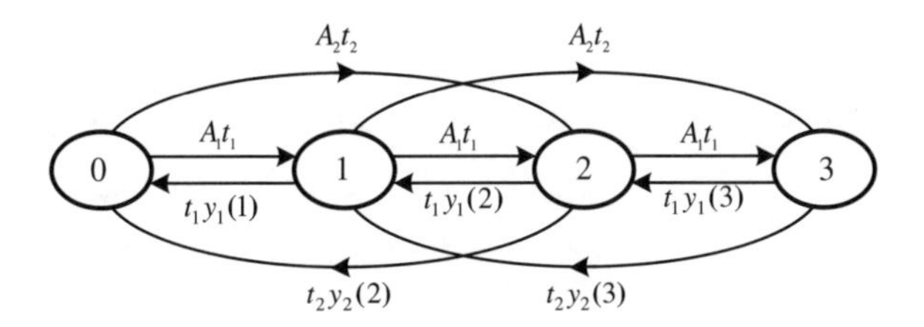

Figure 7.18 The Markov process in the full-availability group without reservation ($V = 3$, $t_1 = 1$, $t_2 = 2$).

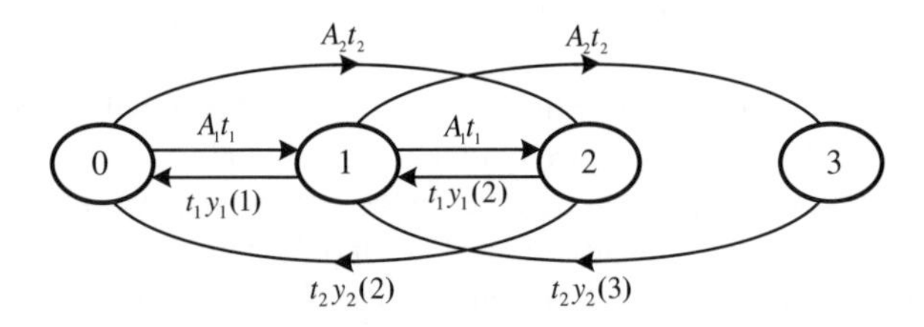

Figure 7.19 The Markov process in the full-availability group with reservation ($V = 3, t_1 = 1, t_2 = 2,$ $Q = 1$).

assumption that a reservation mechanism has been introduced into the group with a threshold ensuring the equalization of the blocking probability: $Q = V - t_2 = 1$. It is noticeable that the introduction of the reservation threshold is followed by the elimination of call and service streams of all classes within the reservation space. In accordance with Figure 7.19, the blocking probability for the calls of the first class is equal to $[P_2]_3 + [P_3]_3$, because in these states the calls of the first class will not be serviced. The calls of the second class will not be serviced in states "2" and "3" either, which means that the blocking probabilities for class 1 and 2 are the same.

The generalized Equation (7.67) – forming a basis for the full-availability group with reservation model – is an approximate distribution. Numerous simulation studies (for instance [9]) have confirmed high accuracy of the full-availability group with reservation distribution and its high potential applicability in practical engineering. Furthermore, other works such as [37, 38] and [35], analyze the influence of differences of average service times and occupancy distributions on the accuracy of the full-availability group with reservation model. As a result of the conducted studies it has been proved that the influence of the average service time and the occupancy distribution on the accuracy of full-availability group with reservation is, from the point of view of engineering applications, insignificant and thus can be omitted [9].

7.7.5 Modified Model of the Full-Availability Group with Reservation

The algorithm for modeling groups with reservation on the basis of a one-dimensional Markov chain presented in the previous section is an approximate algorithm. In order to improve the accuracy of calculations many publications, for example [5], present models of reservation systems based on the multi-dimensional Markov processes. In [39] a simple modification to the full-availability group with reservation distribution is proposed that increases the accuracy

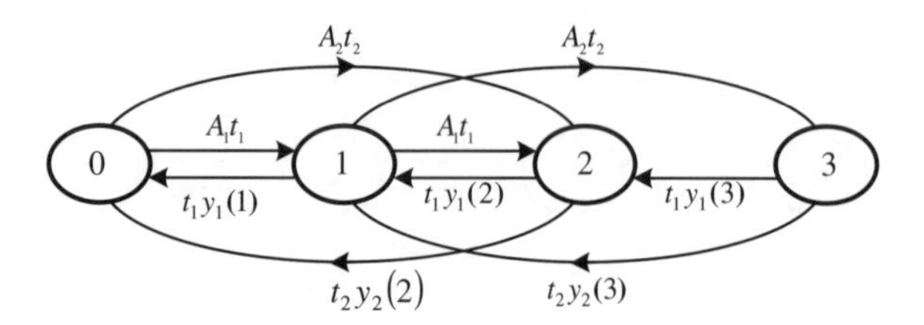

Figure 7.20 Modified diagram of the Markov process in the full-availability group with reservation ($V = 3, t_1 = 1, t_2 = 2, Q = 1$).

of blocking probability calculations in the full-availability group with reservation by more than twofold.

In the previous section the model of the full-availability group with reservation, proposed in [33, 35], is discussed. The model does not take into account the service of call streams of younger traffic classes in the reservation space. In the process shown in Figure 7.19, a call of the first class ($t_1 = 1$) in the occupancy state "3" cannot be terminated. In reality, the system in this state services one call of class 1 and one of class 2. The modified and improved model should therefore have, in state "3," a capability of terminating the service of a call of class 1 and thus be capable of ensuring a transition to state "2." The modified diagram of Markov chain that takes into consideration the service stream of class 1 in the reservation space is presented in Figure 7.20.

In the reservation space local statistical equilibrium equations for calls of all classes, with the exception of the class with the maximum demands $t_{\max}$ (class 2 in Figure 7.20), cannot be formulated. Thus, the determination of the occupancy distribution also cannot take into consideration the service streams presented in Figure 7.20. Therefore, such a system does not have a product-form solution and, consequently, cannot be directly described by recursive equations (7.67).

The introduction of the reservation threshold Q, determined by Equation (7.98) in the full-availability group, results in a situation where in the reservation space R (a set of all occupancy states n such that $Q < n \leq V$), call streams of all traffic classes, except traffic demanding the highest number of BBUs, are equal to zero. An example diagram of a one-dimensional Markov chain in the full-availability group is presented in Figure 7.20. In state "1" the group services one call of the first class ($t_1 = 1$). If, while servicing the call, a new call of the second class arrives ($t_2 = 2$), then the system switches to state "3." In this state one call of the first class is still serviced. This fact indicates that service streams of the first class are identical in states "1" and "3," or that the service stream of the first class $y_1(1)$ has been transferred from state "1" to state "3" by a traffic stream of the second class.

Let us generalize our observations by interpreting the conditional transition probability $\sigma_i(n)$. Let us assume that the traffic stream outgoing from state $n - t_i$ is equal to $A_i t_i$ with the probability $\sigma_i(n)$, and equal to zero with the probability $1 - \sigma_i(n)$.

Let us consider the first case, in which we assume that the traffic stream is equal to $A_i t_i$. Then the service stream of class i (average number of calls of class i in state n) is the same as in the full-availability group and can be determined on the basis of Equation (7.48) in the following way:

$$y_i(n) = A_i t_i \left[P_{n-t_i}\right]_V / \left[P_n\right]_V \quad \text{for } t_i \leq n \leq V \tag{7.101}$$

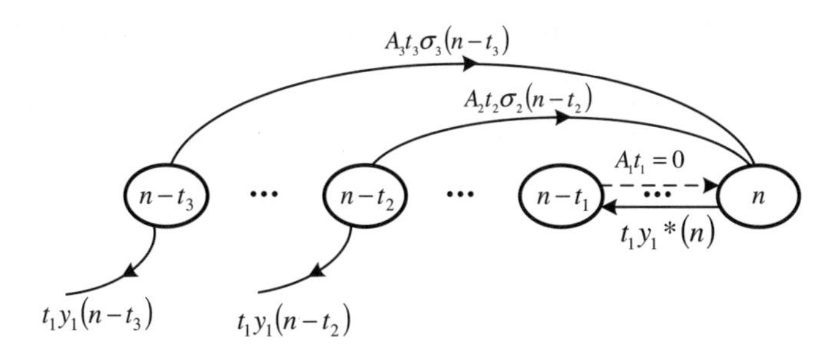

Figure 7.21 Transferring service streams of class 1 to state n ($A_1t_1 = 0$).

Let us now consider the second case, in which we assume that the traffic stream is equal to zero with the probability $1 - \sigma_i(n)$. In this case, the local equilibrium equation of type (7.47) cannot be applied to determine the service stream because the traffic stream is equal to zero – in other words, no traffic stream of class i enters state n. As was illustrated in Figure 7.20, this does not mean, however, that the parameter $y_i(n)$ takes on a value equal to zero because the service stream of class i is transferred to state n by streams that belong to other classes – in other words, which are different from i. This phenomenon is illustrated in Figure 7.21, where $A_1t_1 = 0$. In this situation the service stream of the first class $y_1(n - t_2)$ can be transferred from state $n - t_2$ to state n by a traffic stream of the second class $A_2t_2\sigma_2(n - t_2)$, and similarly, the service stream of the first class $y_1(n - t_3)$ can be transferred from state $n - t_3$ to state n by a traffic stream of the third class $A_3t_3\sigma_3(n - t_3)$. The total service stream of the first class in state n is then the sum of service streams of the first class transferred to state n by traffic streams of other classes (the second and the third in Figure 7.21) with appropriate weights. Weights $w_{i,j}(n)$ determine this part of the total service stream of class i in state n, which has been transferred to the state by the stream of class j. In [39] it is assumed that the weight is directly proportional to the value of the relevant traffic intensity associated with the call stream transferring the service stream of class i to state n:

$$w_{i,j}(n) = A_jt_j\sigma_j(n - t_j) \Bigg/ \sum_{k=1,k \neq i}^{M} A_kt_k\sigma_k(n - t_k) \tag{7.102}$$

Following the above reasoning, the modified value of the service stream $y_i^*(n)$ of class i in state n is determined:

$$y_i^*(n) = \sigma_i(n - t_i)y_i(n) + [1 - \sigma_i(n - t_i)] \left[\sum_{j=1,j \neq 1}^{M} y_i^*(n - t_j)w_{i,j}(n) \right] \tag{7.103}$$

Summing up, in order to evaluate modified values of service streams in a multi-service system with bandwidth reservation, the prerequisite is to calculate – based on Equations (7.39) and (7.48) – the occupancy distribution and the values of service streams in the full-availability group. Then, after determining appropriate weight coefficients (7.102), it is possible with Equation (7.103) to determine modified values of parameters $y_i^*(n)$. The value $y_i^*(n)$ depends on values $y_i^*(n-t_j)$ calculated earlier, so the calculation is done recursively starting with $n=0$.

After the calculation of reverse transitions rates in the multi-service system, it is possible to calculate the occupancy distribution of the system by modifying Equation (7.67) in the following way:

$$n^* [P_n]_V = \sum_{i=1}^{M} A_i t_i \sigma_i(n-t_i) \left[P_{n-t_i}\right]_V \tag{7.104}$$

The parameter n^* defines the total service stream in state n, which can be determined as follows:

$$n^* = \sum_{i=1}^{M} t_i y_i^*(n) \tag{7.105}$$

Equations (7.104) and (7.105) determine the modified model of the multi-service system. The model will be then applied to determine the occupancy distribution of the group with bandwidth reservation.

Let us consider again the full-availability group with the reservation threshold (7.98) that allows blocking probabilities of calls of all classes to be equalized. The values of parameter $\sigma_i(n)$ in the group are determined by Equation (7.99). Note that if the parameter $\sigma_i(n-t_i)$ in Equation (7.103) is equal to unity, then $y_i^*(n) = y_i(n)$. In the reverse case, when $\sigma_i(n-t_i)=0$, Equation (7.103) takes the following form:

$$y_i^*(n) = \sum_{j=1, j \neq i}^{M} y_i^*(n-t_j) w_{i,j}(n) \tag{7.106}$$

The diagram of the one-dimensional Markov process for the group with reservation is constructed in such a way (Figure 7.17) that the service stream of class i calls is linked with a given state n if the condition $n \leq Q + t_i$ is satisfied. In the reverse case, in other words for $n > Q + t_i$, service streams do not exist. Hence, the modified values of service streams in the group with reservation take on the following values:

$$y_i^*(n) = \begin{cases} y_i(n) & \text{for } n \leq Q + t_i \\ \sum_{j=1, j \neq i}^{M} y_i^*(n-t_j) w_{i,j}(n) & \text{for } Q + t_i < n \leq V \end{cases} \tag{7.107}$$

where $y_i(n)$ is the service stream of calls of class i in the full-availability group without reservation, and $y_i^*(n)$ is the service stream in the group with reservation.

The formulas derived above allow us to determine parameters of a modified one-dimensional Markov chain that approximates the service process in the full-availability group with reservation. The formulas can be used to determine the equalized blocking probability with the application of the following sequence of operations:

- determination of the occupancy distribution in the full-availability group without reservation (Formula (7.39));
- determination of service streams $y_i(n)$ in the full-availability group without reservation (Equation (7.101));

- calculation of service streams of particular classes $y_i^*(n)$ in the full-availability group with reservation (Equation (7.102) and (7.107));
- determination of total values of service streams n^* in particular states of the group with reservation (Equation (7.105));
- determination of the modified occupancy distribution $[P_n]_V$ in the full-availability group with reservation (Equation (7.104)) and the equalized blocking probability (Equation (7.100)). Equations (7.104), (7.106) and (7.107) determine the modified distribution in the full-availability group with reservation.

In [39] the results of analytical calculations are compared with the results of simulations of various groups with reservation that have been offered a mixture of call streams with different proportions. The simulation studies involved indicate the high accuracy of the modified full-availability group with reservation distribution, which is at least twofold higher than that of the full-availability group with reservation distribution (relative error comparison). In conclusion, one can state that the inclusion of service streams in the reservation space into the calculations increases the accuracy of blocking probability calculations to such a degree that the obtained results can be regarded as precise. Simulation studies have confirmed the correctness of all the theoretical assumptions adopted for the modified full-availability group with reservation distribution.

7.7.6 Comments

The models of the full-availability groups with dynamic reservation presented in this chapter can be generalized after taking into consideration the service of PCT1 and PCT2 traffic streams. In this case, the occupancy distribution expressed by the dependence (7.104) will change and take on the following form:

$$n\,[P_n]_V = \sum_{i=1}^{M_1} A_{1,i} t_{1,i} \sigma_{1,i}(n - t_{1,i}) \left[P_{n-t_{1,i}}\right]_V + \sum_{j=1}^{M_2} [N_{2,j} - y_{2,j}(n - t_{2,j})] \alpha_{2,j} t_{2,j} \sigma_{2,j}(n - t_{2,j}) \left[P_{n-t_{2,j}}\right]_V \qquad (7.108)$$

The parameters $\sigma_{1,i}$ and $\sigma_{2,j}$ will be determined on the basis of the dependence (7.99) for traffic streams PCT1 and PCT2, respectively, as appropriate. The method for determination of blocking and loss probabilities (Equation (7.100)) will also be changed. To determine the blocking probability the dependencies (7.80) and (7.81) will be used, whereas to calculate the loss probability Equations (7.80) and (7.82) will be applied, with the dependence (7.99) taken into account.

The model of the full-availability group with reservation presented in this chapter can be applied to, for example, modeling of the phenomenon of soft handover in the UMTS network (Section 13.4.3). The model can be also used in the evaluation of traffic capacity of other interfaces in the backbone network in which the dynamic resources reservation mechanism is used.

7.8 Full-Availability Group with Threshold Mechanism

In this section we will present models of full-availability groups with multi-rate traffic in which traffic parameters can change depending on the load of the system. For example, an increase in the load exceeding a given occupancy state Q in such a system can effect in a change in the number of demanded BBUs with a call of a given class. Below, we will consider the following models:

- single-threshold model (Section 7.8.1);
- multi-threshold model (Section 7.8.2).

7.8.1 Single-Threshold Models

7.8.1.1 Assumptions

Let us consider a single-threshold model [40, 41]. It is a full-availability group with multi-rate traffic of the type PCT1 (all traffic streams are Erlang streams), with the capacity of V BBUs, in which the so-called *threshold* Q (identical for all M classes) is introduced. This means that in all states n, such that: $0 \leq n \leq Q$ (the so-called *pre-threshold area*) the traffic stream of class j ($j \in \{1, \ldots, M\}$) is defined by the parameters: $\{\lambda_j, t_{j,0}, \mu_{j,0}\}$, whereas for all states n older than Q, in other words such that $Q < n \leq V$ (the so-called *post-threshold area*) this stream is defined by the parameters: $\{\lambda_j, t_{j,1}, \mu_{j,1}\}$. Generally, in the models provided by the literature it is assumed that $t_{j,0} > t_{j,1}$ and $\mu_{j,0}^{-1} < \mu_{j,1}^{-1}$. This situation occurs in the case of calls belonging to the so-called elastic traffic classes [42]. In the case of adaptive traffic, the parameter μ_j does not undergo any changes, i.e. $\mu_{j,0} = \mu_{j,1} = \mu_j$ [42].

The model of the single-threshold system considered here is presented in Figure 7.22, where the parameters of the offered and service streams in the pre-threshold ($0 \leq n \leq Q$) and post-threshold ($Q \leq n \leq V$) area were designated.

7.8.1.2 Occupancy Distribution of the Basic Single-Threshold Model

The service processes occurring in the single-threshold system are not reversible because, depending on the occupancy state of a group, the parameters of the offered traffic stream change. So we take the view that Assumption 1 and 2 (Section 7.4.4) are satisfied. For these reasons it is possible for all streams to be treated independently and to analyze the process on the basis of the

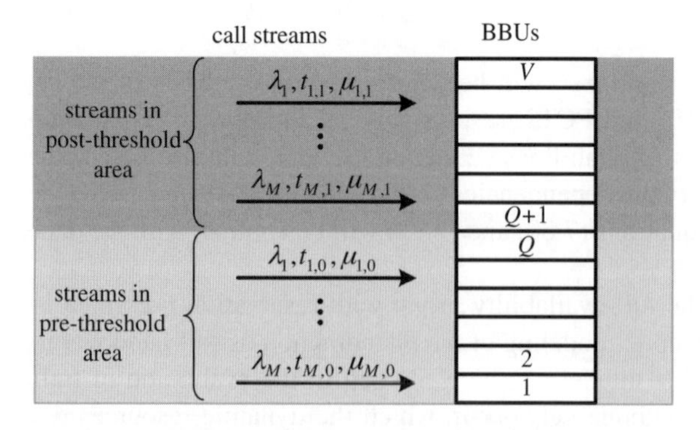

Figure 7.22 Full-availability group with single-threshold mechanism.

local equilibrium equations. This, in turn, means that it is possible to approximate the occupancy distribution in the single-threshold system by the generalized Kaufman–Roberts distribution (Section 7.4.5). With this in mind, let us consider first the local equilibrium equations at the macrostate level (Equation (7.65), Section 7.4.5) in the pre-threshold area ($0 \leq n \leq Q$):

$$x_i \mu_{i,0}[p(x_1, \ldots, x_i, \ldots, x_M)]_V$$
$$= \lambda_i \sigma_{i,0}(x_1, \ldots, x_i - 1, \ldots, x_M)[p(x_1, \ldots, x_i - 1, \ldots, x_M)]_V \quad (7.109)$$

Assumption 1 (Section 7.4.4) enables us to rewrite Equation (7.109) in the following form:

$$x_i \mu_{i,0}[p(x_1, \ldots, x_i, \ldots, x_M)]_V$$
$$= \lambda_i \sigma_{i,0}(n - t_{i,0})[p(x_1, \ldots, x_i - 1, \ldots, x_M)]_V \quad (7.110)$$

where:

$$n = \sum_{i=1}^{M} x_i t_{i,0} \quad (7.111)$$

while the conditional transition coefficient $\sigma_{i,0}(n)$ determines the pre-threshold area in which Equation (7.110) is still valid, and the stream of offered traffic is described by the parameters $\{\lambda_i, t_{i,0}, \mu_{i,0}\}$:

$$\sigma_{i,0}(n) = \begin{cases} 1 & \text{for } n \leq Q \\ 0 & \text{for } n > Q \end{cases} \quad (7.112)$$

We have assumed that we treat all call streams that are offered to the group in exactly the same way as independent streams, so for microstate $\{x_1, x_2, \ldots, x_M\}$ we can add up all M equations of the type (7.110), which results in the following equation

$$[p(x_1, \ldots, x_i, \ldots, x_M)]_V \sum_{i=1}^{M} x_i t_{i,0}$$
$$= \sum_{i=1}^{M} A_{i,0} t_{i,0} \sigma_{i,0}(n - t_{i,0})[p(x_1, \ldots, x_i - 1, \ldots, x_M)]_V \quad (7.113)$$

where $A_{i,0}$ is the traffic intensity of class i, offered to the group in the pre-threshold area $0 \leq n \leq Q$:

$$A_{i,0} = \frac{\lambda_i}{\mu_{i,0}} \quad (7.114)$$

After taking into account the definition of the macrostate (Section 7.3.4, Equation (7.32)) and the application of the procedure that is exactly the same as with the case of the full-availability group (Section 7.3.5, Equation (7.113)) can be finally written as:

$$n\,[P_n]_V = \sum_{i=1}^{M} A_{i,0} t_{i,0} \sigma_{i,0}(n - t_{i,0}) \left[P_{n-t_{i,0}}\right]_V \quad (7.115)$$

Applying the same procedure for the post-threshold area we obtain:

$$n\left[P_n\right]_V = \sum_{i=1}^{M} A_{i,1} t_{i,1} \sigma_{i,1}(n - t_{i,1}) \left[P_{n-t_{i,1}}\right]_V \tag{7.116}$$

where:

$$\sigma_{i,1}(n) = \begin{cases} 0 & \text{for } n \leq Q \\ 1 & \text{for } n > Q \end{cases} \tag{7.117}$$

whereas the traffic intensity $A_{i,1}$, offered to the group in the post-threshold area $Q < n \leq V$, is equal to:

$$A_{i,1} = \frac{\lambda_i}{\mu_{i,1}} \tag{7.118}$$

As state areas: pre-threshold and post-threshold are separable, we can thus add up Equations (7.115) and (7.116), the result of which will be the occupancy distribution in the single-threshold system:

$$\begin{aligned} n\left[P_n\right]_V &= \sum_{i=1}^{M} A_{i,0} t_{i,0} \sigma_{i,0}(n - t_{i,0}) \left[P_{n-t_{i,0}}\right]_V \\ &+ \sum_{i=1}^{M} A_{i,1} t_{i,1} \sigma_{i,1}(n - t_{i,1}) \left[P_{n-t_{i,1}}\right]_V \end{aligned} \tag{7.119}$$

Blocking Probability in the Single-Threshold Model

In the case of the single-threshold system considered here blocking and loss probabilities are identical because all streams of offered traffic classes are, by assumption, PCT1 streams.

In order to determine the blocking (loss) probability let us consider the following two boundary cases:

1. $Q > V - t_{i,1}$;
2. $Q \leq V - t_{i,1}$.

If $Q > V - t_{i,1}$, then the value Q has no influence upon the service process. The system behaves like a full-availability group with multi-rate traffic and is described by the Kaufman–Roberts distribution (Section 7.3.5, Equation (7.39)). A change in the number of BBUs demanded has no influence on the system's performance because the capacity of the post-threshold area is still lower than $t_{i,1}$, i.e. the number of BBUs required to set up a connection. The blocking probability in this case occurs when the group does not have a sufficient number of free BBUs $t_{i,0}$ to set up a connection of class i.

If $Q \leq V - t_{i,1}$, then congestion in the single-threshold system ensues when the group has no $t_{i,1}$ free BBUs for a connection of class i to be effected in the post-threshold area.

Summing up the above considerations, the blocking (loss) probability for a call of class i in the single-threshold system depends on the value of the adopted threshold and can be written

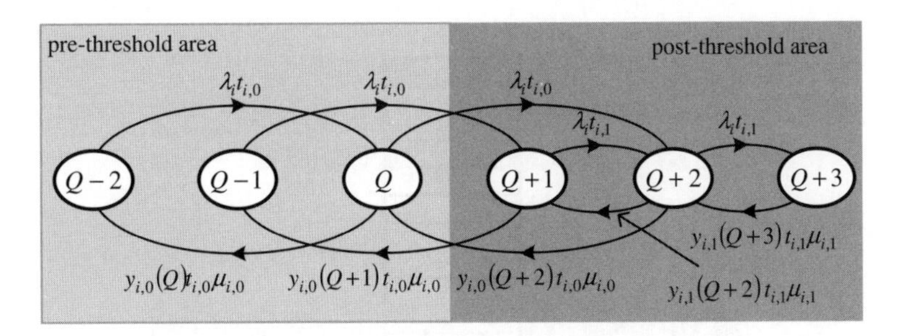

Figure 7.23 A fragment of the Markov process diagram in the single-threshold system.

in the following form:

$$E_i = B_i = \begin{cases} \sum_{n=V-t_{i,0}+1}^{V} [P_n]_V & \text{for } Q > V - t_{i,1} \\ \sum_{n=V-t_{i,1}+1}^{V} [P_n]_V & \text{for } Q \leq V - t_{i,1} \end{cases} \tag{7.120}$$

7.8.1.3 Interpretation of the Recurrent Notation of the Occupancy Distribution in the Single-Threshold System

Figure 7.23 shows a graphical interpretation of the occupancy distribution (7.119) in the single-threshold system, in other words a fragment of a one-dimensional diagram of the Markov process in the single-threshold system with multi-rate traffic. For better clarity, the figure shows only one class of offered traffic, described by the parameters: $\{\lambda_i, t_{i,0} = 2, \mu_{i,0}\}$ in the pre-threshold area, and the parameters: $\{\lambda_i, t_{i,1} = 1, \mu_{i,1}\}$ in the post-threshold area.

The symbols $y_{i,0}(n)$ i $y_{i,1}(n)$ denote (Figure 7.23) – for each state n – the values of service streams of class i in the pre-threshold and post-threshold areas, respectively. These values correspond to the number of serviced calls of class i related to offered traffic in the pre-threshold and post-threshold areas, respectively. The parameters $y_{i,k}(n)$ can be determined on the basis of Equation (7.70) from Section 7.4.7, which, for the system under consideration, will be rewritten in the following form:

$$y_{i,k}(n) = \begin{cases} A_{i,0}\, \sigma_{i,0}(n - t_{i,0}) \left[P_{n-t_{i,0}}\right]_V / [P_n]_V & \text{for } k = 0 \text{ and } n \leq Q + t_{i,0} \\ \\ A_{i,1}\, \sigma_{i,1}(n - t_{i,1}) \left[P_{n-t_{i,1}}\right]_V / [P_n]_V & \text{for } k = 1 \text{ and } n > Q + t_{i,1} \end{cases} \tag{7.121}$$

Let us explain now the domains of determinacy of the parameter $y_{i,k}$ adopted in Equation (7.121). Figure 7.23 shows clearly that the pre-threshold call, i.e. the one that requires $t_{i,0}$ BBUs, can be serviced in, at the maximum, state $n = Q + t_{i,0}$ because the threshold Q is the oldest state in which the system can accept a call with such requirements for service. In turn, it is only in state $Q + 1$ that the group can accept calls with post-threshold requirements, in other words, those demanding $t_{i,1}$ BBUs for service. Thus, the youngest state in which a post-threshold call can be serviced is state $n = Q + 1 + t_{i,1}$.

7.8.2 Multi-Threshold Models

The multi-threshold model [27, 41, 43] is a generalization of the single-threshold model. In the multi-threshold model we assume that, for each call class, a set of thresholds is introduced

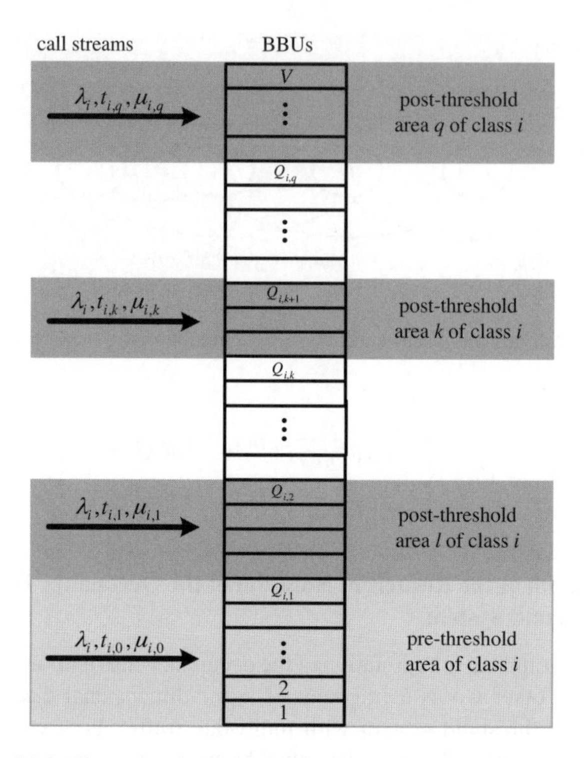

Figure 7.24 Example of a multi-threshold system with one class of calls.

individually. For example, for class i the set of adopted thresholds will be written in the following way $\{Q_{i,1}, Q_{i,2}, \ldots, Q_{i,q}\}$, where the first subscript indicates the class of a call, while the second the number of the threshold, with the initial assumption that: $\{Q_{i,1} \leq Q_{i,2} \leq \ldots \leq Q_{i,q}\}$.

Figure 7.24 shows a fragment of a multi-threshold system for class i. The figure shows the stream of offered traffic of class i in the pre-threshold area ($0 \leq n \leq Q_{i,1}$), in post-threshold area 1 ($Q_{i,1} < n \leq Q_{i,2}$), post-threshold area k ($Q_{i,k} < n \leq Q_{i,k+1}$) and post-threshold area q ($Q_{i,q} < n \leq V$). Notice that the threshold area k is the interthreshold area bordered by thresholds $Q_{i,k}$ and $Q_{i,k+1}$.

The operation of the multi-threshold system can be presented in the following way: in each post-threshold area k the traffic stream of class i is offered, described by its own set of parameters $\{\lambda_i, t_{i,k}, \mu_{i,k}\}$, while $t_{i,0} > t_{i,1} > \ldots > t_{i,k} > \ldots > t_{i,q}$ and $\mu_{i,0}^{-1} < \mu_{i,1}^{-1} < \ldots < \mu_{i,k}^{-1} < \ldots < \mu_{i,q}^{-1}$. This means that as the load of the group increases, the number of the BBUs admitted to calls of particular classes decreases and, simultaneously, the mean service time of the calls increases.

The multi-threshold model is an extension of the single-threshold model, hence the more important characteristics of the system can be determined on the basis of Equations (7.119)–(7.120) with the number of thresholds increased to q_i:

- In such a case, the occupancy distribution can be written in the following way:

$$n\left[P_n\right]_V = \sum_{i=1}^{M}\sum_{k=0}^{q} A_{i,k}t_{i,k}\sigma_{i,k}(n-t_{i,k})\left[P_{n-t_{i,k}}\right]_V \tag{7.122}$$

where for each pair $\{i, k\}$, such that: $1 \leq i \leq M$ and $0 \leq k \leq q$, we have:

$$\sigma_{i,k}(n) = \begin{cases} 1 & \text{for } Q_{i,k} < n \leq Q_{i,k+1} \\ 0 & \text{for the remaining } n \end{cases} \tag{7.123}$$

In Equation (7.123) we assume $Q_{i,0} = 0$.
Let us consider an interval of states for which the conditional transition coefficient $\sigma_{i,k}$ is defined (Equation (7.123)). This interval is determined by the range of the post-threshold area k of class i, where: $Q_{i,k} < n \leq Q_{i,k+1}$. In states belonging to the area the number of demanded bandwidth units for a call of class i is equal to $t_{i,k}$, whereas the intensity of offered traffic is: $A_{i,k} = \lambda_i/\mu_{i,k}$.
- Blocking probability in the multi-threshold model can be expressed by the formula:

$$E_i = \sum_{k=0}^{q} E_{i,k} \tag{7.124}$$

where $E_{i,k}$ is the blocking probability of calls of class i in the threshold area k. This probability is determined in the following way [27]:

$$E_{i,k} = \begin{cases} 0 & \text{for } \begin{cases} V - t_{i,k} \geq Q_{i,k+1} \\ V - t_{i,k} > Q_{i,k} \end{cases} & \text{(a)} \\ \sum\limits_{n=V-t_{i,k}+1}^{Q_{i,k+1}} \left[P_n\right]_V & \text{for } \begin{cases} V - t_{i,k} < Q_{i,k+1} \\ V - t_{i,k} > Q_{i,k} \end{cases} & \text{(b)} \\ \sum\limits_{n=Q_k+1}^{Q_{i,k+1}} \left[P_n\right]_V & \text{for } \begin{cases} V - t_{i,k} < Q_{i,k+1} \\ V - t_{i,k} \leq Q_{i,k} \end{cases} & \text{(c)} \end{cases} \tag{7.125}$$

where $Q_{i,q+1} = V$.
Equation (7.124) introduces the notion of the blocking probability of calls of class i in the post-threshold area k, denoted with the symbol $E_{i,k}$. The value of this probability depends on the mutual location of the post-threshold area k of class i and reference state $n = V - t_{i,k}$. The reference state is the oldest state of the multi-threshold system in which a call that requires $t_{i,k}$ BBUs can be serviced. Three conceivable instances of location of the post-threshold area and the reference state are presented in Figure 7.25.

If the post-threshold area is located below the reference state (or, at the most, the upper limit of the post-threshold area overlaps with the reference state), then the system has always $t_{i,k}$ free BBUs necessary to set up a connection of class i in the post-threshold area k. This

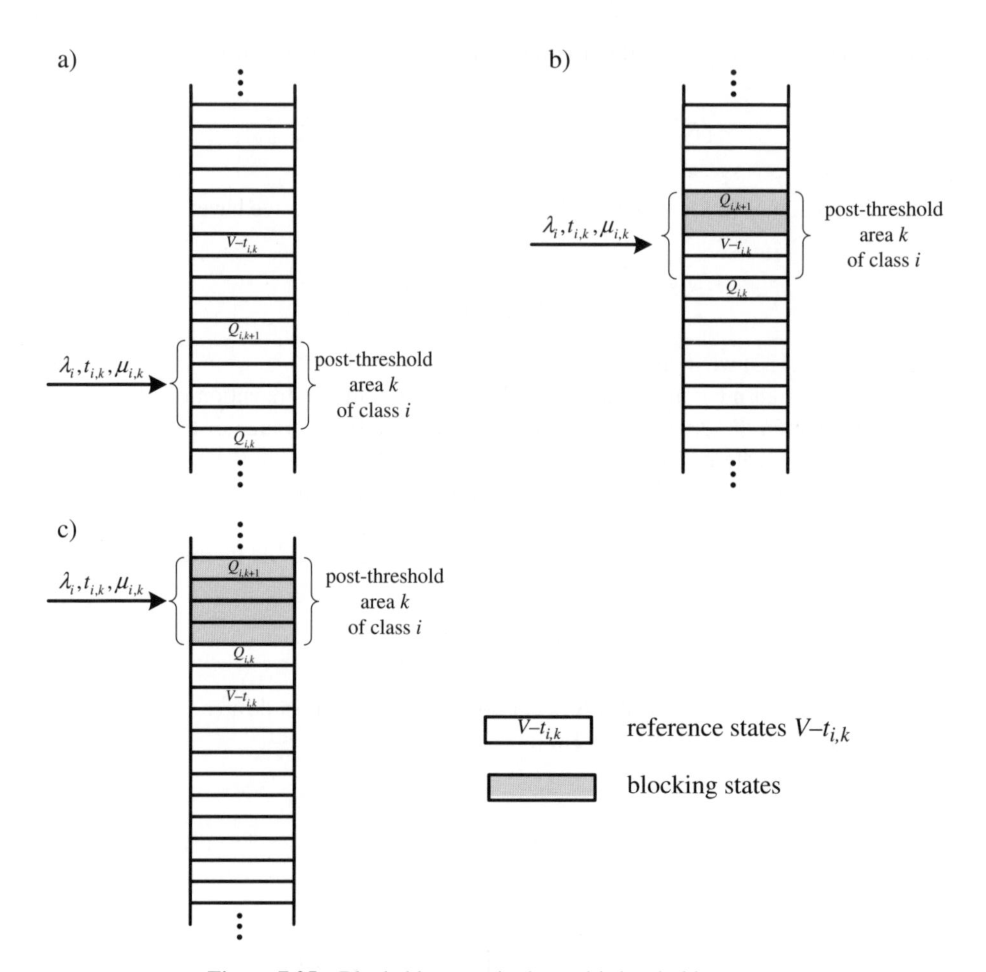

Figure 7.25 Blockable states in the multi-threshold system.

means that, with the same mutual location of the post-threshold area and the reference state (Figure 7.25a), the blocking probability is equal to zero, which is expressed by the equation (a) in the system of equations (7.125).

Figure 7.25b shows a case in which the post-threshold area k of class i includes the reference state. All states of this area above the reference state (in which a class i call can still be serviced) are blocking states. The blocking probability in this particular case will be then determined by the equation (b) in the system of equations (7.125). If the post-threshold area k of class i is located above the reference state, then all states of this area will be blockable states (Figure 7.25c). In this case, the blocking probability will be determined by the equation (c) in the system of equations (7.125).

It should be stressed that until now the subject literature [40, 41, 44] has only discussed the first case widely (Figure 7.25a). It means that each state of the post-threshold area (except the last post-threshold area q) can service appropriate calls. With such assumptions, blockable states in the multi-threshold system can occur exclusively in the last threshold area and the blocking probability will be described by the following equation:

$$E_i = \sum_{n=V-t_{i,q}+1}^{V} [P_n]_V \tag{7.126}$$

- Service streams in the multi-threshold system – the number of class i calls – after Equation (7.121) can be rewritten in the following way:

$$y_{i,k}(n) = A_{i,k}\,\sigma_{i,k}(n-t_{i,k})\left[P_{n-t_{i,k}}\right]_V \Big/ [P_n]_V$$
$$\text{for} \quad 0 \le k \le q \text{ and } Q_{i,k}+t_{i,k} < n \le Q_{i,k+1}+t_{i,k} \tag{7.127}$$

Let us consider a call of class i offered in the post-threshold area k – in other words, a call that requires $t_{i,k}$ BBUs. The youngest state in which the group can admit calls with such requirements is state $n = Q_{i,k}+1$. Therefore, the youngest state (boundary state) in which a call demanding $t_{i,k}$ BBUs can be serviced is state $n = Q_{i,k}+1+t_{i,k}$. The last state of the post-threshold area k of class i in which a new call can be accepted is the state $n = Q_{i,k+1}$. Thus, the oldest state (boundary state) in which a call demanding $t_{i,k}$ BBUs can still be serviced is the state $n = Q_{i,k+1}+t_{i,k}$.

7.8.3 *Comments for Single-Threshold and Multi-Threshold Systems*

The single-threshold and the multi-threshold models can be generalized by taking into consideration the service of the mixture of PCT1 and PCT2 traffic streams. In this case, occupancy distributions expressed by the dependencies (7.119) and (7.122) will change. In the case of a single-threshold system, the occupancy distribution will take on the following form:

$$\begin{aligned} n\,[P_n]_V &= \sum_{i=1}^{M_1}\left(A_{i,0}t_{i,0}\sigma_{i,0}(n-t_{i,0})\left[P_{n-t_{i,0}}\right]_V + A_{i,1}t_{i,1}\sigma_{i,1}(n-t_{i,1})\left[P_{n-t_{i,1}}\right]_V\right) \\ &+ \sum_{j=1}^{M_2}\Big(A_{j,0}(n-t_{j,0})t_{j,0}\sigma_{j,0}(n-t_{j,0})\left[P_{n-t_{j,0}}\right]_V \\ &+ A_{j,1}(n-t_{j,1})t_{j,1}\sigma_{j,1}(n-t_{j,1})\left[P_{n-t_{j,1}}\right]_V\Big) \end{aligned} \tag{7.128}$$

where M_1 is the number of PCT1 traffic classes and M_2 is the number of PCT2 traffic classes.

Appropriately, the occupancy distribution for a multi-threshold system will take on, in this particular case, the form shown below:

$$\begin{aligned} n\,[P_n]_V &= \sum_{i=1}^{M_1}\sum_{k=0}^{q_1} A_{i,k}t_{i,k}\sigma_{i,k}(n-t_{i,k})\left[P_{n-t_{i,k}}\right]_V \\ &+ \sum_{j=1}^{M_2}\sum_{k=0}^{q_2} A_{j,k}(n-t_{j,k})t_{j,k}\sigma_{j,k}(n-t_{j,k})\left[P_{n-t_{j,k}}\right]_V \end{aligned} \tag{7.129}$$

where for each pair $\{i, k\}$, such that: $1 \leq i \leq M_1$ and $0 \leq k \leq q_1$, we have:

$$\sigma_{i,k}(n) = \begin{cases} 1 & \text{for } Q_{i,k} < n \leq Q_{i,k+1} \\ 0 & \text{for the remaining } n \end{cases} \tag{7.130}$$

and where for each pair $\{j, k\}$, such that: $1 \leq j \leq M_2$ and $0 \leq k \leq q_2$, we have:

$$\sigma_{j,k}(n) = \begin{cases} 1 & \text{for } Q_{j,k} < n \leq Q_{j,k+1} \\ 0 & \text{for the remaining } n \end{cases} \tag{7.131}$$

In Equation (7.129) parameters q_1 and q_2 designate numbers of threshold for each PCT1 and PCT2 call stream, respectively. The blocking probability in both models will be determined on the basis of the expressions (7.120) and (7.125).

Models of systems with one or many thresholds for a given traffic class can be used in the call admission control function. In the case of cellular networks, threshold models can be applied for modeling and dimensioning radio interfaces with both hard and soft capacity.

7.9 Full-Availability Group with Compression Mechanism

This section discusses a full-availability group model with the compression mechanism [45–48]. The presented model takes into account a possibility of a change in parameters of serviced calls when a new call appears in the system. Such a model is called the adaptive bit rate model or a model with traffic compression [48]. In the model under discussion it is assumed that the compression mechanism can be introduced to all, or selected, traffic classes serviced in the system.

7.9.1 Description of the Model

Let us consider a model in which a change in the number of BBUs demanded depends on a traffic class that undergoes the compression mechanism. The model assumes that the system services simultaneously a mixture of different multi-rate traffic classes, and that these classes are divided into two sets: classes in which calls can change their demands during service time, and classes that do not change their demands during service time. A measure of a possible change of the number of BBUs is a *maximum compression coefficient* and it determines the ratio of the maximum number of BBUs to the minimum number of BBUs to be accepted for service of calls of a given class. This definition can be written in the following way:

$$K_{j,\max} = \frac{t_{j,\max}}{t_{j,\min}} \tag{7.132}$$

where $t_{j,\max}$ oraz $t_{j,\min}$ denotes respectively the maximum and the minimum number of BBUs accepted to service calls of class j.

In the proposed model we assumed that the compression mechanism was introduced to a full-availability group for M_k traffic classes and that M_{nk} traffic classes did not undergo compression. With the case of a system that services a mixture of traffic streams that undergo and do not undergo compression, it is convenient to rewrite the occupancy distribution in the full-availability group (7.39) after separating both types of traffic in the following way:[2]

$$n\,[P_n]_V = \sum_{i=1}^{M_{nk}} A_i t_i \left[P_{n-t_i}\right]_V + \sum_{j=1}^{M_k} A_j t_{j,\min} \left[P_{n-t_{j,\min}}\right]_V \tag{7.133}$$

where $t_{j,\min}$ is the minimum number of demanded BBUs by a call of class j that belongs to the classes that undergo compression.

The blocking (loss) probability in the full-availability group determined by the distribution (7.133) will, in the case of the system considered here, take on the following form:

$$E_i = B_i = \begin{cases} \sum\limits_{n=V-t_i+1}^{V} [P_n]_V & \text{if class } i \text{ call belongs to } M_{nk} \text{ classes} \\ \sum\limits_{n=V-t_{i,\min}+1}^{V} [P_n]_V & \text{if class } i \text{ call belongs to } M_k \text{ classes} \end{cases} \tag{7.134}$$

In Equation (7.133) and Equation (7.134) the model is characterized by the parameter $t_{i,\min}$, which is the minimum number of BBUs that are demanded by a call of class i under the conditions of maximum compression. Such an approach is necessary to determine the blocking probability in a system with compression because blocking states will occur only in the conditions of maximum compression. The maximum compression defines such occupancy states in the system in which further reduction in the number of BBUs that belong to serviced calls is not possible.

In order to determine carried traffic in a system with compression it is necessary to estimate the number and the type of calls serviced in a given state of the system. For this purpose, Equation (7.48) (Section 7.3.9) will be used. This formula allows us to determine the average number of calls of class i serviced in the occupancy state n. This parameter, determined under the assumption of maximum compression, can be written as follows:

$$y_i(n) = \begin{cases} \dfrac{A_i \left[P_{n-t_i}\right]_V}{[P_n]_V} & \text{if class } i \text{ call belongs to } M_{nk} \text{ classes} \\ \dfrac{A_i \left[P_{n-t_{i,\min}}\right]_V}{[P_n]_V} & \text{if class } i \text{ call belongs to } M_k \text{ classes} \end{cases} \tag{7.135}$$

On the basis of Equation (7.135) it is possible to determine total average carried traffic serviced in state n, under the assumption of the maximum compression:

$$Y^{\max}(n) = Y_{nk}(n) + Y_k^{\max}(n) = \sum_{i=1}^{M_{nk}} y_i(n) t_i + \sum_{j=1}^{M_k} y_j(n) t_{j,\min} \tag{7.136}$$

where $Y_k^{\max}(n)$ is the average number of busy BBUs in state n occupied by calls that undergo compression, whereas $Y_{nk}(n)$ is the average number of busy BBUs in state n occupied by calls without compression.

[2]Further on in the chapter the terms “a set of classes with applicable compression” and “a class with applicable compression” will, for convenience, be simplified to “a set of classes with compression” and “a class with compression,” respectively.

The value of parameter $Y_{nk}(n)$ refers to non compressed traffic. This parameter is thus independent of the degree of compression of the remaining calls. The real value of carried traffic in state n, determined in conditions of maximum compression, will depend on the number of free BBUs in the system. We assume that the real system operates in a way that provides maximum use of resources, that is calls of compressed classes always try to occupy free resources and reduce their maximum demands to the least possible degree. Thus, the real value of traffic $Y(n)$, serviced in the system in a given state that corresponds to state n (determined in the conditions of maximum compression) can be expressed as follows:[3]

$$Y(n) = Y_{nk}(n) + Y_k(n) = \sum_{i=1}^{M_{nk}} y_i(n)t_i + \sum_{j=1}^{M_k} y_j(n)t_j(n) \tag{7.137}$$

The parameter $t_j(n)$ in Equation (7.137) determines the real value of the occupied BBUs of class j (undergoing compression) in state n:

$$t_{j,\min} < t_j(n) \leq t_{j,\max} \tag{7.138}$$

The measure of a degree of compression in state n is the compression coefficient $\xi(n)$, which can be defined in the following way:

$$\xi(n) = \frac{t_j(n)}{t_{j,\min}} \tag{7.139}$$

At this point, taking into consideration Equation (7.139), the average number of busy BBU occupied by calls with compression, we are in position to write what follows:

$$Y_k(n) = \sum_{j=1}^{M_k} y_j(n)t_j(n) = \xi(n) \sum_{j=1}^{M_k} y_j(n)t_{j,\min} \tag{7.140}$$

Note that the sum in Equation (7.140) is the average number of busy BBUs in state n occupied by calls that undergo compression (Equation (7.136)). Therefore we can rewrite Formula (7.140) in the following way:

$$Y_k(n) = \xi(n) \cdot Y_k^{\max}(n) \tag{7.141}$$

To determine the parameter $Y_k(n)$ in accordance with Formula (7.141) the knowledge of the compression coefficient $\xi(n)$ is indispensable. This coefficient can be defined as the ratio of potentially available resources to service calls with compression to the resources occupied by these calls in the condition of maximum compression. Therefore, we can write (Figure 7.26):

$$\xi(n) = \frac{V - Y_{nk}(n)}{Y_k^{\max}(n)} = \frac{V - Y_{nk}(n)}{n - Y_{nk}(n)} \tag{7.142}$$

The value of compression coefficient is limited by the maximum compression defined by Equation (7.132) for particular classes. Taking into account this limitation we can modify the

[3] For convenience, from now on we will use the term "state n" instead of "a given state n in the conditions of maximum compression."

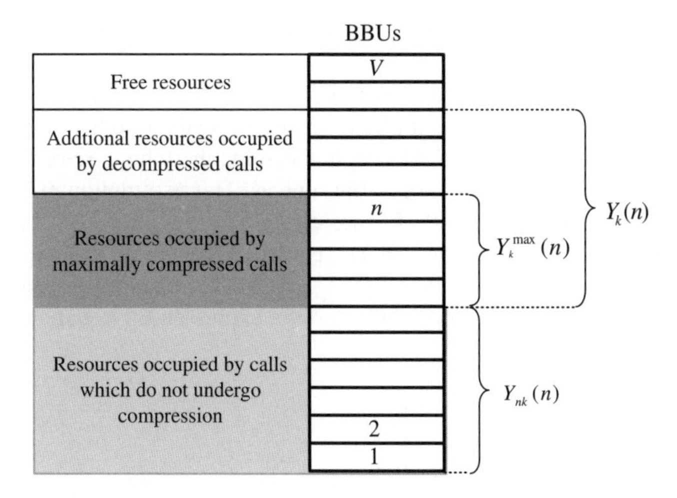

Figure 7.26 Example full-availability group with compression mechanism.

definition of compression coefficient for class j calls in the following way:

$$\xi_j(n) = \begin{cases} K_{j,\max} & \text{for } \quad \xi(n) \geq K_{j,\max} \\ \xi(n) & \text{for } 1 \leq \xi(n) < K_{j,\max} \end{cases} \tag{7.143}$$

The compression coefficient described by Equation (7.143) depends on the class of traffic. This means that each traffic class that undergoes compression can be compressed in state n to a different degree.

Knowing the value of the compression coefficient in each state n, we are in position to determine the average resources occupied by calls with compression:

$$Y_k = \sum_{n=0}^{V} \sum_{j=0}^{M_k} y_j(n) \left[\xi_j(n) t_{j,\min}\right] [P_n]_V \tag{7.144}$$

Notice that the value Y_k in Equation (7.144) is the average carried traffic of calls with compression.

7.9.2 Comments

In this chapter we have used the concept of a system with maximum compression. Appropriate characteristics have been defined for the system, such as the occupancy distribution and the blocking probability. To determine the value of traffic carried in the system, we have introduced the notion of the compression coefficient, which transfers the occupancy state in a system with maximum compression to a given real state. This state is characterized by the same number of serviced calls of individual classes as state n, and a different number of busy BBUs occupied by calls of classes that undergo compression. Thus, in the construction of the model we have used characteristics of the system with maximum compression and the assumption that offered calls always try to occupy as many free BBUS in the system as it is possible. Such an approach simplifies the proposed model as compared to the other models described in the literature of the subject, for example [44] or [46].

The model presented here will be used in Chapter 12 for modeling HSPA traffic in the Iub interface. This interface services traffic from several cells, in other words traffic that is generated by a great number of users. Hence, in modeling the Iub interface it is assumed that traffic is generated by infinite number of traffic sources. This assumption makes it possible to limit the description of the model to PCT1 traffic only.

The model presented in this chapter can be generalized assuming that the system services a mixture of traffic streams of the type PCT1 and PCT2. When this is the case, then the occupancy distribution in the full-availability group, expressed by Equation (7.133), can be written in the following form:

$$\begin{aligned} n\left[P_n\right]_V = & \sum_{i=1}^{M_{1,nk}} A_{1,i} t_{1,i} \left[P_{n-t_{1,i}}\right]_V + \sum_{j=1}^{M_{1,k}} A_{1,j} t_{1,j,\min} \left[P_{n-t_{1,j,\min}}\right]_V \\ & + \sum_{i=1}^{M_{2,nk}} A_{2,i}(n-t_{2,i}) t_{2,i} \left[P_{n-t_{2,i}}\right]_V + \sum_{j=1}^{M_{2,k}} A_{2,j}(n-t_{2,j,\min}) t_{2,j,\min} \left[P_{n-t_{2,j,\min}}\right]_V \end{aligned} \tag{7.145}$$

Knowing that $M_{nk} = M_{1,nk} + M_{2,nk}$ and that $M_k = M_{1,k} + M_{2,k}$, the remaining characteristics of the system, namely the blocking (loss) probability and average resources occupied by compressed traffic, can be determined on the basis of Equation (7.134) and Equation (7.144), respectively.

7.10 Full-Availability Group with Priorities

This chapter presents a model of the full-availability group servicing a mixture of different multi-rate traffic classes with priorities. Solutions to the problem can be found in the literature – for example, in [49–52]. The proposed analytical model assumes that an arrival of a new call with a higher priority can terminate currently serviced calls with lower priority if there are no available free resources. A further assumption in the model is that all classes of calls offered to the system have a defined priority. Class 1 is characterized by the highest priority, whereas class M has been assigned the lowest priority. Traffic offered by classes with lower priority has no influence on the blocking probability of calls with a higher priority [49, 50, 53].

7.10.1 Description of the Basic Model

Let us assume that a given system can be treated as a full-availability group with multi-rate traffic. The occupancy distribution in such a system (without priority) can be then expressed by the recurrent formula (7.39):

$$n\left[P_n\right]_V = \sum_{i=1}^{M} A_i t_i \left[P_{n-t_i}\right]_V \tag{7.146}$$

where M is the number of classes of calls serviced in the system. The blocking probability of calls of class i can be written in the following way:

$$E_i = \sum_{n=V-t_i+1}^{V} [P_n]_V \tag{7.147}$$

The blocking probability for a call of class i in a full-availability group that services M traffic classes can, after some simplifications be presented in the form of the following function:

$$[E_i]_M = f((A_1, t_1), \ldots, (A_i, t_i), \ldots (A_M, t_M)) \tag{7.148}$$

where f is determined on the basis of Equations (7.146) and (7.147).

The two subscripts notation $[x_i]_j$ has been adopted in this chapter. The subscript i denotes the traffic class, whereas subscript j indicates the number of all call classes serviced in the system.

7.10.2 System with Two Priorities

Let us consider the full-availability model with two classes of calls and priorities proposed in [48]. Our assumption is that the first class of calls, in line with the adopted notation, has a higher priority. Accordingly, calls of the second class, namely those with lower priority, have no influence on the service process of calls of the first class (with higher priority). To proceed with an analysis of the system with priorities, three systems will be considered. The systems can be modeled on the basis of the full-availability group with multi-rate traffic. The first system assumes that the full-availability group services only calls of higher priority class. The second system is a system without priorities and services two equivalent traffic classes. Finally, a full-availability group that corresponds to the third system services two classes of calls with priorities.

7.10.2.1 System 1. Full-Availability Group Servicing One Class of Calls

The blocking probability in the first system can be written in the following way:

$$[E_1]_1 = f(A_1, t_1) \tag{7.149}$$

After determining the blocking probability, the total carried traffic, can be determined:

$$[Y]_1 = [Y_1]_1 = A_1 t_1 (1 - [E_1]_1) \tag{7.150}$$

7.10.2.2 System 2. Full-Availability Group Servicing Two Equivalent Classes of Calls

With the instance of the second system, the group services two, equivalent, classes of calls without priorities. Thus, we can write:

$$([E_1]_2, [E_2]_2) = f\,((A_1, t_1), (A_2, t_2)) \tag{7.151}$$

and

$$[Y]_2 = \sum_{k=1}^{2} [Y_k]_2 = \sum_{k=1}^{2} A_k t_k (1 - [E_k]_2) \tag{7.152}$$

In Equation (7.150) and (7.152) the parameter $[Y]_j$ determines the total carried traffic in a system servicing j classes of calls, whereas $[Y_i]_j$ determines carried traffic of class i in a system servicing j classes of calls.

7.10.2.3 System 3. Full-Availability Group Servicing Two Classes of Calls With Priorities

In line with the assumptions made here, the service process of calls with lower priority (class 2) in the full-availability group with two classes of calls and priorities has no influence on the service process of classes with higher priority (class 1). This means that, for class 1, the blocking probability and carried traffic will be the same as in first system that services only the higher priority class of calls:

$$[E_1]_2^P = [E_1]_1 \tag{7.153}$$

and

$$[Y_1]_2^P = [Y_1]_1 \tag{7.154}$$

Index "P" outside the square brackets, in Equation (7.153) and (7.154) indicates a system with priorities.

The operation of the system with priorities assumes that calls with higher priority, in the case of limited resources, "push out" calls with lower priority – that is, they force termination of the service process for calls with lower priority and occupy the resources previously assigned to them. In this mode of operation of the system it can be assumed that the total carried traffic in a system without priorities (System 2) is identical to traffic carried in a system with priorities (the traffic conservation principle). Thus, we can write:

$$[Y]_2^P = [Y]_2 \tag{7.155}$$

or

$$[Y_1]_2^P + [Y_2]_2^P = [Y]_2 \tag{7.156}$$

In Equation (7.156) the total traffic $[Y]_2$ carried in the system without priorities is known (Equation (7.152)). The characteristics of traffic of class 1, with higher priority, are also known (Equation (7.153) and Equation (7.154)). Thus, lower priority traffic can be determined, on the bases of Equations (7.154) and (7.156), in the following way:

$$[Y_2]_2^P = [Y]_2 - [Y]_1 \tag{7.157}$$

Notice that traffic of class 2 in the system with priorities is determined by the difference between the total carried traffic in the system without priorities (System 2) and the total carried traffic in the system with one class of calls (System 1). Taking into consideration the following dependence:

$$[Y_2]_2^P = A_2 t_2 (1 - [E_2]_2^P) \tag{7.158}$$

and substituting Equation (7.158) into Equation (7.157), we can determine the blocking probability of calls of the second class in the system with priorities:

$$[E_2]_2^P = \frac{A_2t_2 - [Y_1]_2 + [Y_1]_1}{A_2t_2} \tag{7.159}$$

Equation (7.159), after taking into consideration Equations (7.150) and (7.152), can be written in the following form:

$$[E_2]_2^P = \frac{A_1t_1([E_1]_2 - [E_1]_1) + A_2t_2[E_2]_2}{A_2t_2} \tag{7.160}$$

Thus, as in the case of class 1 traffic (Equation (7.157)), the blocking probability of class 2 calls (with lower priority) can also be estimated on the basis of the blocking probabilities in Systems 1 and 2.

7.10.3 System with h Priorities

A system with h priorities can be reduced to a $h-1$-stage computational algorithm in which the system with two priorities is considered in each step of the algorithm [48].

In the first step of the algorithm, the lowest priority of calls is considered (traffic A_h) assuming that the remaining traffic classes (traffic $A_1, \ldots, A_{h-1}$) have higher priority and "push out" serviced calls of class h. In the next step, calls of class $h-1$ (i.e. traffic A_{h-1}) has the lowest priority, whereas the remaining classes ($A_1, \ldots, A_{h-2}$) has higher priority and can "push out" serviced calls of class $h-1$. In the following steps calls of class k ($1 < k < h$) has always the lowest priority, while the remaining calls ($A_1, \ldots, A_{k-1}$) has higher priority and "push out" serviced calls of class k. In the last step of the algorithm we consider traffic of two classes of calls (A_1 and A_2) and we can determine relevant characteristics of these two types of traffic directly on the basis of Equations (7.153), (7.154), (7.158), and (7.159).

Performing the consideration, in an identical way as with the system with two priorities, we can thus determine the blocking probability of calls of class k on the basis of the following formula:

$$[E_k]_k^P = \frac{\sum_{i=1}^{k-1}[A_it_i([E_i]_k - [E_i]_{k-1})] + A_kt_k[E_k]_k}{A_kt_k} \quad \text{for } k > 1 \tag{7.161}$$

Thus, to determine the blocking probability for all traffic classes we apply Equation (7.161) to each of them, starting from the class with the lowest priority and finishing with the class with the highest priority.

7.10.4 Comments

This section presented a full-availability group model servicing a mixture of different multi-rate traffic classes, with the assumption that each of the traffic classes is characterized by a different priority.

This model will be used in Chapters 11 and 12 to model Iub interfaces and the radio interface servicing a mixture of different multi-rate traffic classes with priorities. The Iub interface services traffic generated by a great number of users and hence in modeling this interface it is assumed that traffic is generated by an infinite number of traffic sources (PCT1 traffic). In the case of the WCDMA interface one has to assume that traffic is generated by a relatively small number of users (PCT2 traffic). For this reason, the model presented in this chapter generally services a mixture of PCT1 and PCT2 traffic streams. Here, the occupancy distribution in the full-availability group expressed by Equation (7.133) can be rewritten in the following form:

$$n\left[P_n\right]_V = \sum_{i=1}^{M_1} A_{1,i} t_{1,i} \left[P_{n-t_{1,i}}\right]_V + \sum_{i=1}^{M_2} A_{2,i}(n - t_{2,i}) t_{2,i} \left[P_{n-t_{2,i}}\right]_V \tag{7.162}$$

Knowing that M_1 is the number of PCT1 traffic classes and that M_2 indicates the number of PCT2 traffic classes, the blocking probability can be determined in this case on the basis of the dependence (7.161).

References

[1] Akimuru, H. and Kawashima, K. (1999) *Teletraffic: Theory and Application*, Springer.

[2] Beshai, M.E. and Manfield, D.R. (1988) Multichannel Services Performance of Switching Networks. Proceedings of 12th International Teletraffic Congress, Torino, Amsterdam.

[3] Conradt, J. and Buchheister, A. (1985) *Considerations on Loss Probability of Multi-slot Connections*. Proceedings of 11th International Teletraffic Congress, Kyoto. Elsevier North-Holland, Amsterdam.

[4] Iversen, V.B. (1987) The exact evaluation of multi-service loss systems with access control. *Seventh Nordic Teletraffic Seminar (NTS-7)*, August, Lund, pp. 56–61.

[5] Ross, K.W. (1995) *Multi-service Loss Models for Broadband Telecommunication Network*. Springer.

[6] Filipiak, J. (1991) Structured system analysis methodology for design of an ATM network architecture. *IEEE Journal on Selected Areas in Communications*, **7** (8), 1263–73.

[7] Roberts, J.W. (ed.) (1992) *Performance Evaluation and Design of Multi-service Networks, Final Report COST 224*, Commission of the European Communities, Brussels.

[8] Kelly, F.P. (1991) Loss networks. *The Annals of Applied Probability*, **1** (3), 319–78.

[9] Roberts, J.W., Mocci, V. and Virtamo, I. (eds) (1996) *Broadband Network Teletraffic, Final Report of Action COST 242*. Commission of the European Communities, Springer.

[10] Głąbowski, M., Stasiak, M., Wiśniewski, A., and Zwierzykowski, P. (2009) Blocking probability calculation for cellular systems with WCDMA radio interface servicing PCT1 and PCT2 multi-rate traffic. *IEICE Transactions on Communications*, **E92-B(04)**, 1156–65.

[11] Iversen, V.B. (2001) Modelling restricted accessibility for wireless multi-service systems. In Cesana, M. and Fratta, L. (eds), *EuroNGI Workshop*, of *Lecture Notes in Computer Science*, Springer, vol. 3883, pp. 93–102.

[12] Janevski, T. (2003) *Traffic Analysis and Design of Wireless IP Networks*, Artech House.

[13] Iversen, V.B. (1980) The A-formula. *Teleteknik (English ed.)*, **23** (2), 64–79.

[14] Sutton, D.J. (1980) The application of reversible Markov population processes to teletraffic. *Australian Telecommunication Review*, **13** (2), 3–8.

[15] Kelly, F.P. (1979) *Reversibility and Stochastic Networks*, John Wiley & Sons, Ltd.

[16] Dobrushin, R. L. Yu. M. Sukhov, and Fritz. J. A. N. Kolmogorov – the founder of the theory of reversible Markov processes. *Russian Mathematical Surveys*, **43** (6), 157–82.

[17] Brockmeyer, E., Halstrom, H.L., and Jensen, A. (1960) The life and works of A.K. Erlang. *Acta Polytechnika Scandinavia*, **6** (287), 138–55.

[18] Iversen, V.B. (2001) *Teletraffic Engineering Handbook*. June 2001 (draft, published online by International Telecommunication Union). https://www.itu.int/ITU-D/study_groups/SGP_19982002/SG2/StudyQuestions/Question_16/RapporteursGroupDocs/teletraffic.pdf (accessed 19 July 2010).

[19] Aien, J.M. (1978) A multi-user-class, blocked-calls-cleared, demand access model. *IEEE Transactions on Communications*, **26** (3), 378–85.

[20] Kaufman, J.S. (1981) Blocking in a shared resource environment. *IEEE Transactions on Communications*, **29** (10), 1474–81.

[21] Roberts, J.W. (1981) *A Service system with Heterogeneous User Requirements – Application to Multi-service Telecommunications Systems.* Proceedings of Performance of Data Communications Systems and their Applications. North Holland, Amsterdam.
[22] Erlang, A.K. (1917) Solution of some problems in the theory of probabilities of significance in automatic telephone exchanges. *Elektrotechnikeren*, **13**, 5.
[23] Delbrouck, L.E.N. (1983) On the steady-state distribution in a service facility carrying mixtures of traffic with different peakedness factors and capacity requirements. *IEEE Transactions on Communications*, **31** (11), 1209–11.
[24] Kraimeche, B. and Schwartz, M. (1984) Circuit Access Control Strategies in Integrated Digital Networks. *Proceedings of IEEE Infocom*, San Francisco. Institute of Electrical and Electronics Engineers, San Francisco.
[25] Stasiak, M. (1993) Blocking probability in a limited-availability group carrying mixture of different multichannel traffic streams. *Annales des Télécommunications*, **48** (1–2), 71–6.
[26] Głabowski, M. and Stasiak, M. (2004) *An Approximate Model of the Full-availability Group with Multi-rate Traffic and a Finite Source Population. Buchholtz, P., Lehnert, R., and Pióro, M. Proceedings of 3rd Polish-German Teletraffic Symposium* VDE Verlag, September.
[27] Głąbowski, M., Kaliszan, A., and Stasiak, M. (2010) Modeling product-form state-dependent systems with BPP traffic. *Journal of Performance Evaluation*, **67**, 174–97.
[28] Głąbowski, M. and Stasiak, M. (2004) *Multi-rate Model of the Group of Separated Transmission Links of Various Capacities.* De Souzam J.N., Dini, P., and Lorenz, P. (eds) *Telecommunications and Networking - ICT 2004,* 11th International Conference on Telecommunications, Fortaleza, Brazil, August 1-6, 2004, Proceedings, *Lecture Notes in Computer Science*, Springer, vol. 3124, pp. 1101–6.
[29] Delaire, M. and Hebuterne, G. (1977) Call Blocking in Multi-services Systems on One Transmission Link. Proceedings of 5th IFIP Workshop on Performance Modelling and Evaluation of ATM Networks, Ilkley, July IFIP, Bradford.
[30] Karlsson, J.M. (1992) *Loss Performance in Trunk Groups with Different Capacity Demands.* Proceedings of 13th International Teletraffic Congress, Copenhagen. North Holland, Amsterdam.
[31] Lindberger, K. (1987) *Blocking for Multislot Heterogeneous Traffic Streams Offered to a Trunk Group with Reservation.* Proceeding of the 5th ITC Seminar, Lake Como.
[32] Korner, U., Lubacz, J., and Pióro, M. (1989) *Traffic Engineering Problems* in *Multi-service Circuit Switched Networks.* Proceedings of International Teletraffic Congress Specialist Seminar - ITC–12, Adelaide. University of Adelaide, Australia.
[33] Roberts, J.W. (1983) *Teletraffic Models for the Telcom 1 Integrated Services Network.* Proceedings of 10th International Teletraffic Congress, Montreal.
[34] Takagi, K. and Sakita, Y. (1988) *Analysis of Loss Probability Equalised by Trunk Reservation for Mixtures of Several Bandwidth Traffic.* Proceedings of 12th International Teletraffic Congress, Torino, 1988. Elsevier, Amsterdam.
[35] Tran-Gia, P. and Hubner, F. (2006) Analysis of trunk reservation and grade of service balancing mechanisms in multi-service broadband networks, in *Modelling and Performance Evaluation of ATM Technology* (eds G. Pujolle, H.G. Perros, and Y. Takahashi). North-Holland, Amsterdam.
[36] Whitt, W. (1985) Blocking when service is required from several facilities simultaneously. *Bell System Technical Journal*, **64** (8), 1807–56.
[37] Azmodeh, H. and Macfadyen, R.N. (1994) Multi-rate Call Congestion: Fixed Point Models and Trunk Reservation. Proceedings of Teletraffic Symposium, 11th Performance Engineering in Telecommunications Networks, Cambridge, March, Cambridge.
[38] Bean, N.G., Gibbens, R.J., and Zachary, S. (1995) Asymptotic analysis of single resource loss systems in heavy traffic with applications to integrated networks. *Advances in Applied Probability*, **27** (3), 273–92.
[39] Stasiak, M. and Głąbowski, M. (2000) A simple approximation of the link model with reservation by a one-dimensional Markov chain. *Journal of Performance Evaluation*, **41** (2–3), 195–208.
[40] Kaufman, J.S. (1992) Blocking with retrials in a completly shared recource environment. *Journal of Performance Evaluation*, **15**, 99–113.
[41] Moscholios, I.D., Logothetis, M.D., and Kokkinakis, G.K. (2002) Connection-dependent threshold model: a generalization of the Erlang multiple rate loss model. *Journal of Performance Evaluation*, **48** (1–4), 177–200.
[42] Vassilakis, V.G., Moscholios, I.D., and Logothesis, M.D. (2006) *The Extended Connection-dependent Threshold Model for Elastic and Adaptive Traffic.* International Symposium on Communication Systems, Networks and Digital Signal Processing, Patras, Greece, July Institute of Elecctrical and Electronics Engineers, Patras.

[43] Głąbowski, M. (2007) *Continuous Threshold Model for Multi-service Wireless Systems with PCT1 and PCT2 Traffic.* Proceedings of 7th International Symposium on Communications and Information Technologies, Sydney, October Institute of Electrical and Electronics Engineers, Sydney. 10.1109/ISCIT.2007.4392057.

[44] Moscholios, I.D., Logothesis, M.D., and Kokkinakis, G.K. (2005) Call-burst blocking of ON-OFF traffic sources with retrials under the complete sharing policy. *Journal of Performance Evaluation*, **59**, 279–312. DOI: 10.1016/j.peva.2004.06.003.

[45] Kallos, G.A., Vassilakis, V.G., Moscholios, I.D., and Logothetis, M.D. (2006) *Performance Modelling of W-CDMA Networks Supporting Elastic and Adaptive Traffic.* Proceedings of the Fourth International Working Conference on Performance Modelling and Evaluation of Heterogeneous Networks (HET-NETs '06), Ilkley. University of Bradford, Bradford.

[46] Rácz, S., Gerö, B., and Fodor, G. (2002) Flow level performance analysis of a multi-service system supporting elastic and adaptive services. *Journal of Performance Evaluation*, **49** (1–4), 451–69.

[47] Stasiak, M. and Zwierzykowski, P. (2008) Analytical model of the Iub interface carrying HSDPA traffic in the UMTS network, in *Management Enabling the Future Internet for Changing Business and New Computing Services* (eds C.S. Hong, T. Tonouchi, Y. Ma, and C.S. Chao), Springer.

[48] Stasiak, M., Zwierzykowski, P., Wiewióra, J., and Parniewicz, D. (2009) Analytical model of traffic compression in the UMTS network, in *Computer Performance Engineering* (ed. J. Bradley). Springer.

[49] Stasiak, M., Wiewióra, J., and Zwierzykowski, P. (2008) *The Analytical Model of the WCDMA Interface with Priorities in the UMTS Network.* International Symposium on Information Theory and its Applications, ISITA2008, Auckland, New Zealand, 7–10, December.

[50] Stasiak, M., Zwierzykowski, P., Wiewióra, J., and Parniewicz, D. (2008) Approximate model of the WCDMA interface servicing a mixture of multi-rate traffic streams with priorities, in *Computer Performance Engineering* (eds N. Thomas and C. Juiz). Springer. DOI: 10.1007/978-3-540-87412.

[51] Subramaniam, K. and Nilsson, A.A. (2005) *An Analytical Model for Adaptive Call Admission Control Scheme in a Heterogeneous umts-wcdma System.* Proceedings of International Conference on Communications. Institute of Electrical and Electronics Engineers, Seoul.

[52] Subramaniam, K. and Nilsson, A.A. (2005) *Tier-based Analytical Model for Adaptive Call Admission Cntrol Scheme in a umts-wcdma System.* Proceedings of Vehicular Technology Conference. Institute of Electrical and Electronics Engineers, Stockholm.

[53] Katzschner, L. and Scheller, R. (1976) *Probability of Loss of Data Traffics with Different Bit Rates Hunting One Common PCM-channel.* Proceedings of 8th International Teletraffic Congress, Melbourne.

8

Modeling of Systems with Multi-Rate Overflow Traffic

8.1 Introduction

In Chapter 6 we defined the notion of the telecommunications traffic overflow in networks offering call streams that require only one channel (BBU) of the group for service. In Chapter 7, we presented basic issues related to multi-service switching networks.

The aim of the present chapter is to combine these issues. Let us consider, now, on the basis of the information presented in previous chapters, the analytical mechanism that makes it possible to determine traffic characteristics of systems that are offered overflow multi-rate traffic. This chapter presents a method for determining the parameters of traffic that overflows from primary groups (cells) and an approximate method for dimensioning systems servicing multi-service overflow traffic [1–3].

8.2 Single-Service Model of the Full-Availability Group with Overflow Traffic

The algorithm based on the Kaufman–Roberts recursion worked out in Section 7.5.3 has been used to determine the occupancy distribution (Equation (7.79)), the blocking probability (Equations (7.80), (7.81)) and the loss probability (Equation (7.82)) of calls of different traffic classes. It has also been assumed that individual traffic streams are generated both by an infinite number of traffic sources (PCT1 stream) and by a finite number of traffic sources (PCT2 stream). With the instance in which the system services PCT1 stream only, the algorithm for determining traffic characteristics of the system, presented in Section 7.5.3, involves the application of Equations (7.39) and (7.42).

The algorithm worked out in Section 7.5.3 allows us to model systems properly with PCT1 and/or PCT2 streams offered directly to telecommunications groups (the so-called primary groups). However, as we already proved in Section 6.2.3, the call stream that overflows from the primary group is not a Poisson stream. Overflow traffic is therefore described by two parameters: the mean value of overflow traffic R (Equation (6.19)) that overflows to the alternative

Modeling and Dimensioning of Mobile Networks: From GSM to LTE
Maciej Stasiak, Mariusz Głąbowski, Arkadiusz Wiśniewski and Piotr Zwierzykowski

group and its variance σ^2 (Equation (6.20)). The Poisson call stream can also be described by the parameters R and σ^2, although in this case the values of both parameters are equal ($R = \sigma^2$). In the case of the call stream that overflows, the variance is always higher than the mean value. This means that the stream that overflows is more peaked than the stream of offered calls. The measure for this peaked character is the value of the peakedness coefficient $Z = \sigma^2/R$, which for the stream that overflows from the primary group is higher than 1 ($Z > 1$), and for the stream of offered calls is equal to 1 ($Z = 1$).

Bearing in mind the above statements, we may conclude that the Kaufman–Roberts formulas (7.39) and (7.42), in their basic form, worked out with the assumption of the exponential time distribution between calls, cannot be applied to the determination of the blocking probability of the multi-rate traffic in the alternative group. The present chapter aims to present a modification to Equations (7.39) and (7.42) so that a determination of blocking and loss probability of calls belonging to different streams of overflow multi-rate traffic in the alternative group will be possible.

8.2.1 Assumptions of the Model

Let us consider a fragment of the network shown in Figure 8.1. It is assumed in this system that each primary group is offered one call class only. The real links forming a network with multi-rate traffic carry different types of services (voice services, video conferences, and so on) in order to use their resources effectively. The assumption adopted in this chapter attempts at facilitating the understanding of the introduced analytical dependencies. Systems in which primary groups service many traffic classes will be presented in Section 8.4.

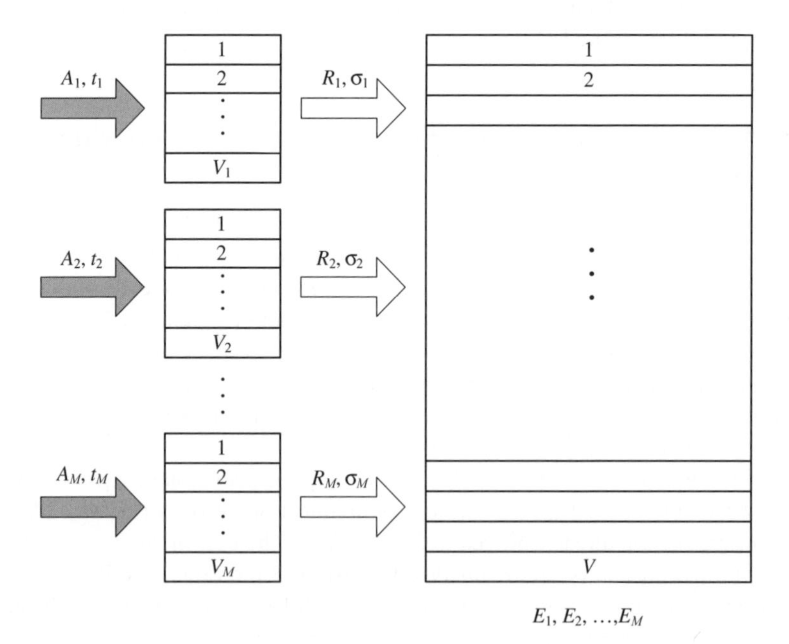

Figure 8.1 A fragment of the telecommunications network with single-service traffic offered to primary groups.

The fragment of the network considered here consists of M primary groups. The group designated by i has a capacity equal to V_i BBUs. Each of the groups is offered a different PCT1 traffic stream characterized by the traffic intensity A_i. The calls of class i demand t_i BBUs to set up a connection.

8.2.2 *Parameters of Overflow Traffic*

In the model with primary and alternative groups, in the case of blocking of primary groups, traffic overflows to an alternative group with the capacity V. The blocking probability in primary groups can be calculated on the basis of Erlang-B formula (5.4). However, it is necessary to note that one call of class i occupies simultaneously t_i BBUs. Therefore, from the point of view of the Erlang model, it is tantamount to a t_i-fold decrease in the capacity of the group with the real capacity of V_i BBUs. What follows is that, before a substitution with Equation (5.4), the group capacity should be divided by the number of BBUs demanded to set up a connection of a given class. Another way to obtain the same values of blocking probabilities is to apply Equations (7.39) and (7.42). They will take into consideration the group with the capacity of V_i which is offered one call stream with Poisson distribution formed by the calls that demand t_i BBUs to set up a connection.

Having the blocking probabilities in primary groups we are in position to calculate the parameters of overflow traffic of each of the classes, i.e the mean value R_i and the variance σ_i^2. For this purpose, the Riordan formulas (6.19) and (6.20) are used. Subsequently, on the basis of the parameters R_i and σ_i^2, we determine the value of peakedness Z_i of individual call streams of overflow traffic: $Z_i = \sigma_i^2 / R_i$.

8.2.3 *Occupancy Distribution and Blocking Probability in the Alternative Group with Multi-Rate Traffic*

The calls overflowing from primary groups form a multi-rate overflow stream offered to an alternative group. Consequently, the alternative group services M call classes. In order to determine blocking probabilities in such the group we apply the analogy to the Hayward method, described in Section 6.4.5 [1–3]. Recall that the method has been proposed to determine the blocking probability in the group servicing an overflow single-service call stream with the mean value R, additionally characterized by the peakedness Z, on the basis of the appropriately modified Erlang-B formula (Equation (6.46)):

$$B = E = E_{\frac{V}{Z}}\left(\frac{R}{Z}\right) \tag{8.1}$$

In the case of a group with multi-rate traffic the identical modification can be applied to the Kaufman–Roberts formulas [1–3]:

$$E_i = F\left(\frac{R_1}{Z_1}, \frac{R_2}{Z_2}, \ldots, \frac{R_i}{Z_i}, \ldots, \frac{R_M}{Z_M}; t_1, t_2, \ldots, t_i, \ldots, t_M; \frac{V}{Z}\right) \tag{8.2}$$

where $F(\cdot)$ denotes the algorithm for determining blocking probability E_i of calls of particular classes on the basis of Equations (7.39) and (7.42), which takes the following form:

$$n\,[P_n]_{V/Z} = \sum_{i=1}^{M} \frac{R_i}{Z_i} t_i \left[P_{n-t_i}\right]_{V/Z} \tag{8.3}$$

$$E_i = \sum_{n=\frac{V}{Z}-t_i+1}^{\frac{V}{Z}} [P_n]_{V/Z} \tag{8.4}$$

Note that in Equations (8.3) and (8.4) we also divide the capacity of the alternative group V by the value of the peakedness coefficient Z, analogously as in the dependence (8.1). The capacity of the alternative group in Equations (8.3) and (8.4) is divided by the so-called overall peakedness coefficient Z. According to Equation (8.2), the alternative group accepts M classes of calls, each of which can have a different value of the coefficient Z_i. What makes a problem, then, is the procedure of determining the value of the overall peakedness coefficient Z in order to normalize the group with the capacity V. The problem can be solved by introducing the weighted average of coefficients Z_i to particular call streams:

$$Z = \sum_{i=1}^{M} Z_i k_i \tag{8.5}$$

where

$$k_i = \frac{R_i t_i}{\sum_{l=1}^{M} R_l t_l} \tag{8.6}$$

It is assumed in Equation (8.5) that the contribution of peakedness of a stream of class i in the overall peakedness coefficient Z is directly proportional to the value of traffic offered to the alternative group by class i. The plausibility of this assumption has been proved by simulation studies [3].

8.3 Dimensioning of Alternative Groups with Multi-Rate Traffic

Let us consider now the problem of dimensioning of alternative groups [3]. Knowing the parameters of multi-rate traffic that overflows to the alternative group, we can determine the capacity of the group in such a way that appropriate quality of service of calls arriving to the system is secured. This problem can be formulated in the following way: *it is necessary to determine the minimum capacity of the alternative group for the assumed values of blocking probabilities for calls of particular classes.*

The basis for the considerations will be Equations (8.3) and (8.4). These formulas provide us with the parameters of the traffic that overflows to the alternative group (R_1, R_2, ..., R_M; Z_1, Z_2, ..., Z_M). Knowing these parameters of the traffic that overflows to the alternative group, we determine the capacity of the group in such a way as not to exceed the assumed values of blocking probabilities E_1, E_2,..., E_M. To achieve that we will be diminishing the capacity of the alternative group V starting from a given value that forms the upper limit V_{UL}. After each reduction in the capacity V, we can determine the current blocking probabilities e_1, e_2, ..., e_M on the basis of Equations (8.2)–(8.4):

$$e_1, e_2, \ldots, e_M = F\left(\frac{R_1}{Z_1}, \frac{R_2}{Z_2}, \ldots, \frac{R_M}{Z_M}; t_1, t_2, \ldots, t_M; \frac{V}{Z}\right) \tag{8.7}$$

The procedure of reducing the capacity V is repeated until we reach the minimum capacity that guarantees the following condition to be fulfilled:

$$e_1 \leq E_1;\; e_2 \leq E_2;\; \ldots; e_M \leq E_M \tag{8.8}$$

As we have stated, the process of decreasing the capacity is carried out from a certain boundary value V_{UL}. In order to perform the algorithm quickly, that is with the minimum number of decremental procedures involved, it is necessary to define the boundary value appropriately. Let us consider this problem in the following way. Let us treat each of the traffic classes forming the overflow multi-rate stream offered to the alternative group V separately. For each of M component call streams with the traffic intensity R_i, we determine the elementary group capacity v_i. This capacity has to be matched in such a way as to secure the quality of service of calls of class i in keeping with the assumed blocking coefficient E_i. The elementary capacities v_i are determined on the basis of the Hayward method:

$$E_i = E_{\frac{v_i}{Z_i}}\left(\frac{R_i}{Z_i}\right) \tag{8.9}$$

From the Erlang tables we read the value v_i/Z_i, which forms the basis for a determination of v_i. The value v_i determined on the basis of Equation (8.9) secures the required quality of service for calls with the blocking probability not exceeding value E_i. Moreover, traffic theory supports us in stating that it is also the alternative group with the capacity of $v_1 + v_2 + \ldots + v_M$, servicing M multi-rate call streams that, at least, ensures the assumed level of service for each of the traffic classes. This is so because, according to the group conservation principle (Section 5.2.8), blocking probabilities of calls in the group formed by a combination of several component groups are lower than blocking coefficients to be found in the component groups treated separately. With reference to the system under consideration, this means that calls are serviced with much lower blocking probabilities than originally assumed. What follows is a need to reduce the capacity of the group servicing overflow traffic.

On the basis of the determined elementary capacities it is possible to calculate the total boundary capacity V_{UL} of the alternative group:

$$\frac{V_{UL}}{Z} = \frac{v_1}{Z_1} + \frac{v_2}{Z_2} + \ldots + \frac{v_M}{Z_M} \tag{8.10}$$

We start the decremental procedure with the initial values v_i that are included in Equations (8.10). Reductions are made for each of the classes recurrently (in cycles), starting, for example, from the youngest class (serviced by a fictitious group v_1), that is the one that demands the lowest number of BBUs to set up a connection. In each step we reduce the capacity of only one component group. After each decrease of one BBU of the successive capacities v_i, we determine the normalized capacity of the transit group V, after the following equation:

$$\frac{V}{Z} = \frac{v_1}{Z_1} + \frac{v_2}{Z_2} + \ldots + \frac{v_M}{Z_M} \tag{8.11}$$

The value obtained (8.11) is then substituted into the dependence (8.7) and the blocking probability of calls of each class is then determined. The operations are carried out until the minimum capacity of the alternative group has been reached that still guarantees condition (8.8) to be satisfied. On finishing the algorithm, we obtain a capacity of the alternative group that secures an appropriate level of service of calls with appropriate blocking probabilities that do not exceed the assumed values E_1, E_2,. . ., E_M.

In conclusion, the algorithm for dimensioning the group that is offered the overflow multi-rate traffic composed of M classes of calls can be presented in the following way:

1. Determination of capacities of elementary groups v_1, v_2,. . .,v_M (Equation (8.9)).
2. Determination of the initial capacity of the group V_{UL} with overflow traffic on the basis of the values v_i calculated in the previous step $(v_1 + v_2 + \ldots + v_M)$ (Equation (8.10)).
3. Controlled cyclic decreases in capacity of elementary groups and determination of blocking probabilities of call until minimal capacity V guaranteeing the satisfaction of condition (8.8) is reached.
4. Determination of the capacity of the alternative group following the summation of capacities of component groups obtained in the previous step $(v_1 + v_2 + \ldots + v_M)$.

8.4 Multi-Service Model of the Full-Availability Group with Overflow Traffic

In previous sections we dealt with the dimensioning process of alternative groups in systems in which primary groups serviced only one call stream. This was a purely theoretical case and its main purpose was to facilitate understanding of the analytical dependencies being introduced. In real systems, primary groups carry multi-rate traffic that is composed of several classes of calls. The assumption that has been used so far allowed us to determine variances of traffic that overflows from primary groups in a simple way, through the application of the Riordan formulas. In the case in which the group carries multi-rate traffic, a direct application of the Riordan formulas is not possible [4–6].

In order to discuss an approximate method for determining variances of different classes of calls with traffic that overflows from groups servicing multi-rate traffic let us consider the telecommunications system presented in Figure 8.2. The system is composed of K primary groups [3]. Each group is offered M call classes; calls of class i demand t_i BBUs to set up a connection. The traffic intensity of the calls of class i offered to group j is $A_{i,j}$ erl. The blocking probability $E_{i,j}$ for calls of class i in the primary group j can be determined on the basis of Equations (7.39) and (7.42). Given the blocking probabilities, the mean value of the intensity of class i traffic that overflows from the group j can be determined:

$$R_{i,j} = A_{i,j}E_{i,j} \tag{8.12}$$

In order to determine the variance of each of call streams a method of decomposition of each of the real groups into M fictitious component groups with the capacities $V_{i,j}$ can be

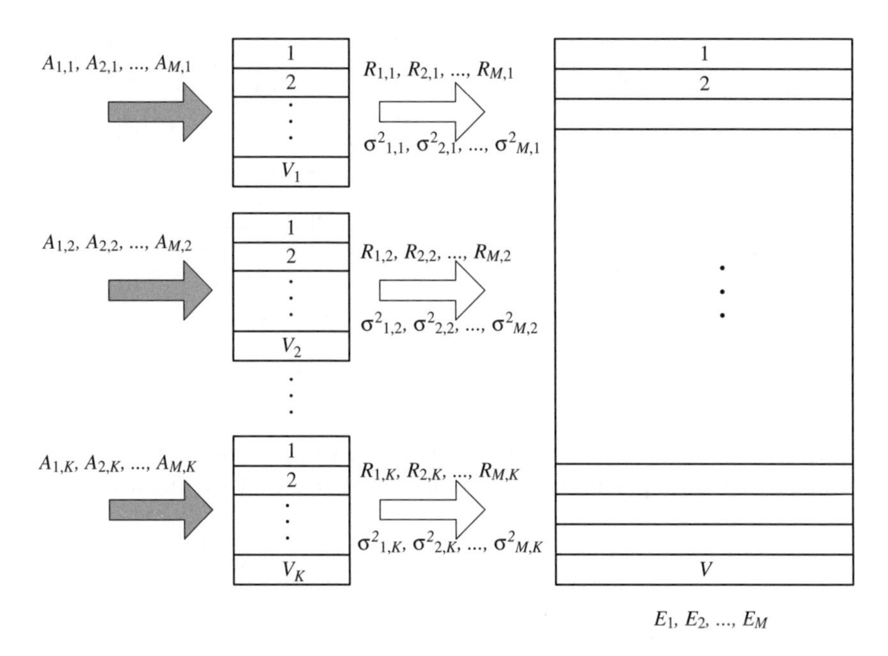

Figure 8.2 A fragment of telecommunications network with multi-service traffic offered to primary groups.

applied [3]. As a result, each fictitious group will be servicing exclusively calls of one class, which will make it possible to apply the Riordan formulas to determine the variance $\sigma_{i,j}^2$ of the traffic of class i that overflows from the group j.

First, let us determine the capacities of the fictitious groups. For this purpose we first determine the carried traffic of class i in group j:

$$Y_{i,j} = A_{i,j}(1 - E_{i,j}) \tag{8.13}$$

Let us notice that the value $Y_{i,j}$ defines the average number of calls of class i serviced in group j. The mean value of the intensity of class i traffic, expressed in BBUs, will therefore be equal to $Y_{i,j}t_i$. The capacity of a fictitious component group $V_{i,j}$ will be defined as this part of the real group V_j, which is not occupied by calls of the remaining classes:

$$V_{i,j} = V_j - \sum_{l=1;l \neq i}^{M} Y_{l,j}t_l \tag{8.14}$$

where V_j is the capacity of the primary group and the sum on the right side of Equation (8.14) determines the number of BBUs occupied by the calls of the remaining classes.[1] The idea of the construction of the fictitious component group is presented in Figure 8.3.

[1] An alternative algorithm for multi-rate links decomposition is proposed in [6].

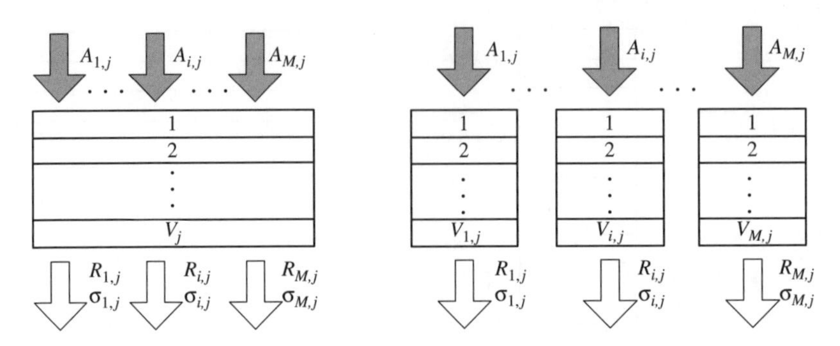

Figure 8.3 Decomposition of primary group with multi-rate traffic.

Knowing the parameters $R_{i,j}$, $A_{i,j}$ and $V_{i,j}$, the variance $\sigma_{i,j}^2$ for individual call streams that overflow to the alternative group can be determined on the basis of the Riordan formula:

$$\sigma_{i,j}^2 = R_{i,j}\left(\frac{A_{i,j}}{\frac{V_{i,j}}{t_i} + 1 - A_{i,j} + R_{i,j}} + 1 - R_{i,j}\right) \tag{8.15}$$

Since individual call streams offered to the system are statistically independent, then the parameters of the total traffic of class i offered to the alternative group will be equal to:

$$R_i = \sum_{j=1}^{K} R_{i,j} \tag{8.16}$$

$$\sigma_i^2 = \sum_{j=1}^{K} \sigma_{i,j}^2 \tag{8.17}$$

At this point we have at our disposal all parameters that characterize M call streams offered to the alternative group. To determine the capacity of the group we can apply the algorithm for dimensioning the alternative group servicing overflow multi-rate traffic described in Section 8.3.

Given the dependencies (8.16) and (8.17), we can also determine the occupancy distribution and blocking probability in the system with overflow multi-rate traffic shown in Figure 8.2. For this purpose, Equations (8.3) and (8.4) will be used, where the overall peakedness coefficient Z is determined after Equation (8.5).

8.5 Comments

The method presented in this chapter can be used to determine the intensity of multi-service overflow traffic between the layers of cellular networks, between cells (sectors) within a network, or between networks representing different technologies/generations (for example, GSM and UMTS). These models are particularly recommended for operators assign a given network a task of servicing a particular type of service (for example, a designation of a GSM network to service speech traffic and an UMTS network to service packet traffic). The determination of overflow traffic, as in the case of the model for single-service overflow traffic, will make it possible to determine the capacity of the cells servicing overflow traffic that will guarantee quality requirements laid down in a technical documentation for a given cellular network – the blocking and loss probability. This model can be also used for optimization of the structure of the existing network by indicating cells with an appropriate capacity for the intensity of overflow traffic.

Note that the method for determining the occupancy distribution and dimensioning of the systems with overflow traffic presented in the chapter can be also applied in systems in which direct groups service PCT2 traffic streams. In this case, the method for the determination of blocking probability $E_{i,j}$ of multi-service PCT2 traffic streams offered to direct groups will be changed – this probability is determined on the basis of Equation (7.81). A determined value of the blocking probability $E_{i,j}$ makes it possible to determine the value of offered traffic $R_{i,j}$ on the basis of Equation (8.12) and allows us to perform a decomposition of the primary group j (with the capacity of V_j BBUs), servicing M traffic classes, on the M groups where each has the capacity of $V_{i,j}$ described by Equation (8.14). Subsequently, we can convert PCT2 traffic stream to the equivalent PCT1 traffic stream according to the method described in Section 6.5, and, in turn, we can determine the parameters of offered overflow traffic $R_{i,j}$ (Equation (8.16)) and $\sigma^2_{i,j}$ (Equation (8.17). Finally, we are in position to determine the occupancy distribution and blocking probability in the system with overflow multi-rate traffic overflowing from primary groups servicing PCT2 traffic on the basis of Equations (8.3) and (8.4).

References

[1] Głąbowski, M. (2008) Modeling systems with multi-service overflow Erlang and Engset traffic streams. *International Journal on Advances in Telecommunications*, **1** (1), 14–26.

[2] Głąbowski, M., Kubasik, K., and Stasiak, M. (2007) *Modeling of Systems with Overflow Multi-rate Traffic*. Proceedings of Third Advanced International Conference on Telecommunications – AICT 2008. DOI 10.1109/AICT.2007.30.

[3] Głąbowski, M., Kubasik, K., and Stasiak, M. (2008) Modeling of systems with overflow multi-rate traffic. *Telecommunication Systems*, **37** (1–3), 85–96. DOI 10.1007/s11235-008-9070-8.

[4] Brandt, A. and Brandt, M. (2001) Approximation for overflow moments of a multi-service link with trunk reservation. *Journal of Performance Evaluation*, **43** (4), 259–68.

[5] Brandt, A. and Brandt, M. (2002) On the moments of the overflow and freed carried traffic for the GI/M/C/0 system. *Methodology and Computing in Applied Probability*, **4**, 69–82.

[6] Huang, Q., Ko, K., and Iversen, V.B. (2007) *An Approximation Method for Multi-service Loss Performance in Hierarchical Networks. Twentieth International Teletraffic Congress, ITC20 2007*, volume 4516 of *Lecture Notes in Computer Science*, DOI 10.1007/978-3-540-72990-7_78.

9

Equivalent Bandwidth

While analyzing the architecture of cellular networks it is noticeable that each of the networks in question is characterized by the radio access network and the backbone network. The core part of the mobile network is constructed through technologies used in telecommunications and computer wide area networks, and the most frequently used technology today is asynchronous transfer mode (ATM). This technology was a standard that was recommended for constructing the backbone network in early versions of the UMTS system (Section 2.2, Figure 2.4) and, in fact, it is still used by many operators. It is envisaged that newer versions of the UMTS system (Figures 2.5 and 2.6) as well as the backbone network in the LTE system (Section 3.2), will be constructed on the basis of the IP protocol (Internet Protocol).

Regardless of the technology used, the backbone network has been and will remain a network with virtual channel switching that services traffic generated by subscribers who use a number of different services (see, for example, Table 2.4). Traffic in such networks can be considered in three time scales (Figure 9.1): packet level, burst level and call level. From the perspective of designing and dimensioning of networks with virtual channel switching, the call level is the most important. It should be remembered, however, that traffic characteristics at the burst and the packet levels have a substantial influence upon characteristics at the call level.

The classification of traffic sources at the burst level involves distinguishing between traffic sources with constant bit rate (CBR) and those with variable bit rate (VBR). For an analytic description, traffic sources with variable bit rate require a higher number of parameters that take into consideration the nature of the changes in the bit rate oscillating between the peak rate and the minimum rate.

The link dimensioning process in networks is based on a determination of loads imposed upon the network by VBR sources at the call level. For this purpose, the concept of so-called *equivalent bandwidth* [1] was introduced. The allocation of the equivalent bandwidth to VBR sources makes it possible to model links of the backbone networks through methods worked out for broadband systems [1, 2]. Therefore, for modeling of present-day cellular networks, the *multi-rate models* [1–3], described in Chapter 7, are used.

The present chapter gives a presentation of the most typical models of traffic sources in networks with multi-rate traffic, that is, interrupted Poisson process (IPP), Markov modulated Poisson process (MMPP), and interrupted Bernoulli process (IBP) [1–3], and provides the

Modeling and Dimensioning of Mobile Networks: From GSM to LTE
Maciej Stasiak, Mariusz Głąbowski, Arkadiusz Wiśniewski and Piotr Zwierzykowski

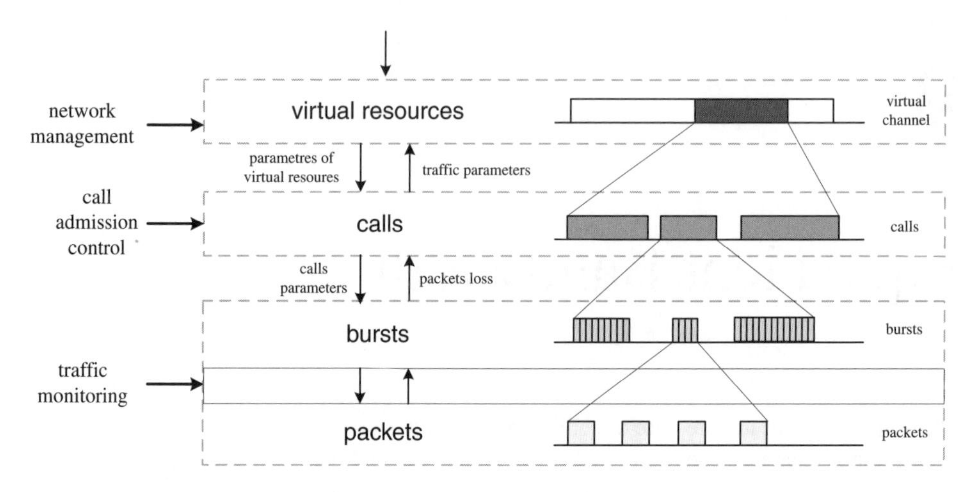

Figure 9.1 Traffic control levels in the present-day cellular network.

reader with comments pertaining to the modeling process for self-similar traffic. Further on in the chapter, methods for a determination of the equivalent bandwidth for VBR sources and a method for the discretization of the bandwidth that allow links of modern cellular networks to be dimensioned through the methods presented in Chapter 7, are presented.

9.1 Interrupted Poisson Process

The first model of traffic sources to be described is the IPP (the interrupted Poisson process) [4]. The source can be alternatively in one of the two following states – active ON and inactive state OFF (Figure 9.2). In the active state, packets, which form a stream with the intensity λ are generated. In an inactive state, packets are not generated. The alternating periods of active and inactive states are switched on and off by the two-state Markov process (Figure 9.2) with the following parameters:

α – intensity of the transition from state ON to state OFF;
β – intensity of the transition from state OFF to state OFF;
$P(\text{ON})$ – ON state probability;
$P(\text{OFF})$ – OFF state probability.

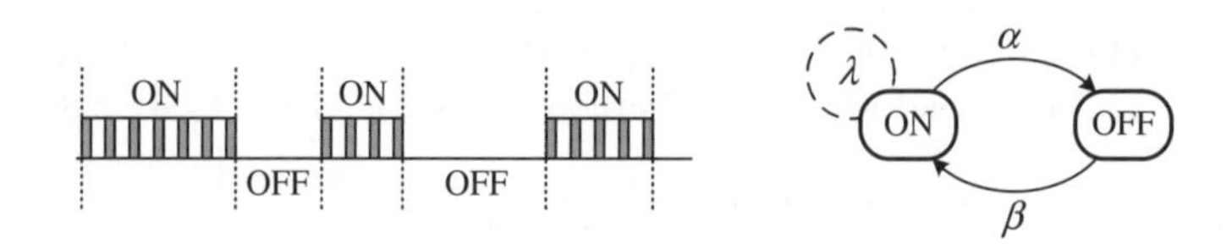

Figure 9.2 Example interrupted Poisson process.

Notice that the average time that the sources spends in states ON and OFF is the inverse of the intensities α and β:

$$\tau_{\mathrm{ON}} = 1/\alpha \tag{9.1}$$

$$\tau_{\mathrm{OFF}} = 1/\beta \tag{9.2}$$

State probabilities can be determined on the basis of the statistical equilibrium equations that, in the case being considered here, take on the following form:

$$\alpha P(\mathrm{ON}) = \beta P(\mathrm{OFF}) \tag{9.3}$$

$$P(\mathrm{ON}) + P(\mathrm{OFF}) = 1 \tag{9.4}$$

The solution of this simple system of Equations (9.3) and (9.4) is:

$$P(\mathrm{ON}) = \frac{\beta}{\beta+\alpha} \qquad P(\mathrm{OFF}) = \frac{\alpha}{\beta+\alpha} \tag{9.5}$$

At this point, the average value m_{IPP} and variance σ^2_{IPP} of the intensity of generated packets can be determined:

$$m_{\mathrm{IPP}} = \lambda P(\mathrm{ON}) = \lambda \frac{\beta}{\beta+\alpha} \tag{9.6}$$

$$\sigma^2_{\mathrm{IPP}} = [\lambda - m_{\mathrm{IPP}}]^2\, P(\mathrm{ON}) + [0 - m_{\mathrm{IPP}}]^2\, P(\mathrm{OFF}) = \lambda^2 \frac{\alpha\beta}{(\beta+\alpha)^2} \tag{9.7}$$

After elementary transformations – with Equations (9.1) and (9.2) taken into consideration – Equations (9.6) and (9.7) can be transformed to the following:

$$m_{\mathrm{IPP}} = \lambda \frac{\tau_{\mathrm{ON}}}{\tau_{\mathrm{ON}} + \tau_{\mathrm{OFF}}} \tag{9.8}$$

$$\sigma^2_{\mathrm{IPP}} = m_{\mathrm{IPP}}(\lambda - m_{\mathrm{IPP}}) \tag{9.9}$$

9.2 Markov Modulated Poisson Process

Let us consider the model of traffic source called Markov modulated Poisson process [5, 6]. The simplest MMPP process can occur in one of the two active states ON_1 and ON_2 (Figure 9.3). In active state ON_1, packets that form a stream with the intensity λ_1 are generated, whereas in active state ON_2 packets with the intensity λ_2 are generated.

Figure 9.3 Example Markov modulated Poisson process.

Active periods ON_1 and ON_2 are switched on and off by the two-state Markov process (Figure 9.3) with the parameters:

α – intensity of the transition from state ON_1 to state ON_2;
β – intensity of the transition from state ON_2 to state ON_1;
$P(ON_1)$ – state probability ON_1;
$P(ON_2)$ – state probability ON_2.

With a similar analysis as in the case of the IPP process it can be proved that the mean value m_{MMPP} and variance σ^2_{MMPP} of the intensities of generated packets can be described by the following formulas:

$$m_{MMPP} = P(ON_1)\lambda_1 + P(ON_2)\lambda_2 = \frac{\beta\lambda_1 + \alpha\lambda_2}{\beta + \alpha} \tag{9.10}$$

$$\begin{aligned}\sigma^2_{MMPP} &= P(ON_1)\,[\lambda_1 - m_{MMPP}]^2 + P(ON_2)\,[\lambda_2 - m_{MMPP}]^2 \\ &= \tfrac{(\lambda_1-\lambda_2)^2}{(\alpha+\beta)^2}\alpha\beta\end{aligned} \tag{9.11}$$

It should be stressed that, in practice, to model real packet streams in broadband networks using ATM technology MMPP processes with a higher number of switched states are also used.

9.3 Interrupted Bernoulli Process

The IBP [7] is a discrete counterpart of the IPP process describing traffic sources in a broadband network that assumes constant time lengths (slots) for packet service.[1]

The IBP occurs in two states – active state ON and inactive state OFF (Figure 9.2). Packets are generated in the active state. Packets create a Bernoulli stream. In the inactive state packets are not generated.

The active and inactive periods are switched on and off by a two-state embedded Markov chain (Figure 9.4) that can be defined by the following parameters:

$P(ON)$ – state probability ON;
p – the probability of an event in which, if a given slot belongs to state ON, then the next slot will also belong to state ON; probability $1 - p$ is a measure of an event in which the next slot will belong to state OFF;

[1] Such an approach can be successfully used in the ATM technology, where every packet (ATM cell) has a constant length.

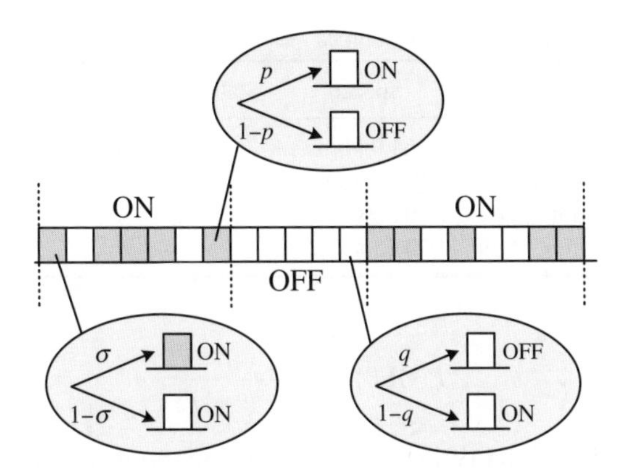

Figure 9.4 Example interrupted Bernoulli process.

σ – the probability of an event in which a given slot in state ON is active; probability $1 - \sigma$ is thus the probability of an event in which a given cell in state ON is not active;

$P(\text{OFF})$ – probability of state OFF;

q – the probability of an event in which if a given slot belongs to state OFF, then in state OFF the next slot will also belong to state OFF; the probability $1 - q$ is a measure of an event in which the next slot will belong to state ON.

The probability of event in which the ON state will terminate after x slots is determined by a geometric distribution and is equal to:

$$P(x) = (1 - p)p^{x-1} \tag{9.12}$$

In a similar way, the probability of an event in which the OFF state terminates after y slots will be determined:

$$P(y) = (1 - q)q^{y-1} \tag{9.13}$$

Thus, on the basis of properties of the geometric distribution (9.12) and (9.13), the average time length (expressed in the number of slots) that a source spends in states ON and OFF can be determined:

$$\tau_{\text{ON}} = 1/(1 - p) \qquad \tau_{\text{OFF}} = 1/(1 - q) \tag{9.14}$$

The intensities α and β of the switching process that switches the source from the ON state to the OFF state and from the OFF state to the ON state are inverses of average time lengths that the system spends in these states:

$$\alpha = 1/\tau_{\text{ON}} = 1 - p \qquad \beta = 1/\tau_{\text{OFF}} = 1 - q \tag{9.15}$$

Taking into consideration (9.5) and (9.15), it is possible to determine the probabilities of the ON and OFF states:

$$P(\text{ON}) = \frac{1 - q}{2 - p - q} \tag{9.16}$$

Table 9.1 Parameters of ON/OFF sources

Parameter	Speech	Stationary images	Data transfer
maximum emission speed	64.0 kbps	2.0 Mbps	10.0 Mbps
average emission speed	22.0 kbps	87.0 kbps	1.0 Mbps
average state duration – ON	0.35 s	0.5 s	0.0017 s
average state duration – OFF	0.65 s	11.0 s	0.013 s

$$P(\mathrm{OFF}) = \frac{1-p}{2-p-q} \tag{9.17}$$

Now, on the basis of (9.16) and (9.17), we are in position to determine the mean value and variance of the intensity of generated packets.

$$m_{\mathrm{IBP}} = \sigma P(\mathrm{ON}) = \frac{\sigma(1-q)}{2-p-q} \tag{9.18}$$

$$\begin{aligned}\sigma_{\mathrm{IBP}}^2 &= [\sigma - m_{\mathrm{IBP}}]^2\, P(\mathrm{ON}) + [0 - m_{\mathrm{IBP}}]^2\, P(\mathrm{OFF}) \\ &= \sigma^2 \tfrac{(1-p)(1-q)}{(2-p-q)^2}\end{aligned} \tag{9.19}$$

It needs stressing that in the theory of traffic much more complex models of traffic sources for the broadband network are considered. A detailed description of these models, along with references, can be found in, for example [8]. Parameters of some of the most typical traffic sources of the type ON/OFF, characteristic for the ATM network, are presented in Table 9.1 [9].

9.4 Comments

The models of traffic sources presented in Sections 9.1 – 9.3 are exemplary models that can be used in equivalent bandwidth determination. An example of the application of the IPP model of traffic source for effective bandwidth calculation is shown in Section 9.6.1.1.

9.5 Self-Similar Traffic

The feature of self-similarity is used to describe objects that are composed of smaller copies of themselves. In mathematics the notion of self-similarity was formulated by the mathematician Benoit Mandelbrot and mainly refers to geometrical figures called fractals [10]. Such figures magnified to any arbitrarily chosen extent reveal parts that are exactly similar to the input object. Self-similarity, which, initially, was analyzed in relation to geometrical figures only, has quickly been transferred into the domain of random processes, because it turns out that many stochastic processes taking place around us have this particular property. The works by the English hydrologist H.E. Hurst are particularly important in this respect. H.E. Hurst, with his proposed rescaled adjusted range (R/S) statistic [11] proved that there is a parameter, later called the Hurst parameter, that defines the level of self-similarity of stochastic processes.

A characteristic feature of the Markovian processes is its property of memorylessness. In such processes, the autocorrelation function disappears in an exponential fashion, which means that long-range dependent events are correlated very poorly. These processes have been called *short-memory processes* or *short-range dependent processes*. There are also processes whose autocorrelation function disappears much slower, which results in an occurrence of long-range correlation between distant events. When this is the case, then we have a *long-memory processes* or *long-range dependent processes*. A slowly disappearing autocorrelation function is characteristic of self-similar processes. However, for a long-memory process to be a self-similar process the autocorrelation function of the process has to manifest its properties regardless of the time scale used (milliseconds, seconds, minutes, hours, ...). Two kinds of self-similar processes are distinguished: exactly self-similar [10] and asymptotically self-similar [12, 13]. A process can be called exactly self-similar if the processes effected by its rescaling are indistinguishable from the initial process. A process that is asymptotically self-similar is, in turn, a process whose autocorrelation functions of the initial process and the rescalled process are in conformity (within an appropriately long time scale) with one another.

In present-day networks telecommunications traffic is characterized by a fairly high irregularity to be observed in different time scales. This irregularity is manifested by the appearance of *bursts* between periods characterized by a lower number of calls. On the basis of long-term observations of packet networks it has been noted that traffic in the networks also reveals features of self-similarity in various time scales (minutes, seconds, and so on). The first long-term studies that indicated the self-similar character of traffic in computer networks were carried out at the beginning of the 1990s on the local Ethernet computer network of Bellcore company. The existence of self-similar traffic was also observed in wide area networks (WANs) [14]. Studies on the self-similarity of traffic in the Internet network are included in [15]. The research work conducted by the authors was based on the analysis of the users' connections to www services (World Wide Web). On the basis of the traffic generated by users, their preferences and the size of data sent by them, it was proved that this traffic has long-range dependent features, that is, the same that are characteristic of self-similar traffic. Subsequent studies revealed that also variable bit rate (VBR) multimedia streams feature similar properties to those of traffic in the Internet network.

These indications that self-similarity occurred in packet networks were followed by numerous studies on its possible causes. There is evidence that reasons for the occurrence of self-similarity could be applications generating traffic streams [15, 16], transport layer protocols [17–20] and human factors [15].

The occurrence of self-similarity, characterized by long-range dependence and slowly diminishing variance, can result in the occurrence of unfavorable traffic phenomena such as greater delays, bigger loads in queues, higher packet loss rate than in the case of networks in which the phenomenon of self-similarity does not exist. Due to the above, over the past few years there has been increased research activity on methods for appropriate analysis of systems with self-similar traffic and on ways of eliminating the causes of self-similarity in packet networks.

The results of an interesting research study on self-similarity in wireless networks are given in [21], where the author analyzes traffic streams transferred from wire networks, for example from the Internet, to wireless networks in terms of the occurrence of self-similarity. It was assumed that traffic from a wire network offered to the gateway connecting wire network to a wireless network would display the self-similarity. On the basis of a typical buffer mechanism and repackaging data at the gate, statistical properties of traffic directed to wireless networks were investigated. The ensuing simulations and analyses found that traffic transferred to the wireless network showed a different degree of self-similarity than the traffic coming from a wire to the wireless networks, including a loss of self-similarity.

In order to determine if the phenomenon of self-similarity should be taken into consideration in the analysis of the network, the Hurst parameter for traffic carried by appropriate links has to be evaluated. There are many methods for determining this parameter, including the R/S statistic and Whittle methods, in other words variance-time plot and periodogram plot [22]. Many works provide us with the values of the Hurst parameter H determined for different telecommunications networks (for example [23, 24]). These works show that the Hurst parameter for telecommunications networks ranges between 0.75 and 0.93.

Ongoing studies on limiting the influence of network mechanisms on the occurrence of the self-similarity phenomenon have not led to its complete elimination. This means that with the introduction of an ever growing range of services involving data transmission in packet networks, including wireless networks, the probability of the occurrence of self-similarity can increase, particularly in those elements of the networks that are responsible for aggregation of many traffic streams.[2]

In order to determine traffic characteristics of packet networks with self-similar call streams it is necessary, then, to apply appropriate analytical models such as fractional Gaussian noise [10] or the fractional autoregressive integrated moving average (FARIMA) model proposed in [12, 13]. These models are characterized by a relatively high level of complexity, which considerably limits their usefulness from the engineering perspective. It should be emphasized here that from among stochastic traffic models, for sources with variable bit rates, self-similar models, those that use distributions with a defined degree of self-similarity, are treated as equivalent to models based on the equivalent bandwidth (Figure 2 in [25]). In models based on the equivalent bandwidth, the parameters of packet level (packet loss rate, delay, intensity of generation of packets) determine the so-called equivalent bandwidth that makes it possible to dimension systems at the call level using multi-rate models. Methods for determining the parameters of the equivalent bandwidth for self-similar traffic streams are worked out in, for example, [26–28]. An example of a method for a determination of the equivalent bandwidth for self-similar traffic is presented in Section 9.6.3.

9.6 Example Methods for Determining Equivalent Bandwidth

The term "equivalent bandwidth" was proposed for the first time in 1988 [29]. The equivalent bandwidth is a method for estimating resources demanded by a given traffic stream with a defined quality of service (QoS) parameter. Formal definition of the equivalent bandwidth was proposed in [30]:

$$C_E(s, t) = \frac{1}{st} \log E[\exp(sA_t)] \tag{9.20}$$

[2]In [31] it is proved that the basic (essential) element of the wireless network whose properties will depend on a degree of self-similarity is the GGSN node (Section 1.2, Figure 1.2).

where s is the so-called spatial parameter (e.g. bits^{-1}), t is so-called time parameter (for example, seconds), and A_t is the process that determines the call arrival process. On the basis of the definition (9.20), a formal representation of the definition of the equivalent bandwidth for defined traffic sources, such as Bernoulli sources in the system without buffers, periodic sources or aggregated traffic streams from multiple sources ON/OFF [30], has also been proposed.

The problem limiting the practical applicability of the definition (9.20) is the need to evaluate the parameters s and t, which should depend on the quality requirements and the parameters of the system [32]. Therefore, a great deal of methods and techniques for the evaluation of the equivalent bandwidth depart from the formal definition. The applicable evaluation methods can be divided according to the way the traffic source is described and according to the kind of system (with or without buffer) for which the bandwidth is determined. Further on in the chapter we will present examples of methods to be applied for systems with losses (the capacity of the buffer can be omitted) and with queuing (systems with buffers).

9.6.1 Methods for Loss Systems

In order to better understand equivalent bandwidth evaluation methods for systems with losses we assume that the multiplexer under consideration is equipped with a small buffer that can absorb only packets from a few sources simultaneously. A further assumption is that the influence of the buffer on the parameters used for the description of traffic sources is omittable.

9.6.1.1 Lindenberger Tidblom Method

Let us consider the Lindenberger Tidblom method used for the evaluation of the equivalent bandwidth for ON/OFF sources (Section 9.1) [1, 2] that are characterized by the following parameters:

- T_{ON}: average duration time of state ON – the state in which the source emits packets;
- T_{OFF}: average duration time of state OFF – the state in which the source does not emit packets;
- m: average packet generation rate (Equation (9.6)):

$$m = \frac{hT_{\mathrm{ON}}}{T_{\mathrm{ON}} + T_{\mathrm{OFF}}}$$

- h: maximum (peak) packet-generation rate;
- σ^2: variance of packet generation intensity (Equation (9.7)):

$$\sigma^2 = m(h - m)$$

According to the Lindenberger Tidblom method, to evaluate the equivalent bandwidth (C_E) the following dependency can be applied:

$$C_E = c_1 m + c_2 \frac{\sigma^2}{C} \tag{9.21}$$

Table 9.2 Example values of the coefficients c_1 and c_2 presented in [2]

Coefficient	$P_{\text{loss}} = 10^{-9}$	$P_{\text{loss}} = 10^{-8}$	$P_{\text{loss}} = 10^{-7}$	$P_{\text{loss}} = 10^{-6}$
c_1	1.18	1.16	1.14	1.12
c_2	63	56	48	41

where C is the capacity of the link, and c_1 and c_2 are coefficients. The value of these coefficients can be determined on the basis of the following heuristic dependencies:

$$c_1 = 1 - \frac{2 \log P_{\text{loss}}}{100} \quad \frac{c_2}{c_1} = -6 \log P_{\text{loss}} \tag{9.22}$$

where P_{loss} is the packet loss probability. Sample values of the coefficients c_1 i c_2 for some values of packet loss probabilities are presented in Table 9.2.

In the case of the packet loss probability $P_{\text{loss}} = 10^{-9}$ assumed for the example network, Equation (9.21) can be rewritten in the following way:

$$C_E = 1.18m + 63\frac{\sigma^2}{C} \tag{9.23}$$

It is recommended that Equation (9.23) should be used when the two following conditions are fulfilled:

$$\frac{h}{C_E} \geq 2, \quad \frac{m}{h} \geq 0.05 \quad 1000 \geq \frac{C}{h} \geq 15 \tag{9.24}$$

If the source does not meet the conditions defined by the dependencies (9.24), then the equivalent bandwidth can be evaluated using the algorithm presented in Figure 9.5. To simplify the

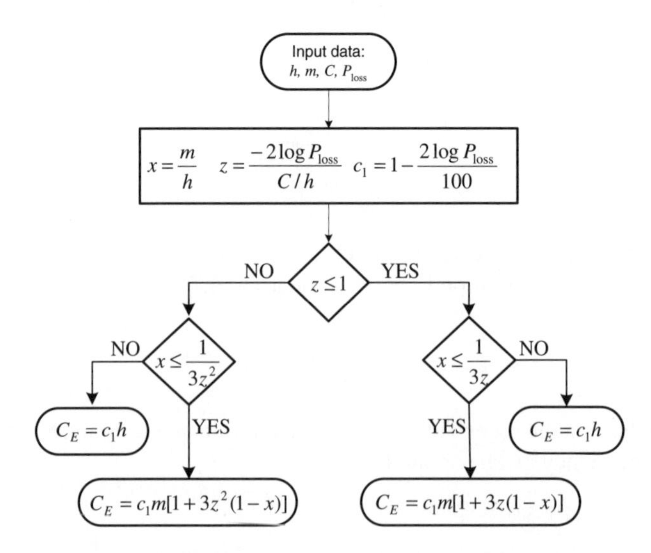

Figure 9.5 Equivalent bandwidth evaluation algorithm for the Lindenberger-Tidblom method.

presentation in the figure, the following coefficients are introduced:

$$z = \frac{-2 \log P_{loss}}{\frac{C}{h}} \qquad x = \frac{m}{h} \tag{9.25}$$

Quotient $\frac{C}{h}$ defines the maximum number of concurrently serviced calls related to a given service that does not contribute to the need for packet buffering. Taking into consideration the coefficients x and z, Equation (9.21) can be rewritten in the following way:

$$C_E = c_1 h x[1 + 3z(1 - x)] \quad \text{for} \quad x \leq \frac{1}{3z} \tag{9.26}$$

For the small value of the fraction $\frac{C}{h}$, use of the following formula is recommended

$$C_E = c_1 h x[1 + 3z^2(1 - x)] \quad \text{for} \quad x \leq \frac{1}{3z^2} \tag{9.27}$$

9.6.1.2 Rate Envelope Multiplexing (REM)

Another commonly used method for bandwidth evaluation is rate envelope multiplexing proposed in [2]. This method assumes that in the description of the aggregated stream the current sum of the intensity of packet generation is used [33]. The advantage of this method is the independence of bandwidth determination from the kind of the burst distribution – mean and peak values are sufficient for the determination of the equivalent bandwidth. By a "burst" we understand a group of packets emitted by sources within a short time interval. The probability that the total traffic stream intensity R will cause an excess in the link capacity C can be determined on the basis of the stream intensity distribution.

Assume that the distribution R can be approximated with a Gauss distribution with the average value m and variance $\sigma^2 = mh$. The probability of exceeding the link capacity can then expressed with the formula [1]:

$$P(R > C) = \frac{1}{\sqrt{2\pi}} \int_{\frac{C_E - m}{\sigma}}^{\infty} \exp\left(\frac{y^2}{2}\right) \mathrm{d}y \tag{9.28}$$

where y is an auxiliary variable. Taking into account the appropriate approximation of the integral in Equation (9.28), proposed in [34], Equation (9.28) can be rewritten as follows [35]:

$$P(R > C) = \varepsilon = \frac{1}{\sqrt{2\pi}} \exp\left(\frac{(C_E - m)^2}{2\sigma^2}\right) \tag{9.29}$$

Equivalent bandwidth is calculated by solving Equation (9.29):

$$C_E = m\left(1 + \sqrt{-2\ln\varepsilon - ln2\pi} \cdot \sqrt{\frac{h}{m}}\right) \qquad \varepsilon < \frac{1}{\sqrt{2\pi}} \tag{9.30}$$

The approximation of the Gauss distribution used above is not accurate enough, but the probability ε obtained by Equation (9.29) corresponds to the probability of queuing $P(Q)$ – a delay in packets delivery ($P(Q > 0)$). This probability is higher than the probability of exceeding the link capacity and can be treated as the upper bound of the probability of exceeding link capacity ($P(R > C)$). This means that the bandwidth determined according to the dependency (9.30) will also be overestimated.

9.6.2 Methods for Queuing Systems

Bandwidth evaluation methods for systems with queuing require a description of the source that includes time parameters. To better understand methods for equivalent bandwidth evaluation in systems with queuing let us assume that the multiplexer under consideration has a buffer with a size that is large enough to accommodate compensation for unevenness occurring during the arrival process of cell bursts. One of the simplest methods for bandwidth evaluation in aggregated streams for systems with queuing is the method proposed in [23, 34, 36]. In [34] it is accepted that the distribution of the queue length is determined with the assumption of the exponential distribution of the size of bursts. Assuming that traffic is generated by superposition of ON/OFF sources, in which duration of ON and OFF states has an exponential distribution, we can determine the distribution of the length of the queue in an infinitely long buffer by providing solutions to appropriate differential equations. If the buffer is appropriately large, then the probability of no space in the buffer $P(Q > x)$ can be approximated by a single exponential equation [35]:

$$P(Q > x) = P(Q > 0) \cdot \exp\left(-\left(1 - \frac{m}{C_E}\right)\frac{x}{b}\right) \tag{9.31}$$

where $\frac{m}{C}$ denotes the average load of the system, while b is the average number of packets in a burst, and x represents the size of the buffer. A determination of the probability of the occurrence of the queue ($P(Q > 0)$) is a complex matter so in [34] a simplifying assumption was proposed: $P(Q > 0) = 1$. This corresponds to approximating $P(Q > x)$ by the conditional probability $P(Q > x|Q > 0)$, which leads to:

$$P(Q > x) = \varepsilon = \exp\left(-\left(1 - \frac{m}{C_E}\right)\frac{x}{b}\right) \tag{9.32}$$

The effective bandwidth can be obtained by solving Equality (9.32) for the assigned packet loss probability:

$$C_E = \frac{m}{1 + \frac{b}{x}\ln\varepsilon}, \quad \frac{x}{b} > -\ln\varepsilon \tag{9.33}$$

9.6.3 Determination of the Equivalent Bandwidth for Self-Similar Traffic

The Fractional Brownian Motion model (FBM) is one of the basic models that allows us to take into consideration the phenomenon of self-similarity. It was proposed in [28]. In this model, the process of cumulated call arrival can be described with the formula:

$$A_t = mt + \sqrt{m\sigma^2} Z_t \tag{9.34}$$

Table 9.3 Parameters for FBM traffic

Parameter	Bellcore 1	Bellcore 2	ADSL	local ISP
m	2279 kbps	12.3 kbps	10.5 kbps	9.76 kbps
α	262,8 kbps	68.6 kbps	440 kbps	38 kbps
H	0.78	0.86	0.915	0.88

Table presents example values of the parameters used in Equation (9.36) [23].

where m denotes the average stream intensity, σ^2 is the variance determined on the basis of measurement and A_t is a parameter that defines the call arrival process (Equation (9.20)). The variable Z_t represents normalized FBM with the Hurst parameter within the interval [0.5, 1).

In [28, 37] an approximation is proposed of the distribution of the length of the queue for the system with infinite buffer to which streams described by the FMB traffic model arrive. In [34] it was assumed that the distribution of the length of the queue could be described by the Weibull distribution:

$$P(Q > x) = \varepsilon = \exp\left(-\frac{(C_E - m)^{2H} x^{2-2H}}{2 \cdot \kappa(H)\sigma^2 m}\right) \tag{9.35}$$

where H is the Hurst parameter and $\kappa(H) = H^H(1 - H)^{1-H}$.

The probability (9.35) forms a basis for the determination of the equivalent bandwidth [28]:

$$C_E = m + \left(\kappa(H)\sqrt{-2\ln\varepsilon}\right)^{\frac{1}{H}} (\sigma^2)^{\frac{1}{2H}} x^{\frac{H-1}{H}} m^{\frac{1}{2H}} \tag{9.36}$$

The parameters used in Equation (9.36) can be determined on the basis of traffic measurements [38]. Example values obtained on the basis of the measurements are presented in Table 9.3. The first two columns correspond to the measurements that were obtained for the ethernet network by Bellcore [28], whereas the values in columns 3 and 4 were obtained as the result of the measurements of traffic related to http protocol in the access network and in the network of a local internet service provider respectively [35]. Equation (9.36) can be also derived from the definition provided in [30].

9.7 Bandwidth Discretization

The concept of the equivalent bandwidth is related to the concept of the basic bandwidth unit (BBU) for traffic sources offered in the backbone (broadband) network [1]. The allocation of the equivalent bandwidth packets to traffic sources with variable bit rate, in other words given constant bit rates that can be a multiple of a given basic bit rate ΔC, at a given network level, makes it possible to evaluate the traffic characteristics of the network and nodes in a cellular network with the application of multi-rate models [1].

Let us consider a link in a cellular network characterized by the following parameters:

C – total capacity of the available bandwidth in the considered system;
C_i – equivalent bandwidth for sources of class i;
ΔC – basic bandwidth unit (BBU).

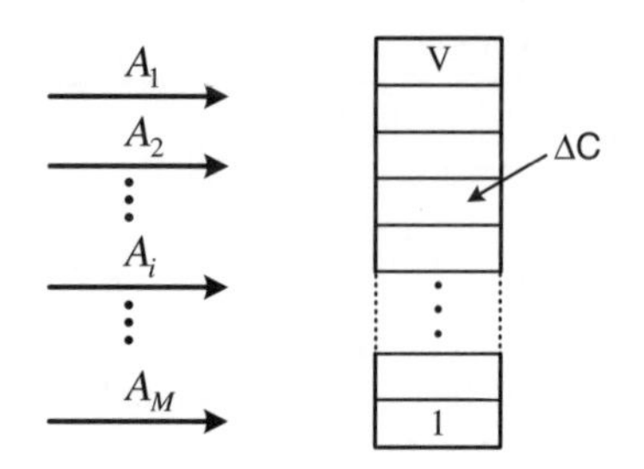

Figure 9.6 Bandwidth discretization.

The basic bandwidth unit can be determined as the greatest common divisor of all equivalent bandwidths required for carrying out calls that are offered to the system:

$$\Delta C = \mathrm{GCD}(C_1, \ldots, C_M) \tag{9.37}$$

After determining the value of BBU for a given link, it is possible to determine the number of BBUs demanded by each call of class i:

$$t_i = \lceil C_i/\Delta C \rceil \tag{9.38}$$

The total link capacity, expressed in basic bandwidth units, is then equal to:

$$V = \lfloor C/\Delta C \rfloor \tag{9.39}$$

The bandwidth is discretized in this way (Figure 9.6) – a link in a cellular network is replaced by an equivalent model of the link with multi-rate traffic. Such a replacement enables us to take advantage of the models of multi-rate systems – described in Chapter 7 – in dimensioning links of cellular networks.

9.7.1 Comments

This chapter described examples of models of traffic sources and methods for determining the equivalent bandwidth for different traffic sources, for loss systems and for queuing systems. The methods presented have been chosen from many different, oftens very complex, methods for evaluating the bandwidth available in the literature. Currently, none of the organizations preparing standards for telecommunications and/or computer networks proposes an appropriate method for equivalent bandwidth determination. Recommendations sometimes mention the necessity of using the equivalent bandwidth, but no specific method for its evaluation is indicated. Therefore, while choosing methods for the evaluation of the equivalent bandwidth, we were guided, by our own assessment of their usefulness in the present course book, and also by the frequency of their citations in literature. An application of one of the methods for bandwidth determination will be necessary in dimensioning, for example, the Iub interface in the UMTS network and also other interfaces in the wire part of the mobile networks. The choice of a method will depend on the traffic characteristics offered to the interface.

References

[1] Roberts, J.W. (ed.) (1992) *Performance Evaluation and Design of Multi-service Networks, Final Report COST 224*. Commission of the European Communities.

[2] Roberts, J.W., Mocci, V., and Virtamo, I. (eds) (1996) *Broadband Network Teletraffic, Final Report of Action COST 242*. Commission of the European Communities, Springer.

[3] Akimuru, H. and Kawashima, K. (1993) *Teletraffic: Theory and Application*, Springer.

[4] Kuczura, A. (1973) The interrupted poisson process as an overflow process. *Bell System Technical Journal*, **52** (3), 437–48.

[5] Freed, D.S. and Shepp, L.A. (1982) A Poisson process whose rate is a hidden Markov process. *Advances in Applied Probability*, **14** (1), 21–36.

[6] Heffes, H. and Lucantoni, D. (1986) A Markov modulated characteristic of packetized voice and data traffic and related multiplexer performance. *IEEE Journal on Selected Areas in Communications*, **4** (6), 856–68.

[7] Hong, S., Yamashita, H., and Perros, H. (1991) *Performance Analysis of a Shared Buffer ATM Switch under Bursty Arrivals*. Proceedings of the 13th International Teletraffic Congress, Copenhagen, 19–26 June. North-Holland, Amsterdam.

[8] Saito, H. (1994) *Teletraffic Technologies in ATM Networks*. Artech House.

[9] Onvural, R.O. (1993) *Asynchronous Transfer Mode Networks: Performance Issues*. Artech House.

[10] Mandelbrot, B.B. and Van Ness, J.W. (1968) Fractional brownian motions, fractional noises and applications. *SIAM Review*, **10** (4), 422–37. DOI 10.1137/1010093.

[11] Hurst, H. (1951) Long term storage capacity of reservoirs. *Transactions of the American Society of Civil Engineers*, **116**, 770–99.

[12] Granger, C.W.J. and Joyeux, R. (2001) An introduction to long-memory time series models and fractional differencing, in *Essays in Econometrics: Collected Papers of Clive W.J. Granger*. Harvard University Press.

[13] Hosking, J. (1981) Fractional differencing. *Biometrika*, **68**, 165–76.

[14] Paxon, V. and Floyd, S. (1995) Wide-area traffic: the failure of traffic modeling. August *IEEE/ACM Transactions on Networking*, **3** (3), 226–44.

[15] Crovella, M. and Bestavros, A. (1996) Self-similarity in World Wide Web traffic: evidence and possible causes. *Performance Evaluation Review*, **24**(1), 160–9.

[16] Willinger, W., Tacqu, M.S., Sherman, R., and Wilson, D.V. (1997) *Self-Similarity through High Variability: Statistical Analysis of Ethernet LAN Traffic at the Source Level.* Nineteenth Annual Joint Conference of the IEEE Computer and Communications Societies (INFOCOMM). Institute of Electrical and Electronics Engineers, Tel Aviv.

[17] Feldmann, A., Gilbert, A., and Willinger, W. (1998) Data networks as cascades: investigating the multifractal nature of internet WAN traffic, *ACM SIGCOMM Computer Communication Review*, **28** (4), 42–55.

[18] Figueiredo, D., Liu, B., Misra, V., and Towsley, D. (2000) On autocorrelation structure of TCP traffic. *Computer Networks*, **40** (3), 339–61.

[19] Sikdar, B. and Vastola, K. (2001) *The Effect of TCP on the Self-similarity of Network Traffic*. Proceedings of 35th Conference on Information Sciences and Systems, Baltimore, MD, March.

[20] Veres, A. and Boda, M. (2000) *The chaotic nature of TCP congestion control.* 19th Annual Joint Conference of the IEEE Computer and Communications' Societies (INFOCOMM), Institute of Electrical and Electronics Engineers (IEEE), Tel Aviv, Israel, **3**, 1715–1723.

[21] Yu, J. and Petropulu, A.P. (2004) *Is High Speed Wireless Network Traffic Self-Similar*? IEEE International Conference on Acoustics, Speech and Signal Processing, Montreal, May. Institute of Electrical and Electronics Engineers, Montreal, Quebec.

[22] Fox, R. and Taqqu, M.S. (1986) Large-sample properties of parameter estimates for strongly dependent stationary gaussian time series. *Annals of Statistics*, **14**, 517–32.

[23] Bodamer, S. and Charzinski, J. (2000) *Evaluation of Effective Bandwidth Schemes for Self-similar Traffic*. Proceedings of 13th ITC Specialist Seminar on IP Measurement, Modelling and Management, Monterey, CA, September.

[24] Leland, W.E., Taqq, M.S., Willinger, W., and Wilson, D.V. (1993) On the self-similar nature of Ethernet traffic. *ACM SIGCOMM Computer Communication Review*, **23** (4), 183–93.

[25] Wu, D. (2005) QoS provisioning in wireless networks. *Wireless Communications and Mobile Computing*, **5**, 957–69.

[26] Bodamer, S. and Charzinski, J. (2000) *Evaluation of Effective Bandwidth Schemes for Self-similar Traffic*. Proceedings of 13th ITC Specialist Seminar on IP Measurement, Modelling and Management, Monterey, CA, September.

[27] Fonseca, N., Mayor, G., and Neto, C. (2000) On the equivalent bandwidth of self-similar sources. *ACM Transactions on Modeling and Computer Simulation*, **10** (2), 104–24. DOI 10.1145/364996.365003.

[28] Norros, I. (1995) On the use on fractional brownian motion in the theory of connectionless networks. *Journal on Selected Areas in Communications*, **13** (6), 953–62.

[29] Hui, J.Y. (1988) Resource allocation in broadband networks. *Journal on Selected Areas in Communications*, **6** (9), 1598–608.

[30] Kelly, F.P. (1996) Notes on effective bandwidth. Technical report, University of Cambridge.

[31] Technical Report on Project COST 290 (2004) COST 290: Traffic and QoS Management in Wireless Multimedia Networks (Wi-QoST). Report ID TD(04)003.

[32] Gibbens, R.J. and Teh, Y.C. (1999) *Critical Time and Space Scales for Statistical Multiplexing. 16th International Teletraffic Congress (ITC 16)*, Edinburgh, UK, June. Elsevier, Amsterdam.

[33] Neame, T.D., Zuckerman, M., Addie, R.G. (1999) A Practical Approach for Multimedia Traffic Modeling. *5th International Conference on Broadband Communications*, Hong Kong, November Kluwer, Hong Kong.

[34] Gurein, R., Ahmadi, H., and Naghshineh, M. (1991) Equivalent capacity and its application to bandwidth allocation in high-speed networks. *Journal on Selected Areas in Communications*, **9** (7), 968–81.

[35] Charzinski, J. (2000) *Internet Traffic Measurment and Characterisation Results*. 13th International Symposium on Services and Local Access (ISSLS'2000). Stockholm, Sweden, June.

[36] Anick, D., Mitra, D., and Sondhi, M. (1982) Stochastic theory of a data-handling system with multiple sources. *Bell Systems Technical Journal*, **61** (8), 1871–94.

[37] Norros, I. (1994) A storage model with self-similar input. *Queueing Systems*, **16** (2), 387–96.

[38] Patel, A. and Wiliamson, C. (1997) Statistical multiplexing of self-similar traffic: Theoretical and simulation results. Technical report, University of Saskatchewan, Department of Computer Science.

10

Models of the Nodes in the Packet Network

10.1 Introduction

10.1.1 Parameters of the Queuing System

A queuing system is a system that is designed for servicing any call (demands) arriving to the system in random (or not) moments of time. Both the traditional telephone exchange and the IP router are examples of the queuing system. Queuing theory is concerned with stochastic processes taking place in queuing systems. Its main objective is to work out analytical methods enabling the identification and determination of parameters describing queuing, for example the length of the queue, the number of servers, and so on, as well as finding the dependencies between these parameters and efficiency characteristics of queuing systems (for example, the average waiting time of calls in the queue, or the average number of occupied servers).

Figure 10.1 presents a basic model of the queuing system. The figure shows the server, input stream, output stream and the queue. Any device (or a system of devices) that services calls is called the server. When the server is occupied, calls wait in the queue. The input stream is the sequence of calls arriving to the system to be serviced, whereas the output stream is the sequence of calls leaving the system. Calls are serviced in the system by the server or a wider set of servers. In order to characterize the queuing system it is necessary to determine the statistical distributions defining time intervals between subsequent calls arriving to the system and the distribution of the service time that describes the time distribution required for servicing subsequent calls.

10.1.2 Classification of Queuing Systems

The classification of the queuing systems can be based on different criteria. The most commonly used classifications are presented below:

Modeling and Dimensioning of Mobile Networks: From GSM to LTE
Maciej Stasiak, Mariusz Głąbowski, Arkadiusz Wiśniewski and Piotr Zwierzykowski

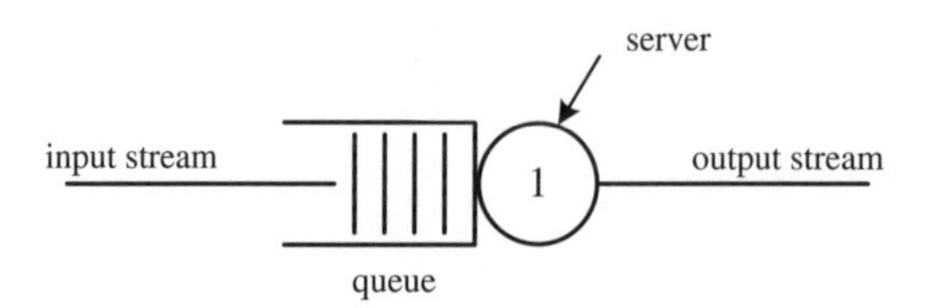

Figure 10.1 A simple queuing system.

1. According to the number of servers:
   - *Single-server* – the number of servers is limited to just one server. A single transmission line can be considered as example here.
   - *Multi-server* – the number of servers is greater than one. An example of this would be a group composed of a few transmission lines jointly servicing a certain call stream.
2. According to whether queuing occurs or whether losses occur:
   - *Systems with losses* – calls arriving to the system when all servers are occupied leave the queuing system and are lost (no space in the queue).
   - *Systems with queuing* – a call that arrives at the system while all servers are occupied does not leave the queuing system but is directed to the queue in which it waits until any of the servers is released.
3. According to the length of the queue (the number of calls in the queue):
   - *Limited*
     - with respect to the number of items in the queue;
     - with respect to the time spent by a call in the queue – "systems with impatient customers".
   - *Unlimited.*
4. According to the discipline of service:
   - Systems without priorities:
     - *FIFO* – First In First Out;
     - *LIFO* – Last In First Out;
     - *SIRO* – Service In Random Order.
   - Systems with priorities:
     - pre-emptive systems;
     - nonpre-emptive systems.

10.1.3 Kendall's Notation

In queuing theory, Kendall's notation [1] is the standard system used to describe and classify the queuing model that a queuing system corresponds to. A queue is described in shorthand notation by $A/B/N/K/S$. The letters stand for the following:

A – time distribution between the arrival of subsequent calls;
B – occupancy distribution of servers;
N – number of servers;
K – capacity of the queue (number of places in the queue);
S – number of traffic sources.

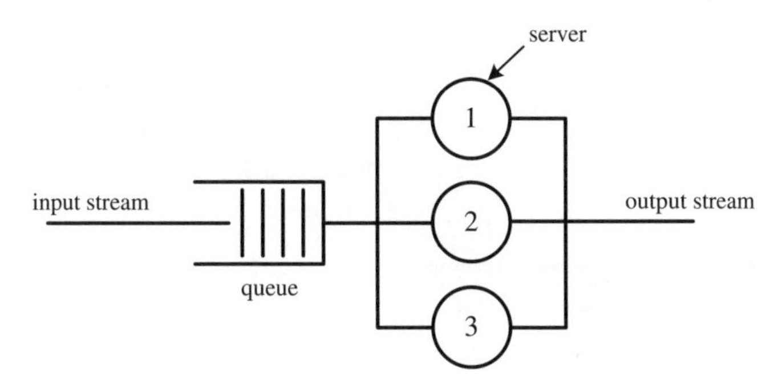

Figure 10.2 The $M/M/3$ queuing system.

It is assumed that if the capacity of the queue and/or the number of traffic sources is infinite, then these values are usually omitted – that is, the symbol ∞ in corresponding positions of Kendall's notation is usually not introduced.

In the notation codes (A and B), the arrival process and the service time distribution, the letter notation describes the kind of applied distribution in the system. The most commonly-used distributions with their notations are the following:

M – exponential distribution;
D – deterministic distribution (fixed inter-arrival time);
E_k – an Erlang distribution with k as the number of exponentially distributed random variables;
G – arbitrary distribution.

For example, the M/M/3 system denotes a queuing system with three servers, infinite queue, exponential service time and call arrival distribution (Figure 10.2).

Further on in this chapter we will consider basic queuing systems [1] in which a call that is not admitted for service because of a resource occupancy waits for the resources to be released in accordance with the queuing discipline. We will consider the systems of the types: M/M/1, M/M/1/N, M/M/m, M/M/m/n, M/G/1, M/D/1, M/G/R PS. We will also discuss the most widely and commonly used methods of admission control to shared resources. Before that, we will derive the so-called *Little's law*, which makes it possible to connect the intensity of arriving calls, average waiting time in the queue (system), and the average number of calls in the queue (system).

10.2 Little's Law

Now we introduce an essential formula for queuing systems. The formula combines the most important characteristics of such systems as the average number of calls in the system (or in the queue), total time delay for calls in the system (or the queue), and average call stream intensity of calls offered to the system. The average total time delay for calls in the system is the sum of the average call waiting time in the queue and the average service time. Similarly, the average number of calls in the system is composed of the average number of calls in the queue and the average number of serviced calls.

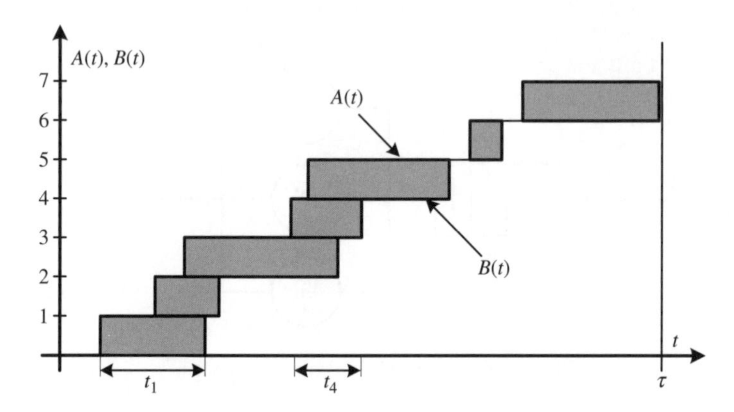

Figure 10.3 Call arrival and service process in the queuing system.

We adopt the following notation for the above-mentioned parameters:

L – average number of calls in the system;
W – average waiting time of a call in the system;
Q – average number of calls in the queue;
T – average waiting time of a call in the queue.

Let us derive Little's law using a diagram like that presented in Figure 10.3 (the same approach appear e.g. in [1–3].

Let us consider now the call arrival and service process presented in Figure 10.3. The following notation has been applied:

$A(t)$ – number of calls arriving to the system in time $(0, t)$;
$B(t)$ – number of calls leaving the system in time $(0, t)$;
t_i – time spent by a call with the number i in the system;
τ – observation period.

If a system is in a steady state, then the average number of calls arriving at the system and the average number of calls departing from the system is equal. Both streams have the same intensity, equal to λ. The functions $A(t)$ and $B(t)$, determining the number of calls arriving in and leaving the system, are discrete functions. Thus, at a given moment, t, their difference $Z(t) = A(t) - B(t)$ determines the number of calls in the system. If the values $A(t)$ and $B(t)$ are equal (merging into one line in Figure 10.3) then it means that the system has no calls.

Let us consider the time interval τ and calculate the average number L of calls in the system. This number can be determined as an integral of function $Z(t)$ divided by the length of the time interval τ:

$$L = \frac{1}{\tau} \int_0^{\tau} Z(t) \mathrm{d}t \tag{10.1}$$

Thus calculated integral is the darkened area in Figure 10.3. This figure is composed of rectangles, the height of each of which is equal to unity, the base being equal to the total time delay of a call in the system. Towards the end of the considered time segment τ some of the rectangles only partially enter the darkened area. However, with an appropriately long time τ this fact will have no significance. Thus, we can write the following:

$$L = \frac{1}{\tau} \int_0^{\tau} Z(t) \mathrm{d}t = \frac{1}{\tau} \sum_i t_i \tag{10.2}$$

Multiplying and dividing the right side of Equation (10.2) by intensity λ, we get:

$$L = \frac{1}{\tau\lambda} \left(\sum_i t_i \right) \lambda \tag{10.3}$$

The value $\tau\lambda$ is the average number of calls that arrived to the system within time τ. Thus, if we divide the sum of all times t_i into an average number of calls $\tau\lambda$, then we will obtain the average total time delay in the system:

$$W = \frac{1}{\tau\lambda} \left(\sum_i t_i \right) \tag{10.4}$$

Substituting value (10.4) in Equation (10.3) we eventually get Little's formula:

$$L = \lambda W \tag{10.5}$$

Equation (10.5) expresses Little's principle, which states that *the average number of calls in the system is equal to the product of the intensity of calls and the average time a call spends in the system.*

It is possible to derive Little's so-called second law, which refers to the queue only: *the average number of calls in the queue is equal to the product of the intensity of calls and the average time a call spends in the queue*:

$$Q = \lambda T \tag{10.6}$$

Little's second principle is derived in the same way as the first principle. For this purpose, to derive Equation (10.6) instead of using function $B(t)$ from Figure 10.3, we can consider function $C(t)$ – the number of calls departing from the queue (and not the system) within time $(0, t)$. It is further assumed that if a call is not directed to the queue (is serviced instantly), then the call is in the queue for a time interval equal to zero.

Little's formulas are very important in traffic theory. They refer to any queuing system regardless of:

- type of the call stream offered to the system;
- service time distribution;
- discipline of service.

10.3 Model of the M/M/1 System with Single-Server and Infinite Queue

10.3.1 Assumptions of the Model

Let us consider the system consisting of a single-server, presented in Figure 10.4, which services a given call stream according to the following assumptions:

- the server is available for any arbitrary call if it is not occupied;
- arriving calls create the Poisson stream with the intensity λ;
- service time is determined by the exponential distribution with the parameter μ (call stream intensity);
- a call not admitted for service waits in the queue under the FIFO (first in first out) discipline; the queue is infinite, which means that an infinite number of calls can wait for service in the queue.

10.3.2 Diagram of the Service Process

The assumptions presented above define the birth-and-death process shown in Figure 10.5. System states are determined according to the number of calls waiting for service and those that are serviced:

- state "0" – the server is free;
- state "1" – the server is occupied (one call is serviced) no calls wait in the queue;
- state "2" – one call is serviced and one call waits in the queue;

...

- state "n" – one call is serviced and $n-1$ calls wait in the queue.

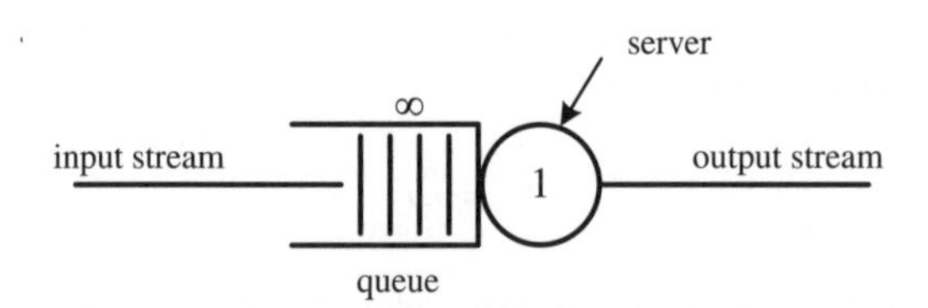

Figure 10.4 M/M/1 queuing system.

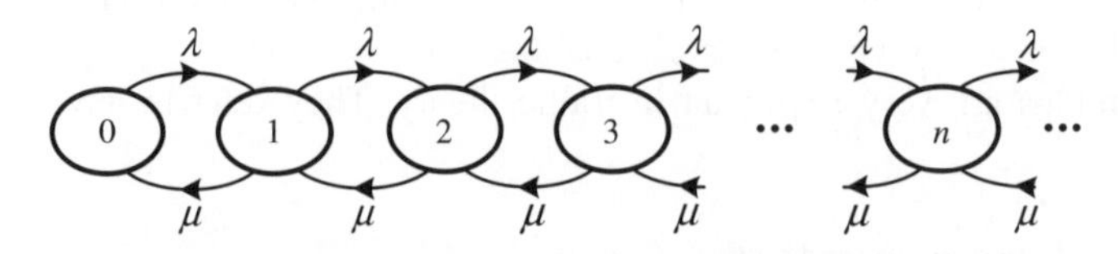

Figure 10.5 Markov process equilibrium in the M/M/1 system.

10.3.3 State Equations

The process, a diagram of which is presented in Figure 10.5, is the birth-and-death process, so in order to determine the occupancy distribution we can use the state equations directly, which for the considered system will take on the following form:

$$\begin{cases} \lambda[P_0]_\infty = \mu[P_1]_\infty \\ \lambda[P_1]_\infty = \mu[P_2]_\infty \\ \quad \ldots \\ \lambda[P_{n-1}]_\infty = \mu[P_n]_\infty \\ \quad \ldots \\ \sum_{k=0}^{\infty} [P_k]_\infty = 1 \end{cases} \tag{10.7}$$

The solution to systems of equations like (10.7) is given by the application of Equations (4.45) and (4.46), derived in Chapter 4 for the generalized birth-and-death process. In the case under consideration here we obtain:

$$[P_k]_\infty = A^k [P_0]_\infty \tag{10.8}$$

$$[P_0]_\infty = \left[1 + A + A^2 + \ldots + A^k + \ldots\right]^{-1} \tag{10.9}$$

where A is the average traffic offered to the system:

$$A = \frac{\lambda}{\mu} \tag{10.10}$$

Note that the sum in Equation (10.9) is the sum of a geometric series, which for $A < 1$ is convergent. Summing the series, Equations (10.8) and (10.9) can eventually be written in the following form:

$$[P_0]_\infty = \left[1 + A + A^2 + \ldots + A^k + \ldots\right]^{-1} = \left[\frac{1}{1-A}\right]^{-1} = (1 - A) \tag{10.11}$$

$$[P_k]_\infty = A^k [P_0]_\infty = A^k (1 - A) \tag{10.12}$$

We pointed out above that the geometrical series (10.11) is convergent if the average traffic offered to the system is less than unity. For $A > 1$, the series is divergent, which means that the length of the queue tends towards infinity. Let us discuss a boundary case when $A = 1$ erl. According to the definition of traffic intensity (Chapter 4), the value $A = 1$ means that within the time interval equal to the average service time of a call, one call will arrive to the system. The system can serve this type of traffic only when the call stream offered to the system will be regular, whereas the service time will be constant and equal to the time intervals between subsequent calls. In such an "ideal" system, the queue will be empty, while the service station will always be occupied. However, it is enough to just "spoil" the regularity of the call stream and the service stream a little, introducing random mechanisms to these streams, and the queue will instantly grow and tend towards infinity.

10.3.4 Characteristics of the M/M/1 System

In this chapter we will determine the most important traffic characteristics for both the whole M/M/1 system and the queue in the M/M/1 system itself.

10.3.4.1 Average Number of Calls in the System

Taking into consideration Equation (10.12), the average number of calls in the system can be determined as the expected value of the discrete variable:

$$L = \sum_{k=1}^{\infty} k[P_k]_{\infty} = \sum_{k=1}^{\infty} kA^k(1-A) = A(1-A)\sum_{k=1}^{\infty} kA^{k-1} \tag{10.13}$$

The sum in Equation (10.13) is indexed from $k = 1$, because the zero component for $k = 0$ is equal to zero. Note now that the expression kA^{k-1} under the sum sign in Equation (10.13) is a derivative of A^k with respect to A. Considering this, Equation (10.13) can be rewritten in the following form:

$$L = A(1-A)\sum_{k=1}^{\infty} kA^{k-1} = A(1-A)\sum_{k=1}^{\infty} \frac{\mathrm{d}A^k}{\mathrm{d}A} \tag{10.14}$$

Changing places of the summation and differentiation operations in Equation (10.14) we obtain:

$$L = A(1-A)\sum_{k=1}^{\infty} \frac{\mathrm{d}A^k}{\mathrm{d}A} = A(1-A)\frac{\mathrm{d}\sum_{k=1}^{\infty} A^k}{\mathrm{d}A} \tag{10.15}$$

Since in Equation (10.15) there is a geometric series under the sum sign, then we can write:

$$\frac{\mathrm{d}\sum_{k=1}^{\infty} A^k}{\mathrm{d}A} = \frac{\mathrm{d}\left(\frac{A}{1-A}\right)}{\mathrm{d}A} = \frac{1}{(1-A)^2} \tag{10.16}$$

Substituting the result (10.16) into Equation (10.15), we finally obtain the formula defining the average number of calls in the system:

$$L = \frac{A}{1-A} \tag{10.17}$$

Analyzing Equation (10.17) it is noticeable that the average number of calls in the system tends towards infinity when the traffic intensity tends towards unity.

10.3.4.2 Variance of the Average Number of Calls in the System

Let us determine, now, the variance σ_L^2 of the number of calls in the system calculated in accordance with the formula for the discrete random variable:

$$\begin{aligned}\sigma_L^2 &= \sum_{k=1}^{\infty} k^2[P_k]_\infty - L^2 = \sum_{k=1}^{\infty} k^2(1-A)A^k - L^2 \\ &= (1-A)A\sum_{k=1}^{\infty} k^2 A^{k-1} - L^2\end{aligned} \tag{10.18}$$

We transform the sum in Equation (10.18):

$$\begin{aligned}\sum_{k=1}^{\infty} k^2 A^{k-1} &= \sum_{k=1}^{\infty} k\frac{\mathrm{d}A^k}{\mathrm{d}A} = \sum_{k=1}^{\infty} \frac{\mathrm{d}(kA^k)}{\mathrm{d}A} = \frac{\mathrm{d}}{\mathrm{d}A}\sum_{k=1}^{\infty} kA^k \\ &= \frac{\mathrm{d}}{\mathrm{d}A}\left[A(1-A)^{-2}\right] = \frac{1+A}{(1-A)^3}\end{aligned} \tag{10.19}$$

Substituting L, defined by Equation (10.17), and the result of the transformation (10.19) into Equation (10.18), we obtain:

$$\sigma_L^2 = \frac{A}{1-A^2} \tag{10.20}$$

10.3.4.3 Average Waiting Time of a Call in the System

Knowing the average number of calls in the system, expressed by Equation (10.17), we can determine – by the application of Little's formula (10.5) – the average time W a call spends in the system:

$$W = \frac{A}{\lambda(1-A)} \tag{10.21}$$

10.3.4.4 Average Number of Calls in the Queue

In order to determine the average number of calls in the queue, we apply the following reasoning. The number of calls is equal to the difference between the number of calls in the system and the number of calls that are currently serviced. This means that, on the basis of the theorem of the sum of average values, the average number of calls in the queue Q is also equal to the difference between the average value of the number of calls in the system L and the average value Y of calls that are serviced:

$$Q = L - Y \tag{10.22}$$

The number of serviced calls can be equal to unity (if the server is occupied) or to zero (if the server is not occupied). The average value Y of this random variable is equal to the occupancy probability of the server. This probability is, in turn, equal to unity minus the probability of an event in which the service station is not occupied. Thus, on the basis of (10.11), we obtain:

$$Y = 1 - [P_0]_\infty = A \tag{10.23}$$

Substituting Equations (10.17) and (10.23) into (10.22), we get the average value of the number of calls in the queue:

$$Q = \frac{A}{1-A} - A = \frac{A^2}{1-A} \tag{10.24}$$

10.3.4.5 Average Waiting Time of a Call in the Queue

Average time a call spends in the queue can be determined on the basis of Little's second formula (10.6):

$$T = \frac{A^2}{\lambda(1-A)} \tag{10.25}$$

10.3.5 Tail Probability

An important issue is also a calculation of the so-called tail probability [4] – the probability that the number of calls in the queue is higher than the required number. It often happens, in engineering calculations, that the queuing system is approximated with the help of the M/M/1 model, despite the limitation of the number of possible calls in the system to a certain finite value. A limited queue size causes a call, when it arrives at the fully occupied system, to be lost.

The probability of a loss of a call can be interpreted as a probability of the occurrence of a defined occupancy state. This means that, for a known value of the buffer k, the loss probability – in a real system with constrained capacity of the buffer – will be equal to the probability that in the M/M/1 system there are more than k calls:

$$\Pr[N > k] = 1 - \Pr[N \leqslant k] = 1 - \sum_{v=0}^{k} [P_v]_\infty = 1 - [P_0]_\infty \frac{1 - A^{k+1}}{1-A} = A^{k+1} \tag{10.26}$$

10.4 Model of the M/M/1/N-1 System with Single-Server and Limited Queue Size

10.4.1 Assumptions for the Model

The following assumptions underpin the model of the M/M/1/N-1 system, presented in Figure 10.6:

- one server;
- server is available for any call if it is not occupied;
- arriving calls form the Poisson stream with the intensity λ;
- service time is defined by the exponential distribution with the parameter μ (intensity of the service stream);
- a call that is not admitted for service waits in the queue under the FIFO discipline (first in first out); queue size is limited to $N-1$ places; if there are just N calls in the system, which means that one call is serviced and the remaining $N-1$ calls are in the queue, then each successive call arriving to the system will be lost because of blocking.

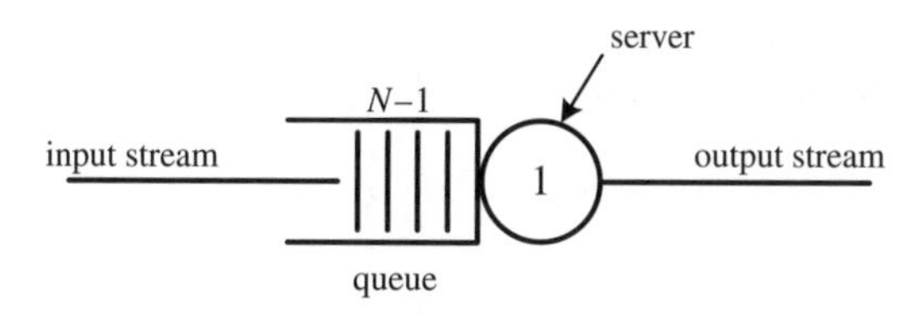

Figure 10.6 M/M/1/N-1 queuing system.

10.4.2 Diagram of the Service Process

Figure 10.7 shows a diagram of Markovian birth-and-death process in the M/M/1/N-1 system with one server and limited queue size. This diagram differs from the diagram of the $M/M/1$ system with one server and unlimited queue size (Figure 10.5) by the finite number of possible states that the system can be in.

The states of the system are determined according to the number of calls waiting for service and calls that are being serviced:

- state "0" – the server is idle;
- state "1" – the server is occupied (one call is being serviced), no calls in the queue;
- state "2" – one call is being serviced and one is in the queue;

...

- state "N" – one call is being serviced and $N-1$ calls are in the queue.

10.4.3 State Equations

The system of equilibrium balance equations (Chapter 4), corresponding to the diagram in Figure 10.7, will take on the following form:

$$\begin{cases} \lambda[P_0]_N = \mu[P_1]_N \\ \lambda[P_1]_N = \mu[P_2]_N \\ \quad\cdots \\ \lambda[P_{N-1}]_N = \mu[P_N]_N \\ \quad\cdots \\ \sum_{k=0}^{N}[P_k]_N = 1 \end{cases} \tag{10.27}$$

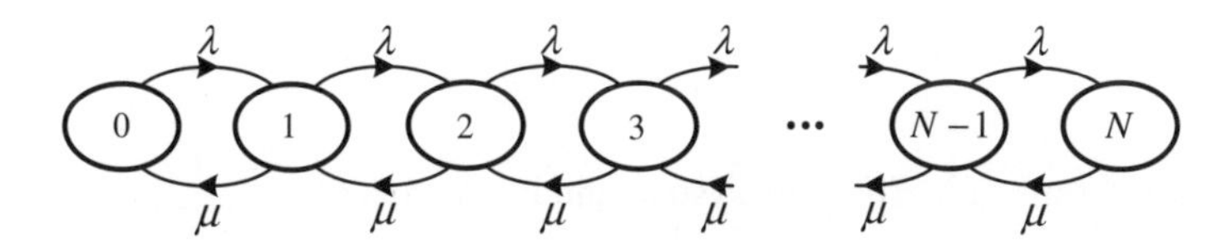

Figure 10.7 Markov process equilibrium (birth-and-death process) in the M/M/1/N-1 system.

On the basis of Equations (4.44) and (4.46), the solution to the system of equations (10.27) leads to the following occupancy distribution in the system:

$$[P_k]_N = A^k [P_0]_N \tag{10.28}$$

$$[P_0]_N = \left[1 + A + A^2 + \ldots + A^N\right]^{-1} = \frac{1 - A}{1 - A^{N+1}} \tag{10.29}$$

where $A = \lambda/\mu$ is the average traffic offered to the system. The numerator in Equation (10.29) defining $[P_0]_N$, is the same as Equation (10.11) for $[P_0]_N$ in the M/M/1 system with an infinite queue. The denominator in Equation (10.29) appears because in the M/M/1/N-1 system the space of the states has been "truncated" and reduced to a finite number of states.

10.4.3.1 The Remaining Traffic Characteristics of the M/M/1/N-1 System

- The blocking probability of the system – the probability of the state $[P_N]_N$. On the basis of Equations (10.28) and (10.29), the blocking probability of the M/M/1/N-1 system is equal to:

$$E = [P_N]_N = A^N \frac{1 - A}{1 - A^{N+1}} \tag{10.30}$$

- The average number of occupied servers Y (probability that a server is occupied):

$$Y = 1 - [P_0]_N \tag{10.31}$$

- The average number of calls in the queue:

$$Q = \sum_{i=2}^{N} i[P_i]_N \tag{10.32}$$

Substituting Equation (10.28) and (10.29) into Equation (10.32) we obtain:

$$Q = \frac{A^2\left[1 - A^{N-1}(N - (N-1)A)\right]}{\left(1 - A^{N+1}\right)(1 - A)} \tag{10.33}$$

- Average number of calls in the system:

$$L = Q + Y \tag{10.34}$$

- Average waiting time of a call in the system and in the queue.

$$W = L/\lambda; \quad T = Q/\lambda \tag{10.35}$$

10.5 Model of the M/M/m System with m Servers and Infinite Queue Size

10.5.1 Assumptions of the Model

Let us consider, now, a multi-server queuing system with infinite queue size. In this model we assume that:

- the system has m servers;
- the queue in the system is infinite;
- the system is offered a Poisson call stream with the intensity λ;
- service time of one call has an exponential character with the parameter μ;
- calls in the blocking state (all servers occupied) wait in the queue under the FIFO service discipline.

Figure 10.8 is a block diagram of the system under consideration. In the literature this model is often called the *Erlang model with queuing*.

10.5.2 Diagram of the Service Process

A diagram of the birth-and-death process for this system is presented in Figure 10.9. States in the system are defined according to the number of calls waiting for service and serviced, i.e.:

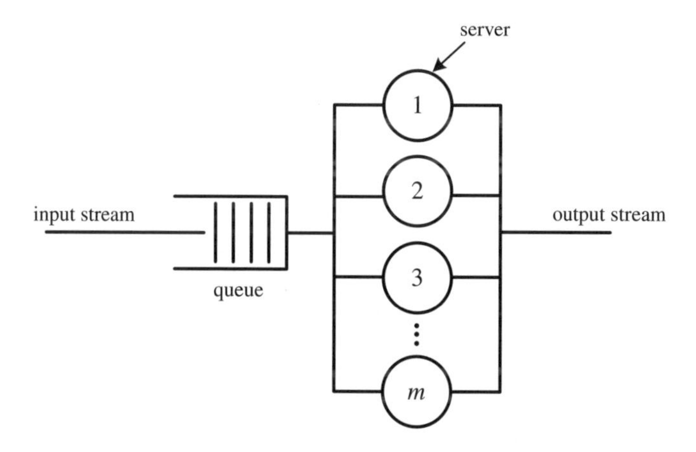

Figure 10.8 M/M/m queuing system.

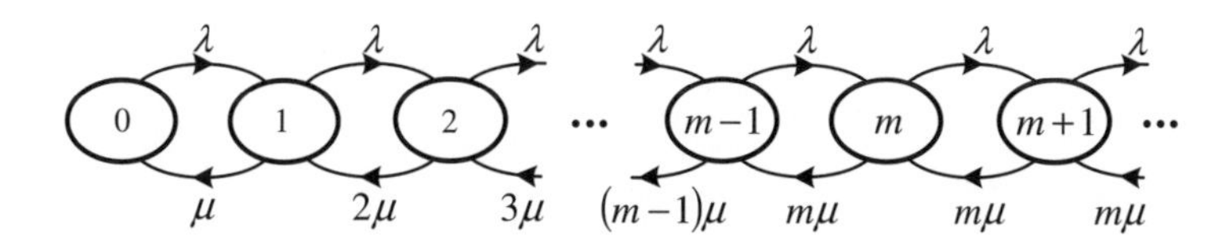

Figure 10.9 Markov process equilibrium (birth-and-death diagram) in the $M/M/m$ system.

- state "0" – no calls, queuing system free;
- state "1" – one server occupied, no calls in the queue;

...

- state "k" – k servers occupied ($1 \leq k \leq m$), no calls in the queue;

...

- state "m" – all m–servers occupied, no calls in the queue;
- state "$m+1$" – all m servers occupied, one call in the in the queue;

...

- state "$m+r$" – all m servers occupied, r calls in the queue.

Note that, as with the M/M/1 model (with infinite queue), the prerequisite for "stabilizing" the length of the queue is $A/m = \chi < 1$, where $A = \lambda/\mu$ (this means that the intensity of traffic offered to one server should be smaller than 1 erl). In the case when $A/m > 1$, the length of the queue tends limitlessly towards infinity.

10.5.3 State Equations

In order to construct the state equations, let us first determine the intensities of the service stream. It is noticeable that, for the occupancy state n lower or equal to the number m of servers, the intensity of the service stream increases and for each occupancy state is $n\mu$, whereas for the occupancy states $n > m$, the intensity of the traffic stream is constant and is $m\mu$. Local equilibrium equations, then, have the following form:

$$\lambda[P_{n-1}]_\infty = n\mu[P_n]_\infty \text{ for } 1 \leq n < m \tag{10.36}$$

$$\lambda[P_{n-1}]_\infty = m\mu[P_n]_\infty \text{ for } n \geq m \tag{10.37}$$

On the basis of the above system of equations we can determine local equilibrium equations for particular (individual) states:

$$[P_1]_\infty = A[P_0]_\infty \tag{10.38}$$

$$[P_2]_\infty = \frac{A^2}{2!}[P_0]_\infty \tag{10.39}$$

$$[P_m]_\infty = \frac{A^m}{m!}[P_0]_\infty \tag{10.40}$$

$$[P_{m+1}]_\infty = \frac{A}{m}\frac{A^m}{m!}[P_0]_\infty \tag{10.41}$$

$$[P_{m+2}]_\infty = \left(\frac{A}{m}\right)^2\frac{A^m}{m!}[P_0]_\infty \tag{10.42}$$

Using the normative condition $\sum_{n=0}^{\infty} = 1$, we are in position to determine the state probability P_0:

$$[P_0]_\infty = \frac{1}{\sum\limits_{i=0}^{m-1} \frac{A^i}{i!} + \frac{A^m}{m!} \sum\limits_{i=0}^{\infty} \left(\frac{A}{m}\right)^i} \tag{10.43}$$

Note that the sum in Equation (10.43) is the sum of a geometric series, which for $A/m < 1$ is convergent and equal to:

$$\sum_{i=0}^{\infty} \left(\frac{A}{m}\right)^i = \frac{A^m}{m!} \frac{m}{m - A} \tag{10.44}$$

Given the sum of the series $\sum_{i=0}^{\infty} \left(\frac{A}{m}\right)^i$, Equation (10.43) can be rewritten in the following form:

$$[P_0]_\infty = \frac{1}{\sum\limits_{i=0}^{m-1} \frac{A^i}{i!} + \frac{A^m}{m!} \frac{m}{m-A}} \tag{10.45}$$

10.5.4 Occupancy Probability of all Servers

Let us determine, now, for the $M/M/m$ system the probability $E_{2,m}$ that a call arriving at the system will wait in the queue. This probability can be determined in the following way:

$$E_{2,m} = \sum_{i=m}^{\infty} [P_i]_\infty = \frac{A^m}{m!} \sum_{i=0}^{\infty} \left(\frac{A}{m}\right)^i [P_0]_\infty = \frac{A^m}{m!} \frac{m}{m - A} [P_0]_\infty \tag{10.46}$$

Substituting (10.45) into Equation (10.46), we can eventually determine the probability of waiting in the M/M/m system:

$$E_{2,m}(A) = \frac{\frac{A^m}{m!} \frac{m}{m-A}}{\sum\limits_{i=0}^{m-1} \frac{A^i}{i!} + \frac{A^m}{m!} \frac{m}{m-A}} \tag{10.47}$$

This formula in literature is defined as Erlang's second formula or the Erlang's C formula.

10.5.5 Traffic Characteristics of the M/M/m System

The remaining characteristics of the efficiency (effectiveness) of the M/M/m queuing system can be determined using the following dependencies:

- Average number of calls in the queue:

$$Q = \sum_{i=m+1}^{\infty} i[P_i]_\infty = \frac{A^{m+1}[P_0]_\infty}{m \cdot m!(1 - \chi)^2} = \frac{\chi[P_m]_\infty}{(1 - \chi)^2} \tag{10.48}$$

where $\chi = A/m$;

- Average number of calls in the system:

$$L = Q + Y = Q + A \tag{10.49}$$

- Average waiting time for a call in the system:

$$W = \frac{L}{\lambda} \tag{10.50}$$

- Average waiting time for a call in the queue:

$$T = \frac{Q}{\lambda} \tag{10.51}$$

10.6 Model of the M/M/m/N System with Limited Queue Size and Limited Number of Servers

10.6.1 Assumptions of the Model

The assumptions and the numbering of states are the same as in Section 10.5, except that the number N of places in the queue is limited. Steady-state probabilities exist for every value of parameters λ and μ and can be determined with the help of the following equations:

$$[P_0]_{m+N} = \left\{1 + \frac{A}{1!} + \ldots + \frac{A^m}{m!} + \frac{A^{m+1}}{m \cdot m!} \cdot \frac{1-\chi^N}{1-\chi}\right\}^{-1}$$

$$[P_k]_{m+N} = \frac{A^k}{k!}[P_0]_{m+N} \quad (1 \leq k \leq m)$$

$$[P_{m+r}]_{m+N} = \frac{A^{m+r}}{m^r \cdot m!}[P_0]_{m+N} \quad (1 \leq r \leq N) \tag{10.52}$$

where $\chi = A/m = \lambda/(m\mu)$.

10.6.2 Traffic Characteristics of the M/M/m/N System

The introduction of analytical formulas defining traffic efficiency characteristics of a system, is analogous to the case of the M/M/m and M/M/1/N-1 models. We will therefore limit ourselves to just giving the final results below:

- Blocking probability E:

$$E = [P_{m+N}]_{m+N} \tag{10.53}$$

- Value of carried traffic Y:

$$Y = A\left(1 - [P_{m+N}]_{m+N}\right) \tag{10.54}$$

- Average number of calls in the queue:

$$Q = \frac{A^{m+1}[p_0]_{m+N}}{m \cdot m!} \frac{1 - (N+1)\chi^N + N\chi^{N+1}}{(1-\chi)^2} \tag{10.55}$$

- Average number of calls in the system:

$$L = Q + Y \tag{10.56}$$

- Average waiting time of a call in the system:

$$W = \frac{L}{\lambda} \tag{10.57}$$

- Average waiting time of a call in the queue:

$$T = \frac{Q}{\lambda} \tag{10.58}$$

10.7 Model of the M/G/1 System with Single-Server and Infinite Queue Size

Let us consider an example of a non-Markovian system with one server that services a given call stream according to the following assumptions:

- the server is available to any call if it is not occupied;
- arriving calls form the Poisson stream with the intensity λ;
- service time can be determined by any distribution with the average intensity μ;
- a call not admitted for servicing waits in the queue under the FIFO discipline;
- the queue has infinite size.

The M/G/1 system under consideration here is a non-Markovian system because the service time is not determined (defined) by an exponential distribution.

To derive important characteristics of the system we use the embedded Markov chain method [5]. This is based on a choice of moments t_i in a particular process $\upsilon(t)$ in which the values of the trajectory of this process in the selected moments create a Markov chain. Then, using the methods adopted for the analysis of Markov chains, distributions of particular variables $\upsilon(t_i)$ are determined. Eventually, on the basis of the distributions obtained, it is possible to evaluate the distribution of the process $\upsilon(t)$ continues in time. In many cases, the last stage can be omitted since the values $\upsilon(t_i)$ include enough information on the functioning of the system.

In the M/G/1 system's theory, it was proved [2] that the steady-state probabilities, determined on the basis of the embedded Markov chain at moments t_i are equal to the steady state probabilities at any time instant (i.e. $\upsilon(t_i) = \upsilon(t)$), if the moments t_i are chosen at the moments of termination of service of subsequent calls.

In order to determine the basic characteristics of M/G/1 system let us apply the reasoning presented, for example, in [3, 6]. First let us assume that the number of calls arriving at the system when a call i is serviced will be denoted by the symbol z_i, while the number of calls

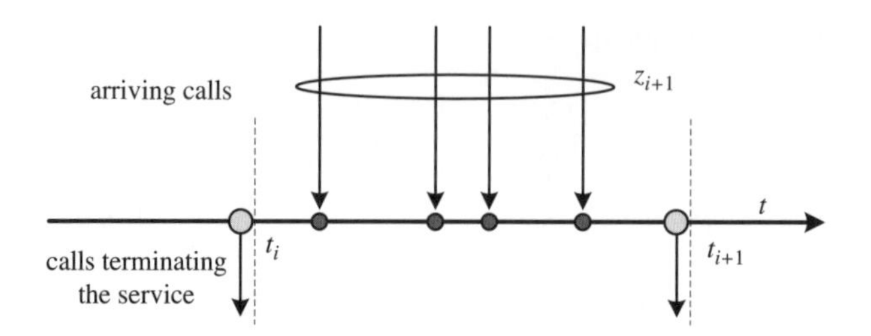

Figure 10.10 Time diagram for a defined embedded Markov chain.

in the system "instantly" after termination of service of the call with the ordinal number $i + 1$ will be denoted by the symbol n_{i+1}. This number is equal to the number of calls in the system after termination of service of the call i minus 1 (termination of service of the call $(i + 1)$) plus the number of calls that have arrived at the system between the moments of termination of service of subsequent calls with the numbers i and $i + 1$. If the number of calls in the system at the moment t_n is equal to zero, then the number of calls in the system at the moment of the termination of service of the call $(i + 1)$ be equal to the number of these calls that have arrived at the empty queue. The first call (with the number $i + 1$) is instantly directed to the server and leaves the system at the moment t_i of the termination of service. This call is therefore not taken into consideration while determining the number of arriving calls z_{i+1}. On the basis of the above reasoning we can thus write:

$$n_{i+1} = \begin{cases} z_{i+1} & \text{for } n_i = 0 \\ n_i - 1 + z_{i+1} & \text{for } n_i > 0 \end{cases} \tag{10.59}$$

Figure 10.10 shows a time diagram determining the embedded Markov chain over the call service process in the M/G/1 system.

We introduce a logical step function δ(n), which will be defined in the following way:

$$\delta(n_i) = \begin{cases} 0 & \text{for } n_i = 0 \\ 1 & \text{for } n_i > 0 \end{cases} \tag{10.60}$$

Taking into consideration the derived and introduced function (10.60), Equation (10.59) can be rewritten into the following form:

$$n_{i+1} = n_i - \delta(n_i) + z_{i+1} \tag{10.61}$$

Equation (10.61) describes the embedded Markov chain over the service process in the M/G/1 system. This chain is defined by the number of calls in the system n_i at moments of termination of the serviced calls.

10.7.1 Pollaczek–Khinchin Formula

Let us square both sides of Equation (10.61) and replace the obtained values of random variables with their average values:

$$E\left[(n_{i+1})^2\right] = E\left[(n_i - \delta(n_i) + z_{i+1})^2\right] \tag{10.62}$$

After performing elementary transformations we obtain:

$$E\left[(n_{i+1})^2\right] = E\left[(n_i)^2\right] + E\left[(\delta(n_i))^2\right] + E\left[(z_{i+1})^2\right] - 2E\left[n_i\delta(n_i)\right] + 2E\left[n_i z_{i+1}\right] - 2E\left[\delta(n_i)z_{i+1}\right] \tag{10.63}$$

Let us consider now all elements of the sum (10.63) one by one. Since the considered process is a Markov chain, then in the state of statistical equilibrium the following equality will take place:

$$E\left[(n_{i+1})\right] = E\left[(n_i)\right] \qquad E\left[(n_{i+1})^2\right] = E\left[(n_i)^2\right] \tag{10.64}$$

From Equation (10.60) for the function $\delta(\mathrm{n})$ it follows that:

$$(\delta(n_i))^2 = \delta(n_i) \tag{10.65}$$

Thus, the following equality has to take place:

$$E\left[(\delta(n_i))^2\right] = E\left[\delta(n_i)\right] \tag{10.66}$$

Let us consider more thoroughly the average value $E\left[\delta(n_i)\right]$. It follows from the definition of the average value that:

$$E\left[\delta(n_i)\right] = \sum_{n=0}^{\infty} \delta(n_i) P(n_i = n) = \sum_{n=1}^{\infty} P(n_i = n) \tag{10.67}$$

and, finally:

$$E\left[\delta(n_i)\right] = 1 - P(n_i = 0) \tag{10.68}$$

Since the considered M/G/1 system is a system with one server, then the number of serviced calls can be equal either to one or to zero. Therefore, the right side of Equation (10.68) expresses the occupancy probability of the server. In systems with one server and infinite queue size, $1 - P(0)$ value is equal to the average number of serviced calls in the system. Let us try to determine this probability applying the reasoning given in [1]. If the system operates within a given period of time t, then it will be occupied within the time $t(1 - P(0))$. The average number of serviced calls during that time is equal to $t\mu(1 - P(0))$, where μ^{-1} is the average service time of a call in the server. Assuming that the system holds up under the statistical equilibrium, the average number of serviced calls will be equal to the average number of arriving calls:

$$\lambda t = t\mu(1 - P(0)) \tag{10.69}$$

Taking into account the result obtained (10.69) and the fact that the intensity of traffic offered to the system A is equal to λ/μ, the equation (10.68) can be eventually rewritten in the following way:

$$E\left[\delta(n_i)\right] = 1 - P(0) = A \tag{10.70}$$

It should be stressed at this point that this reasoning is also valid in more complex systems than the M/G/1 system – it is generally is applicable to systems with arbitrary service time and arbitrary input streams.

Let us consider now the next expressions in Equation (10.63): $2E[n_i\delta(n_i)]$ and $2E\left[\delta(n_i)z_{i+1}\right]$. Since $n_i\delta(n_i) = n_i$, then:

$$2E[n_i\delta(n_i)] = 2E[n_i] \tag{10.71}$$

The number of calls arriving at the system between the moments of termination of call service with the ordinal numbers i and $i+1$ is independent of the number of calls in the queue. This allows us to write the following:

$$E\left[\delta(n_i)z_{i+1}\right] = E[\delta(n_i)]\,E\left[z_{i+1}\right] \tag{10.72}$$

We know that $E[\delta(n_i)] = A$ (Equation (10.70)). Let us now determine the average value $E\left[z_{i+1}\right]$. For this purpose let us replace the random variables in Equation (10.61) with their average values:

$$E\left[n_{i+1}\right] = E[n_i] - E[\delta(n_i)] + E\left[z_{i+1}\right] \tag{10.73}$$

Taking into consideration, in Equation (10.73), the results described by Equations (10.64) and (10.70) we obtain:

$$E\left[z_{i+1}\right] = E[\delta(n_i)] = A \tag{10.74}$$

Now, by substituting Equations (10.70) and (10.74) with Equation (10.72) we can determine the value of the expression $2E\left[\delta(n_i)z_{i+1}\right]$:

$$2E\left[\delta(n_i)z_{i+1}\right] = 2A^2 \tag{10.75}$$

Following similar reasoning with reference to the last expression of Equation (10.63), that is $2E\left[n_i z_{i+1}\right]$, we can state that:

$$2E\left[n_i z_{i+1}\right] = 2E[n_i]\,E\left[z_{i+1}\right] = 2AE[n_i] \tag{10.76}$$

Take into account all the results obtained: (10.64), (10.70), (10.71), (10.74), (10.75) and (10.76) in Equation (10.63). After substitution we obtain:

$$E[n_i] = \frac{A + E\left[(z_{i+1})^2\right] - 2A^2}{2(1-A)} \tag{10.77}$$

In [5] it was proved that the average number of calls in the system, at the moments of termination of service of individual calls, is the same as the average number of calls in any moment of time. Similarly, the average number of calls arriving at the system during the service time of a call i is independent on the choice of the ordinal number of this call. Thus, we can write:

$$E[n_i] = E[n] = L \qquad E\left[z_{i+1}\right] = E[z] \tag{10.78}$$

Now Equation (10.77) can be brought to the following form:

$$L = \frac{A + E\left[(z)^2\right] - 2A^2}{2(1 - A)} \tag{10.79}$$

The result is a simple (algebraic) expression for determining the average number of calls in the system. We try to simplify Equation (10.79) even more by finding the expression for $E\left[(z)^2\right]$. The parameter z is the average number of calls arriving at the system during the service time. Taking into consideration the properties of the variance of the random variable [7] and Equation (10.74), we can write:

$$E\left[(z)^2\right] = \{E\left[z\right]\}^2 + \sigma_z^2 = A^2 + \sigma_z^2 \tag{10.80}$$

where σ_z^2 is the variance of the number of calls arriving to the system during the average service time. To define this parameter we can use the formula for determining the random variable conditioned by another random variable [2, 7]:

$$\sigma_z^2 = E\left[\sigma_{z|\tau}^2\right] + \left[\sigma_{E[z|\tau]}^2\right] \tag{10.81}$$

The parameters $E\left[z|\tau\right]$ and $\sigma_{z|\tau}^2$ are the average value and the variance of the random variable z in relation to the random variable τ. In the case under consideration here, it is the average value and the variance of the number of calls arriving at the system during the service time, with the assumption that the service time is equal to τ. The call stream is a Poisson stream (Section 4.2) with the intensity λ, so:

$$\sigma_{z|\tau}^2 = E\left[z\,|\tau\right] = \lambda\tau \tag{10.82}$$

Now we are in position to determine the components in Equation (10.81):

$$\begin{aligned} E\left[\sigma_{z|\tau}^2\right] &= E\left[\lambda\tau\right] = \lambda E\left[\tau\right] = \lambda\frac{1}{\mu} = A \\ \sigma_{E[z|\tau]}^2 &= \sigma_{\lambda\tau}^2 = \lambda^2\sigma_\tau^2 \end{aligned} \tag{10.83}$$

The parameter σ_τ^2 in Equation (10.83) is the variance of the service time in the system. Taking into consideration the relations (10.74) and (10.83) in Equation (10.80), we obtain:

$$E\left[(z)^2\right] = A + A^2 + \lambda^2\sigma_\tau^2 \tag{10.84}$$

Substituting the expression $E\left[(z)^2\right]$ into Equation (10.79) we obtain the so-called Pollaczek-Khinchin formula:

$$L = \frac{2A - A^2 + \lambda^2\sigma_\tau^2}{2(1 - A)} \tag{10.85}$$

This formula can be also transformed into the following form:

$$L = A + \frac{A^2 + \lambda^2\sigma_\tau^2}{2(1 - A)} \tag{10.86}$$

Equation (10.85) was derived independently by Pollaczek [8] and Khinchin [9]. The formula was derived with the application of a far more complex methods than the one mentioned in this chapter – the embedded Markov chains method [5]. The value of the Pollaczek–Khinchin formula is based on the fact that it expresses the average number of calls in the system as a function of the average value of the call intensity and the average value and variance of the service time.

10.7.2 Characteristics of the M/G/1 System

Using Little's law (Equations (10.5) and (10.6)) we can now determine all important characteristics of the M/G/1 system. The average time spent by a call in the system is equal to:

$$W = \frac{1}{\mu} + \frac{A^2 + \lambda^2\sigma_\tau^2}{2\lambda(1-A)} \tag{10.87}$$

As $1/\mu$ in Equation (10.87) is the average service time, the average waiting time of a call in the queue is therefore equal to:

$$T = \frac{A^2 + \lambda^2\sigma_\tau^2}{2\lambda(1-A)} \tag{10.88}$$

The average number of calls in the queue:

$$Q = \frac{A^2 + \lambda^2\sigma_\tau^2}{2(1-A)} \tag{10.89}$$

Note that Equations (10.87) and (10.88) indicate the obvious dependency between average time spent by a call in the system, in the queue, and in the server:

$$W = T + 1/\mu \tag{10.90}$$

Let us try to write the Pollaczek–Khinchin Equation (10.85) in an algebraic form that would be easier to remember. To achieve that let us put Equation (10.85) in the expression $A/(1-A)$ before the bracket:

$$L = \frac{A}{1-A}\left[1 - \frac{A}{2}\left(1 - \mu^2\sigma_\tau^2\right)\right] \tag{10.91}$$

Using Little's formula (10.5), the average time spent by a call in the system can be written as follows:

$$W = \frac{L}{\lambda} = \frac{1/\mu}{1-A}\left[1 - \frac{A}{2}\left(1 - \mu^2\sigma_\tau^2\right)\right] \tag{10.92}$$

Now substituting (10.92) into Equation (10.90) we obtain the following form of the Pollaczek–Khinchin formula:

$$T = \frac{\lambda E\left[\tau^2\right]}{2(1-A)} \tag{10.93}$$

where $E\left[\tau^2\right]$ is determined by the following formula:

$$E\left[\tau^2\right] = \sigma_\tau^2 + 1/\mu^2 \tag{10.94}$$

Let us compare the Pollaczek–Khinczin formula with the results for L and W in the M/M/1 system (Section 10.3.4, Equations (10.17) and (10.21)). We identify that the average number of calls in the M/G/1 system is equal to the average number of calls in the M/M/1 system multiplied by a certain correction factor. This factor is the expression within the square brackets in Equations (10.91), (10.92) and depends on the service time variance. If the service time variance is equal to $1/\mu^2$ (exponential distribution – the M/M/1 model) then, as was to be expected, Equations (10.91), (10.92) reduce to Equations (10.17) and (10.21). For $\sigma_\tau^2 > 1/\mu^2$, the average number of calls in the system L and the average time spent by a call in the system W increase when compared with the values of these parameters in the M/M/1 system. For $\sigma_\tau^2 < 1/\mu^2$, the parameter values L and W decrease when compared with the M/M/1 system. Equation (10.93) is often written in the following form:

$$T = T_0/(1 - A) \tag{10.95}$$

where:

$$T_0 = \frac{\lambda E\left[\tau^2\right]}{2} \tag{10.96}$$

The parameter T_0 is the average value of the so-called residual service time. If during the service time of a given call we randomly choose the moment t, then the time measured from the moment t until the termination of service of this call is called the residual service time. The average value of this time is expressed by Equation (10.96). Derivation of the formula determining the residual service time as well as its parameters can be found in, for example, [1].

10.8 M/D/1 System

Let us use Pollaczek–Khinczin formulas to determine the parameters of the M/D/1 system that is characterized by the following assumptions:

- the server is available for any arbitrary call if it is not occupied;
- arriving calls form a Poisson stream with the intensity λ;
- service time is fixed for each call and is equal to $1/\mu$;
- a call that is not accepted for service waits in a queue under the FIFO (first in first out) discipline;
- the queue is infinite.

Since the service time is constant, then its variance must be equal to zero: $\sigma_\tau^2 = 0$. Taking the zero variance into consideration in Equations (10.91) and (10.92) we obtain the following:

$$L = \frac{A}{1 - A}\left[1 - \frac{A}{2}\right] \tag{10.97}$$

$$W = \frac{1/\mu}{1 - A}\left[1 - \frac{A}{2}\right] \tag{10.98}$$

The M/D/1 system is often used for the analysis of servers and nodes in packet networks.

10.9 Queuing Systems with One Server and Nonpre-Emptive Priorities

Services offered to customers in modern telecommunications and telecommunications networks are often divided into various classes. This division enables the introduction of different priorities for calls of particular traffic classes and, as a result, the use of a varied queue discipline of the buffers of packet networks.

Usually, in queuing systems, two types of priorities are distinguished: pre-emptive priorities with service interruptions and nonpre-emptive priorities without service interruptions.

In the nonpre-emptive system, if a call with a higher priority arrives at the system, then the call waits until the current call has been serviced even if the serviced call has a lower priority. After the termination of the service of the current call, the next call to be serviced is the call with the highest priority.

In the pre-emptive system, if a call with higher priority arrives at the system then the system interrupts the service that is currently performed and initiates servicing of the call with higher priority. After the termination of the service with higher priority, the system commences servicing the interrupted call.

In this section we will consider a system with nonpre-emptive priorities, with the following assumptions:

- one server is available for any arbitrary call if only it is not occupied;
- the system is offered M classes of Poisson call streams with intensities $\lambda_1, \lambda_2, \ldots, \lambda_M$;
- the service time of calls of particular classes have arbitrary distributions with average service intensities respectively equal to: $\mu_1, \mu_2, \ldots, \mu_M$;
- a call that is not accepted for service waits in the queue under the FIFO discipline with nonpre-emptive priorities;
- the queue size is infinite;
- the call stream with ordinal number i has a higher priority than the call stream with the number $i + 1$; thus, the stream with the intensity λ_1 has the highest priority, whereas the stream with the intensity λ_M has the lowest priority.

Let us calculate, now, the average waiting time in the queue of a class k call that arrives at the system at a given moment of time t_0. The waiting time in the queue of this call can be expressed with the help of the three components [1]:

$$t_k = \tau_0 + \sum_{i=1}^{k} \omega_i + \sum_{i=1}^{k-1} \omega_i' \tag{10.99}$$

where:

τ_0 – random variable – the residual time needed for terminating currently going on service;

ω_i – random variable – service time of calls of class i, with the priority higher or equal to k, that were currently in the queue at the moment of the arrival of the considered call of class k; according to the adopted assumptions for the system, an arriving call must wait for the termination of the service of these calls;

$\omega_i^{'}$ – random variable – service time for class i calls, with the priority higher than k (but not equal to k), which arrived at the system during the waiting of the class k call; according to the adopted assumptions for the system, an arriving class k call must wait until the calls have been serviced.

Let us swap the random variables in Equation (10.99) with their mean values:

$$T_k = E[t_k] = E[\tau_0] + \sum_{i=1}^{k} E[\omega_i] + \sum_{i=1}^{k-1} E\left[\omega_i^{'}\right] \tag{10.100}$$

The average value of the service time of all calls of class i present in the system at the moment t_0, that is the value $E[\omega_i]$, can be expressed in the form of the product of the average number of calls of class i in the queue and the average service time of a call of this class:

$$E[\omega_i] = Q_i/\mu_i \tag{10.101}$$

Alternatively, the average number of calls in the queue, according to Little's second law, can be expressed by the average waiting time T_i spent by a call of class i in the queue:

$$Q_i = \lambda_i T_i \tag{10.102}$$

Taking into consideration Equation (10.102), Equation (10.101) can be rewritten:

$$E[\omega_i] = A_i T_i \tag{10.103}$$

Let us consider now the parameter $E\left[\omega_i^{'}\right]$ in Equation (10.100), i.e. the average service time of class i calls that arrive at the system within the waiting time t_k of a call of the considered class k. Let us denote the average number of these calls as $Q_i^{'}$. Therefore

$$E\left[\omega_i^{'}\right] = Q_i^{'}/\mu_i \tag{10.104}$$

The average number of calls of class i arriving to the system during the waiting time of class class k call is equal to:

$$Q_i^{'} = \lambda_i T_k \tag{10.105}$$

Hence, taking into consideration (10.105), Equation (10.104) will take on the following form:

$$E\left[\omega_i^{'}\right] = A_i T_k \tag{10.106}$$

Let us now consider the first component of Equation (10.100), which is $E[\tau_0]$. This parameter expresses the average value of the residual service time of the call serviced currently (Figure 10.11). It is possible to demonstrate that this parameter is independent of the way the queue is serviced (it is independent of the service discipline or whether there are priorities) [1]. Hence, to evaluate the average value of the residual service time one can use the result obtained

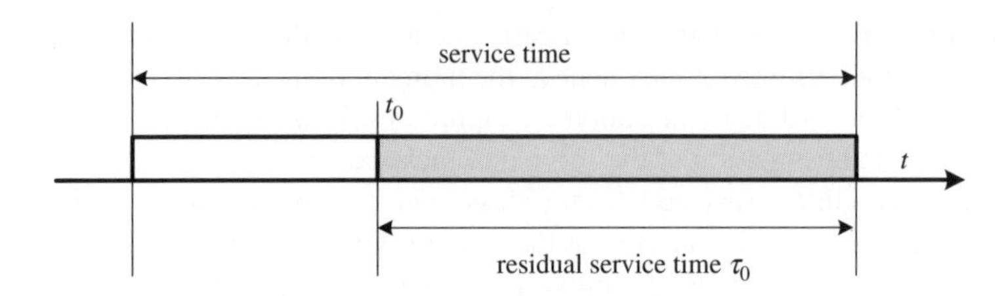

Figure 10.11 Residual service time of currently serviced call.

for the M/G/1 system with the FIFO queue that associates the average value of the residual service time with the second moment of the service time (Equation (10.96)).

Assume that the M/G/1 system with the FIFO queue services k call streams (without priorities) so that the resultant intensity λ is equal to:

$$\lambda = \lambda_1 + \lambda_2 + \ldots + \lambda_k \tag{10.107}$$

The arriving call streams are statistically independent, so the second moment of the service time of the resultant call stream can be expressed as a weighted sum of the second moments of individual classes:

$$E\left[\tau^2\right] = \sum_{i=1}^{k} \frac{\lambda_i}{\lambda} E\left[\tau_i^2\right] \tag{10.108}$$

Taking into consideration (10.108) in Equation (10.96), for the average value of the residual service time we get:

$$T_0 = E\left[\tau_0\right] = \frac{\lambda E\left[\tau^2\right]}{2} = \sum_{i=1}^{k} \frac{\lambda_i E\left[\tau_i^2\right]}{2} \tag{10.109}$$

We can use the results obtained for $E\left[\omega_i\right]$ (Equation (10.103)), $E\left[\omega_i^{'}\right]$ (Equation (10.106)) and $E\left[\tau_0\right]$ (Equation (10.109)) to transform Equation (10.100):

$$T_k = T_0 + \sum_{i=1}^{k} A_i T_i + \sum_{i=1}^{k-1} A_i T_k \tag{10.110}$$

Equation (10.110) is a recursive equation that can be solved starting from the call class with the highest priority (with the index equal to unity). Following this operation we get:

$$T_k = \frac{T_0}{\left(1 - A_k^*\right)\left(1 - A_{k-1}^*\right)} \tag{10.111}$$

where:

$$A_k^* = \sum_{i=1}^{k} A_i \tag{10.112}$$

The parameter T_0 in Equation (10.111) is determined by Equation (10.109). When servicing two classes of calls with priorities we obtain:

$$T_1 = \frac{T_0}{(1 - A_1)} \tag{10.113}$$

$$T_2 = \frac{T_0}{(1 - A_1)(1 - A)} \tag{10.114}$$

$$A = A_1 + A_2 \tag{10.115}$$

10.10 The M/G/R PS Model – Model of Buffers in the UMTS System

Let us consider a direct application of processor-sharing methods [1] in the case of packet networks with separated traffic streams of particular service classes. One of the most commonly used and most general queuing models in the analysis and dimensioning of packet networks is the M/G/R PS model, where PS means *processor sharing*.

10.10.1 Assumptions of the Model

The following assumptions have been adopted for the M/G/R PS model being considered here:

- call streams have Poisson distribution;
- service time distribution is arbitrary;
- resources of the system are shared fairly between all streams offered to the system, i.e. each of the streams occupied the same amount of given resources (such as bandwidth);
- all offered streams are serviced quasi-simultaneously through an allocation of a brief period of time (time slot) to each of the streams and by servicing them in a sequential way;
- the number of servers is R.

The number of servers in the model being considered here, denoted by the letter R, needs a brief explanation. We assume that, in the case when a single call stream fully uses the available resources (for example, the available bandwidth of a link), then the system comes down to a M/G/1 PS system. In the case when a single call stream of a given class is not capable of making a full use of the resources of the server, then – from the point of view of the call stream – the system can be treated as a system with R servers, where

$$R = V/r_{max} \tag{10.116}$$

and V is the capacity of the server (link), whereas r_{max} is the maximum bit rate of the traffic stream. A typical example of such a system is multiplexion of streams of many users with defined capacities of access links r_{max}, in a link of the framework network, with higher capacity, equal to V.

In handling the service of individual streams, the M/G/R PS model corresponds to the operation of the mechanisms implemented in transmission control protocol (TCP). Like the mechanism of "processor-sharing." TCP also aims at ensuring equal access to a shared transmission channel. The similarity of the operation of the M/G/R PS model and the most-commonly-used protocol of the transmission layer results in the fact that currently the M/G/R PS model is widely used in dimensioning packet networks, including the UMTS network [10–12].

Following [11], the average time spent by a call in the M/G/R PS system can be expressed in relation to the so-called *delay factor* f_R, defined in the following way:

$$f_R = 1 + \frac{E_{2,R}(R\rho)}{R(1-\rho)} \tag{10.117}$$

where ρ denotes the use (offered traffic per server) of the link with the capacity V (Equation (10.120)), servicing aggregated traffic streams, whereas E_2 denotes the Erlang C formula. Having at our disposal the value of the coefficient f_R, we can determine the average waiting time for a task with the size x in the system (for example, transfer time for a file with the size x) [10]:

$$W(x) = \frac{x}{r} f_R = \frac{x}{r}\left[1 + \frac{E_{2,R}(R\rho)}{R(1-\rho)}\right] \tag{10.118}$$

where:

- Intensity of offered traffic:

$$A = R\rho \tag{10.119}$$

- Intensity of traffic offered to the link ρ is expressed by the dependence:

$$\rho = \sum_{j=1}^{N} \lambda_j x_j \Big/ V \tag{10.120}$$

  - N is the number of users;
  - λ is the intensity of users' calls;
  - x is the average time length of, for example, data file transfer.
- Erlang's C formula is expressed by the dependence (10.47) which can be rewritten in the following form:

$$E_2(R, A) = \frac{\frac{A^R}{R!}\frac{R}{R-A}}{\sum_{i=0}^{R-1}\frac{A^i}{i!} + \frac{A^R}{R!}\frac{R}{R-A}} \tag{10.121}$$

In [10] it was proved that the results of the delay value calculated on the basis of Equation (10.118) are – in the case of long sessions (for example, a transfer of a huge file) – understated. The reason for this in the model is that the delay in access links are ommitted. This assumption is right, however, only with insignificant load in access links, in other words for $\rho \leq 5\%$. Hence, the present authors introduced a modification to Equation (10.118) that improves its accuracy and precision in relation to large loads in access links.

From [10], the average transmission time in the access link can be determined on the basis of the M/G/1 PS model (a stream can entirely use the link capacity offered by the access link):

$$W_a(x) = \frac{x}{r} f_{R_a} = \frac{1}{r} \frac{x}{1 - \rho_a} \tag{10.122}$$

where ρ_a determines the load of the access link with capacity r:

$$\rho_a = \lambda x / r \tag{10.123}$$

Given the dependence (10.122), we can determine the total value of the delay factor that takes into account not only delay in the access link but also in the link servicing aggregated streams [10]:

$$f_{\text{total}} = \frac{E_2(R, R\rho)}{R(1 - \rho)} + \frac{1}{1 - \rho_a} \tag{10.124}$$

and, subsequently, the total transmission delay for a file with the size x in a network consisting of a defined number of access links and one link servicing aggregated traffic streams from access links:

$$W(x) = \frac{x}{r} f_{\text{total}} \tag{10.125}$$

10.10.2 System Dimensioning based on the M/G/R PS Model

The presented M/G/R PS model can be used in dimensioning packet networks with requested acceptable delay values. In accordance with [10], the dimensioning process is an iterative process and can be expressed in the following form:

1. Determination of the initial value of the aggregated link capacity $V_k = r$.
2. Determination, on the basis of Equation (10.125) of the transmission delay.
3. Do the obtained delay values exceed the required threshold values?
   3.1. YES – increase the capacity and go to Step 2.
   3.2. NO – the required capacity of the aggregated link has been reached.
4. Terminate calculation.

This algorithm is not complicated and can be easily implemented in software.

References

[1] Kleinrock, L. (1975) *Queuing Systems*, John Wiley and Sons, Inc., vol. 1.
[2] Gross, D. and Harris, C.M. (1985) *Fundamentals of Queuing Theory*, John Wiley & Sons, Inc.
[3] Robertazzi, T.G. (1990) *Computer Networks and Systems: Queuing Theory and Performance Evaluation*, Springer-Verlag.
[4] Korilis, Y.A. (2003) Networking theory and fundamentals. Internet, http://www.seas.upenn.edu/t~com501/ (accessed 19 July 2010).
[5] Kendall, D.G. (1951) Some problems in the theory of queues. *Journal of the Royal Statistical Society, Series B*, **B-13** (2), 151–185.
[6] Chan, W. (2000) *Performance Analysis of Telecommunications and Local Area Networks*. Kluwer Academic Publishers.

[7] Feller, W. (1961) *An Introduction to Probability Therory and its Applications*, John Wiley & Sons, Inc.
[8] Pollaczek, F. (1930) Uber eine Aufgabe der Wahrscheinlichkeitstheorie. *I-II Math. Zeitschrift*, **32**, 64–100.
[9] Khinchin, A. (1932) Matematicheskaja teoria stacionarnoj ocheredi. *Matematicheskij sbornik*, **39** (4), 73–84.
[10] Rodrigo, M. and Pleich, R. (2002) *Validation and Extension of the M/G/R Processor Sharing to Dimension Elastic Traffic in TCP/IP Networks*. Proceedings of 2nd Polish-German Teletraffic Symposium (9th Polish Teletraffic Symposium), Gdańsk. Gdańsk University of Technology, Gdańsk.
[11] Key, P. and Smith, D. (1999) *Teletraffic Engineering in a Competitive World*. Proceedings of the International Teletraffic Congress, ITC-16, Edinburgh, 7–11 June. Elsevier, Amsterdam.
[12] Riedl, A., Bauschert, T., Perske, M., and Probst, A. (2000) *Investigation of the M/G/R Processor Sharing Model for Dimensioning of IP Access Networks with Elastic Traffic*. Proceedings of 1st Polish-German Teletraffic Symposium, Dresden. VDE Verlag GMBHm Berlin.

Part III

Application of Analytical Models for Mobile Networks

11

Modeling and Dimensioning of the Radio Interface

This chapter presents some examples of the application of the methods and models discussed in Chapters 5, 6 and 7 for modeling the radio interface in the cellular network in everyday engineering practice. The examples allow us to analyze the radio interface of a single cell in a single-service and multi-service network. The phenomenon of soft capacity and its influence on methods applied in the analysis of the radio interface is also taken into consideration. The chapter also presents models of the interface servicing HSPA traffic (Section 2.8) and traffic with priorities for services and for groups of users.

11.1 Modeling of Resource Allocation in the Radio Interface

A single cell of the mobile system can be treated as a full-availability group with constant or variable capacity, depending on a possible influence of the environment upon the load of the radio interface. In the first part of the book GSM, UMTS and LTE standards were presented. Radio interface in the GSM system can be treated as an example of a hard capacity system, whereas the radio interface in the UMTS or in the LTE system is an example of a soft capacity system.

11.1.1 Hard and Soft Capacity of the Mobile System

In the GSM system, the maximum number of subscribers serviced by one cell is unequivocally determined and depends exclusively on the number of frequency channels used. The GSM system was originally to take on voice services only, so any traffic models can be constructed on the basis of the full-availability group (Chapter 6). It is worth emphasizing that the GSM system can also be considered under special circumstances as a system with soft capacity. Such a situation can occur, for instance, when the operator of the mobile network has only assigned a few channels (frequencies) in a particular area and the mobile network in this area consists of many small base stations fully equipped with transceivers. In this case preparation

Modeling and Dimensioning of Mobile Networks: From GSM to LTE
Maciej Stasiak, Mariusz Głąbowski, Arkadiusz Wiśniewski and Piotr Zwierzykowski

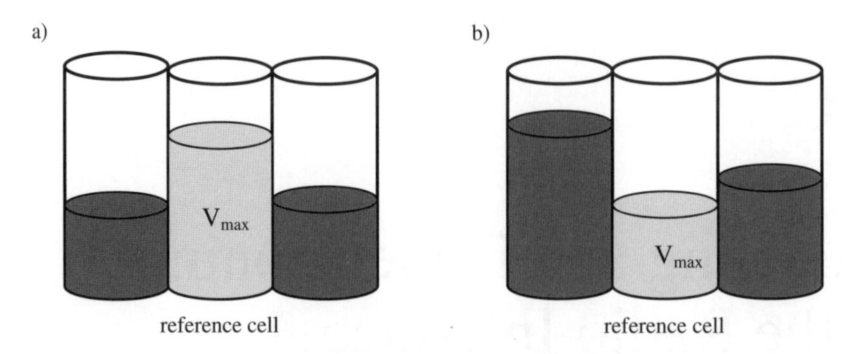

Figure 11.1 The phenomenon of soft capacity.

of a frequency plan for the area under consideration, without taking the appropriate ratio of the signal level to interference into account, is not possible. The traffic load of the radio interface in neighboring cells therefore has an influence on the possibility of servicing new calls in the discussed cell. The UMTS and LTE systems can also serve as examples of a soft capacity system. Soft capacity indicates the possibility of changing the capacity of the cell, depending on external influence, in which important element is the degree of load in neighboring cells. The idea of the dependence of the system capacity on the loads of neighboring cells is illustrated in Figure 11.1, which shows the reference cell in the form of a cylindrical vessel surrounded by two neighboring cells. The height of the liquid in the vessel reflects the load of a cell. In Figure 11.1a relatively small load of the neighboring cells allows to maintain a high maximum capacity of the reference cell. An increase in the load in neighboring cells causes, in turn, a reduction in the maximum capacity of the reference cell (Figure 11.1b).

11.1.2 Resource Allocation in Mobile Systems with Hard Capacity

In traditional systems (in which noise and interference limitations are not taken into account), resource allocation consists in adding bit rates related to calls of particular traffic classes.

$$\sum_{i=1}^{M} N_i r_i \leq V_{[\mathrm{kbps}]} \tag{11.1}$$

where M is the number of traffic classes offered to the system, r_i is the required bit rate of the traffic source of class i, while N_i is the maximum number of traffic sources of class i that can be serviced by the system.

The capacity of a system can be also expressed in the number of allocation units, the so-called basic bandwidth units (BBUs). In the case of a multi-service mobile system (the so-called multi-rate system) servicing many traffic classes with different bit rates, it is conventionally assumed that the value of the BBU should be equal to the greatest common divisor (GCD) of the resources demanded by particular call streams [1, 2]. In the case of the radio interface of a cellular system with hard capacity, we can write:

$$r_{\mathrm{BBU}} = \mathrm{GCD}(r_1, r_2, \ldots, r_M) \tag{11.2}$$

The interface capacity can then be expressed in BBUs:

$$V = \left\lfloor \frac{V_{[\text{kbps}]}}{r_{\text{BBU}}} \right\rfloor \tag{11.3}$$

Similarly, we can express the number of BBUs required by a call of a given class as follows:

$$t_i = \left\lceil \frac{r_i}{r_{\text{BBU}}} \right\rceil \tag{11.4}$$

If all traffic sources require the same bit rate r, then the capacity of the system can be expressed by the number of allocation units those of BBUs, for example:

$$V = \left\lfloor \frac{V_{[\text{kbps}]}}{r} \right\rfloor \tag{11.5}$$

11.1.3 Resource Allocation in Mobile Systems with Soft Capacity

The UMTS system can be considered as an example of a soft capacity system.[1] The radio interface of an isolated single cell in the UMTS network has a large theoretical link capacity. The available interface capacity is constrained by the accepted level of the interference in the frequency channel. In every mobile system with a spread signal spectrum, the capacity of the radio interface is limited as the result of the occurrence of a few types of interference: cochannel interference within a cell (coming from concurrent users of a frequency channel within the cell), outer co-channel interference within a cell (from the concurrent users of the frequency channel working within adjacent cells), adjacent channels interference (from the adjacent frequency channels of the same operator or other mobile operators) and all possible noise and interference from other sources. For the same reason, a growth in the load in the radio interface is accompanied by a growth in interference generated by other serviced users (in this cell or in other cells). To secure an acceptable level of service, it is necessary to limit the number of occupied resources by serviced traffic sources. In [3] it was estimated that the maximum usage of resources of the radio interface in the UMTS network without lowering the quality of service is about 50–80%. Soft capacity of the radio interface is also known as the interference limited capacity.

The example cellular system with soft capacity (UMTS) services multi-rate traffic that can be composed of a few classes, each of which demands a certain bit rate to service its own call. Therefore, in the analysis of the radio interface it is necessary to consider the class of a call and the bit rate required by the call of this class. Summing up, the radio interface in the UMTS system can be considered as a discrete multi-service network.[2]

In the radio interface of the UMTS system the accurate signal reception in the receiver is possible only when the ratio of energy per bit E_b to noise spectral density N_0 is appropriate. This means that a too low value of E_b/N_0 will cause the receiver to be unable to decode the

[1] System LTE, presented in Chapter 3, is also an example of the soft capacity system. In this chapter we choose UMTS system as an example of the soft capacity system based on their popularity in cellular networks.

[2] In the following analysis of the radio interface we will use L_{BBU}, defined in next part of this chapter, as the basic bandwidth unit (Equation (11.13)).

Table 11.1 Examples of R99 services and load factors [5]

	Service			
Parameters	Speech (Voice)	Videotelephony	Data	Data
W (Mcps)	3.84	3.84	3.84	3.84
r_i (kbps)	12.2	64	144	384
ν_i	0.67	1	1	1
E_b/N_0 (db)	4	2	1.5	1
L_i	0.005	0.026	0.050	0.112

received signal, while a too high value E_b/N_0 will be perceived by other users of the same radio channel as interference.

The ratio of the energy per bit in relation to noise spectral density, for a given traffic source of class i, can be expressed as follows [4]:

$$\left(\frac{E_b}{N_0}\right)_i = \frac{W}{\nu_i r_i} \cdot \frac{P_i}{I_{\text{total}} - P_i} \tag{11.6}$$

where P_i is the average signal power received from a traffic source of class i, I_{total} is total power of the received signal in the base station with thermal noise, W is bit rate of the spread signal (the so-called chip rate),[3] r_i represents link capacity of the data signal from the traffic source of class i in kbps and ν_i is activity factor of the traffic source of class i, which denotes the percentage of occupancy time of resources in which the source is active, in other words it transmits a signal with the link capacity r_i.

Equation (11.6) can be converted in such a way as to get the average power of the received signal from the traffic source of class i:

$$P_i = \frac{I_{\text{total}}}{1 + \frac{W}{\left(\frac{E_b}{N_0}\right)_i r_i \nu_i}} = I_{\text{total}} \cdot L_i \longrightarrow L_i = \left(1 + \frac{W}{\left(\frac{E_b}{N_0}\right)_i r_i \nu_i}\right)^{-1} \tag{11.7}$$

where L_i is the load factor, associated with a class i call. Sample loads of the radio interface by calls of different R99 traffic classes are shown in Table 11.1.

It should be noticed, that in the HSUPA technology changes at the required E_b/N_0 level, in relation to R99, appears. The following factors have influence on the value of E_b/N_0 in HSUPA: outer loop power control, target block error (BLER), transmit time interval (TTI) i.e. transmit time of each block of data in HSUPA, transport block size (TBS) i.e. the number of bits transmitted in each "transport block" and the number of HARQ transmission. Therefore in the modeling process we assumed that the load factor for the HSPA traffic can be determined by a simulation procedure [6]. Sample values of load factors for an exemplary HSUPA traffic stream (service) are shown in Table 11.2.

[3] W in the UMTS system is conventionally equal to 3.84 Mcps (Table 11.1).

Table 11.2 HSUPA services and load factors [6]

Parameters	Service		
	Service 1	Service 2	Service 3
W (Mcps)	3.84	3.84	3.84
r_i (kbps)	54.72	800.12	82.1
v_i	1	1	1
E_b/N_0 (db)	4.84	4.55	3.74
L_i	0.0416	0.3726	0.0481

11.1.3.1 The Uplink Direction

Let us note that the load coefficient is dimensionless and defines the fraction of a possible interface load. The coefficient also shows nonlinear dependency between the percentage load of the interface and the load factor for a single source of a given traffic class. Therefore based on the known load factors of single traffic sources, it is possible to determine the total load η_{UL} for the uplink direction:

$$\eta_{\mathrm{UL}} = \sum_{i=1}^{M} N_i L_i \tag{11.8}$$

where N_i is the number of serviced traffic sources of class i in the considered uplink direction.

Dependency (11.8) determines the ideal, maximum interface load in the radio interface of one, isolated, cell. In real cellular networks, however, traffic generated in other cells that also limits the capacity of the interface of a given cell, has to be taken into consideration. Hence, Equation (11.8) is complemented with the coefficient that takes into account interference from other cells. The total load for the uplink takes on the following form [4]:

$$\eta_{\mathrm{UL}} = \left(1 + \delta_j\right) \sum_{i=1}^{M} N_i L_i \tag{11.9}$$

where δ_j is defined as the ratio of the interference from other cells to the cell's own interference for j-th user.[4]

It should be emphasized that intercellular interference can also be taken into consideration using the so-called fixed-point method, presented in Section 13.2.1. With this solution, the total load for the uplink direction does not take into account the coefficient δ_j.

With the increase in the load of the radio interface, the level of interference generated in the system also increases. The increase in noise is defined as the ratio of the total power of the signal received in the system I_{total} to thermal noise P_N. The ratio is defined as the noise multiplication coefficient w. It can be proved that [4]:

$$w = \frac{I_{\mathrm{total}}}{P_N} = \frac{1}{1 - \eta_{\mathrm{UL}}} \tag{11.10}$$

[4]This coefficient, in the case of the uplink direction, is determined in the receiver of the base station [3].

and in decibel scale:

$$w_{[db]} = -10\log_{10}(1 - \eta_{\mathrm{UL}}) \tag{11.11}$$

When the load of the uplink direction increases to unity, the corresponding increase in noise tends to infinity. In [4] it was assumed that the maximum use of the resources of the radio interface, without a reduction in the quality of service, amounts to 50–80% of its theoretical capacity.

In the uplink direction in the HSPA standard (HSUPA – Section 2.8.2), the influence of interference can be taken into consideration either by the application of the coefficient δ_j, or by the application of the fixed-point method presented in Section 13.2.1.[5]

11.1.3.2 The Downlink Direction

The total load for the downlink direction can be written as follows [4]:

$$\eta_{\mathrm{DL}} = \sum_{i=1}^{M}\sum_{j=1}^{N_i}\left(1 - \xi_j + \delta_j\right) L_i \tag{11.12}$$

In Equation (11.12) parameter ξ_j is defined as the orthogonality factor for j-th user and δ_j is defined as the ratio of the interference from other cells to the cell's own interference which is dependent from location of the j-th user in the cell. Parameter ξ_j indicates the level of interference between the users of the same cell or other cells through the application of channel codes on the basis of orthogonal variable spreading factor (OVSF). Thus, the user can be characterized by a different orthogonal code and their mutual correlation is, theoretically, equal to zero [7]. In the book we will use mean values of parameters δ_j nd ξ_j, which depend on the location of mobile users in the cell and these average values will be designated by symbols δ and ξ, respectively.

In the case of the downlink direction in the HSPA standard (HSDPA – Section 2.8.1), the influence of interference can be taken into consideration, in a similar way to HSUPA, either by the application of the parameters δ and ξ, or by the application of the fixed-point method (Section 13.2.1).

11.1.3.3 Allocation Units in the Radio Interface

In systems with soft capacity, the available throughput of a system is different from the theoretical capacity of the isolated cell (Figure 11.2). The capacity of such cell can change from maximum (capacity of the isolated cell which is not exposed to external influences) to a certain minimum capacity, when the influence of the load of the neighboring cells is at its maximum. In the discussed system, the use of the bit rates as a measure for allocation is not very convenient. It is much more convenient in such a situation to measure the state of allocated resources in other units more sophisticated to the physical nature of a mobile system with soft capacity.

Equations (11.9) and (11.12) clearly indicate that the measure of resource allocation in the radio interface can be the percentage of the usage of the radio interface by calls of all traffic classes. Therefore, the radio interface allocation is not based on adding bit rates, but it is

[5] In the fixed-point method, the interference is taken into consideration directly in the computational method.

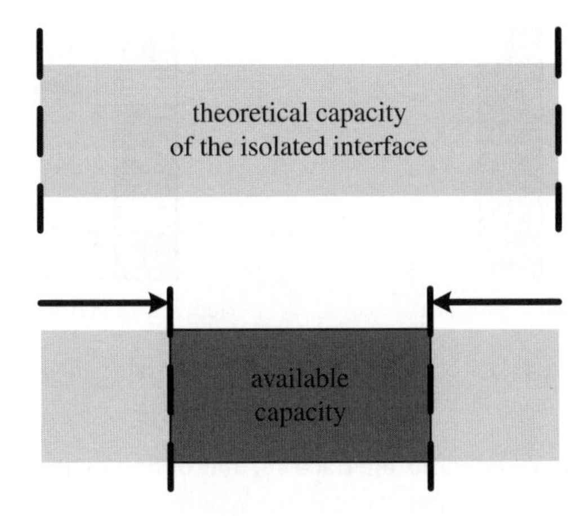

Figure 11.2 Available resources in a system with soft capacity.

based on adding interference load factors of particular calls. A scheme of changing resource allocation, expressed in kbps, into the resource allocation expressed in the percentage of the load of the radio interface is shown in Figure 11.3.

The cellular system with soft capacity considered here services many traffic classes with different load factors. Such a system can be treated as a multi-rate system in which the value of a BBU should be lower than or equal to the greatest common divisor of the resources demanded by individual call streams [1, 2]. In the case of the WCDMA radio interface in the UMTS network, we can write:

$$L_{\text{BBU}} = \text{GCD}(L_1, L_2, \ldots, L_M) \tag{11.13}$$

The capacity of the interface can then be expressed by the number of BBUs:

$$V = \left\lfloor \frac{\eta}{L_{\text{BBU}}} \right\rfloor \tag{11.14}$$

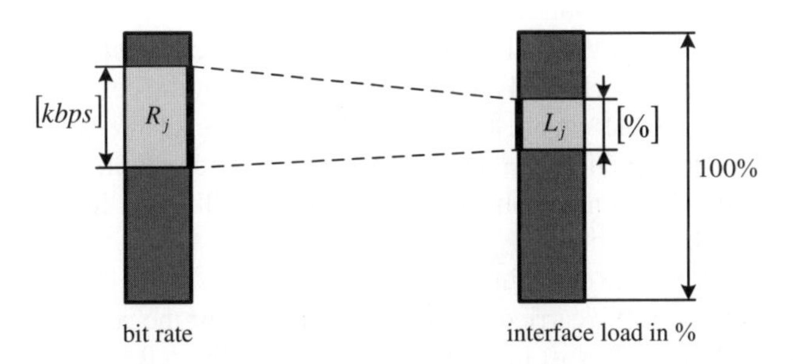

Figure 11.3 Resource allocation in the radio interface.

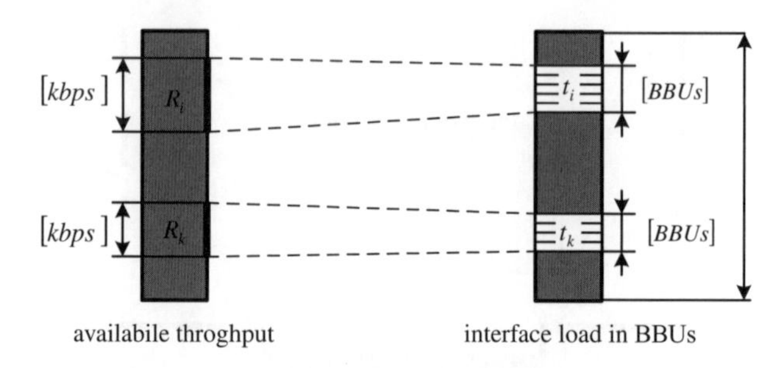

Figure 11.4 Resource allocation in the multi-rate radio interface.

where η is the capacity of the radio interface for the uplink or the downlink direction. In a similar way we can express the number of BBUs required by a call of a given class:

$$t_i = \left\lceil \frac{L_i}{L_{\text{BBU}}} \right\rceil \tag{11.15}$$

Figure 11.4 shows the way of changing resource allocation expressed in kbps, into resource allocation expressed in BBUs, according to the definition (11.15). The figure shows the calls of two classes being serviced and their matching values in the load of the radio interface expressed in kbps and BBUs.

11.2 Cellular System with Hard Capacity Carrying Single-Service Traffic

In this section we shall present models that can be used in determining traffic characteristics of cell radio interfaces in mobile networks with hard capacity. The starting point for this chapter are the models of full-availability groups presented in Chapter 6.

The following notation is used in this chapter:

- E – blocking probability for calls offered to the radio interface of the cell;
- B – loss probability for calls offered to the radio interface of the cell;
- A – intensity of traffic offered to the radio interface of the cell;
- L – interference load factor introduced by one call into the radio interface of the cell (expressed as a percentage);
- η – assumed capacity of the radio interface in the cell, expressed in percentage;
- V_h – number of available channels in the radio interface of the cell (hard capacity);
- V – number of available channels in the radio interface of the cell, equal to V_s (soft capacity) or V_h depending on the model under consideration;
- δ – ratio of the interference from other cells to the cell's own interference in the (uplink/downlink) direction;
- ξ – orthogonality factor (downlink);
- Y_V – intensity of the traffic carried by the radio interface of the cell;
- R – intensity of the traffic lost in the radio interface of the cell;
- y_V – intensity of the traffic carried by one channel of the radio interface of the cell.

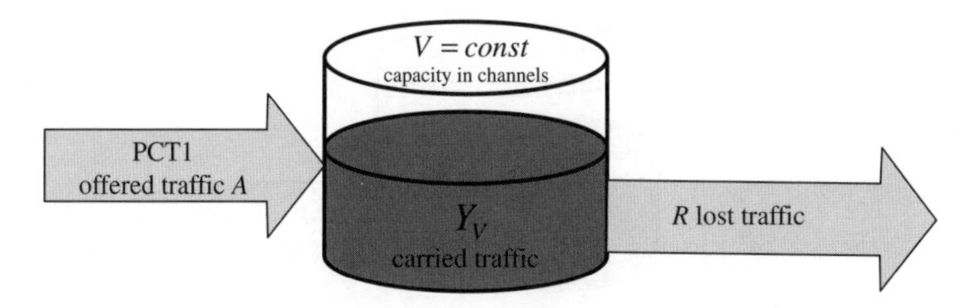

Figure 11.5 Model of the radio interface with hard capacity that services PCT1 traffic.

11.2.1 Erlang Model of the Radio Interface

This model concerns the GSM system and the systems preceding it – those in which the number of access channels in a radio interface of the cell is constant. We also assume that the cell services one class of calls exclusively, which means that each call requires one channel to set up a connection.

Figure 11.5 shows the model of the cell under consideration here. All traffic characteristics – the blocking probability, loss probability, carried and offered traffic – result from the Erlang formula (Section 5.2). In this model we assume that $V = V_h$.

- Blocking (loss) probability:

$$E_V(A) = E = B = \frac{A^V}{V!} \Bigg/ \sum_{i=0}^{V} \frac{A^i}{i!} \tag{11.16}$$

- The intensity of the carried traffic in the cell ((Equation 5.12), Section 5.2.11):

$$Y_V = A[1 - E] \tag{11.17}$$

- Intensity of the traffic carried by one channel of the radio interface in the cell:

$$y_V = Y_V/V \tag{11.18}$$

- Mean value of the intensity of lost traffic R (which is the difference between the intensity of the offered traffic and the intensity of the carried traffic, Equation (5.13)):

$$R = A - Y_V = AE \tag{11.19}$$

11.2.2 Engset Model of the Radio Interface

This model concerns systems in which the number of traffic sources is comparable with the capacity of the radio interface (the number of traffic sources is greater or equal to the capacity of the cell). This can have a substantial influence upon traffic properties of the system. We further assume that the cell services exclusively one class of calls, which means that each call demands one BBU to set up a connection.

The following notation is adopted in this chapter:

γ – intensity of calls generated by one free source;
α – intensity of traffic offered by one free source;

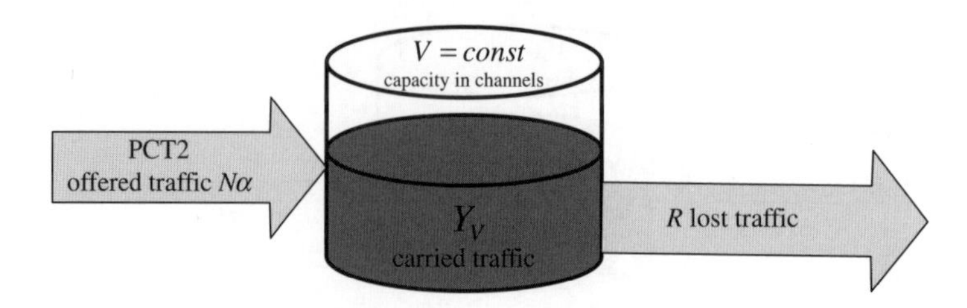

Figure 11.6 Model of the radio interface with hard capacity that services PCT2 traffic.

N – number of traffic sources in the cell;
$E(\alpha, V, N)$ – blocking probability for calls offered to the radio interface of the cell;
$B(\alpha, V, N)$ – loss probability for calls offered to the radio interface of the cell;
$C(\alpha, V, N)$ – traffic loss probability in the radio interface of the cell;
y_N – intensity of traffic carried by one source.

In this model we assume that the number of channels available in the interface is constant and that the cell exclusively services one class of calls, which means that each radio interface call requires one BBU to set up a connection. Figure 11.6 shows the considered model of the interface.

All traffic characteristics – the blocking probability, loss probability, carried and offered traffic – result from the Engset distribution presented in Section 5.3:

- Blocking probability (Section 5.3.5):

$$E(\alpha, V, N) = \binom{N}{V}\alpha^V \Big/ \sum_{j=0}^{V}\binom{N}{j}\alpha^j \tag{11.20}$$

- Loss probability (Section 5.3.6):

$$B(\alpha, V, N) = E(\alpha, V, N-1) \tag{11.21}$$

- Intensity of the traffic offered to the cell:

$$A = Y_N = N\frac{\alpha}{1+\alpha} \tag{11.22}$$

- Intensity of carried traffic in the cell (Section 5.3.10):

$$Y_V = \frac{\gamma N[1 - B(\alpha, V, N)]}{1 + \gamma[1 - B(\alpha, V, N)]} \tag{11.23}$$

or, in the equivalent way:

$$Y_V = \frac{\alpha}{1+\alpha}[N - (N - V)E(\alpha, V, N)] \tag{11.24}$$

- Intensity of traffic carried by one source:

$$y_N = \frac{Y_V}{N} = \frac{\alpha}{1+\alpha}\left[1 - \frac{N-V}{N}E(\alpha, V, N)\right] = \frac{\alpha}{1+\alpha}[1 - C(\alpha, V, N)] \tag{11.25}$$

- Intensity of the traffic carried by one channel of the radio interface in the cell:

$$y_V = \frac{Y_V}{V} = \frac{\alpha}{1+\alpha}\left[\frac{N}{V} - \frac{N-V}{V}E(\alpha, V, N)\right] \tag{11.26}$$

- Mean value of the lost traffic intensity:

$$R = \frac{\alpha}{1+\alpha}(N-V)E(\alpha, V, N) \tag{11.27}$$

Comments on the origin of Equations (11.22) and (11.27) are to be found in Section 5.3.12.

11.3 Cellular System with Soft Capacity Carrying Single-Service Traffic

This chapter presents models of the radio interface in mobile network with soft capacity. The basis for the proposed models of the interface are the models of full-availability groups presented in Chapter 6.

11.3.1 Erlang Model of the Radio Interface

This model applies to the mobile system with soft capacity, in which the number of BBUs offered to the radio interface of the cell is variable and dependable on external circumstances (load of neighboring cells). We assume further that the cell services exclusively one class of calls, which means that each call demands one BBU to set up a connection. Figure 11.7 shows visually the cell model under consideration.

To determine the properties of the system, it is necessary to precede the operation with a determination of the equivalent capacity of the system:

- Equivalent system capacity in BBUs for the uplink:

$$V_{\mathrm{UL}} = \left\lfloor \frac{\eta}{L(1+\delta)} \right\rfloor \tag{11.28}$$

- Equivalent system capacity in BBUs for the downlink direction:

$$V_{\mathrm{DL}} = \left\lfloor \frac{\eta}{L(1-\xi+\delta)} \right\rfloor \tag{11.29}$$

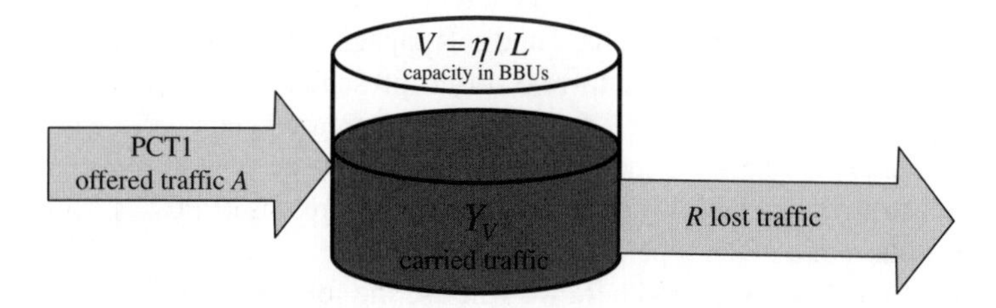

Figure 11.7 Model of the radio interface with soft capacity that services PCT1 traffic.

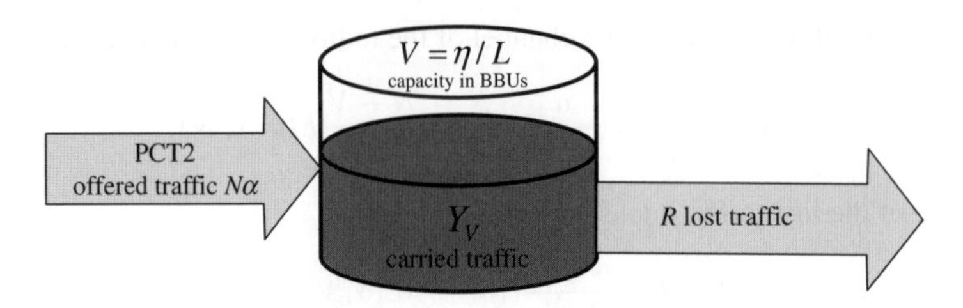

Figure 11.8 Model of the radio interface with soft capacity that services PCT2 traffic.

Following the above, the remaining characteristics of the system can be determined in a similar way as in the previous model, assuming that $V = V_{\mathrm{UL}}$ for the uplink and $V = V_{\mathrm{DL}}$ for the downlink direction.

In the above model, the "softness" of the radio interface is determined by the adopted coefficient δ and orthogonality factor ξ (downlink). Therefore, in this particular case, we model the system with soft capacity by a system with equivalent hard capacity, in which this capacity depends exactly on the adopted values of coefficients δ and ξ.

11.3.2 Engset Model of the Radio Interface

This model concerns radio interface in the cells in the mobile network with soft capacity, in which the number of traffic sources is comparable, though higher, than the capacity of the interface. We assume that the number of channels available in the cell is variable and dependable on external factors (the loads of neighboring cells). Additionally, the standard assumption that each call demands one BBU to set up a connection remains as valid as before.

In this model we proceed exactly as in Section 11.3.1, i.e. we model the system with soft capacity by a system with equivalent hard capacity, in which the capacity depends exactly on the values of the adopted interference coefficient δ and orthogonality factor ξ.

The considered model of the cell is shown in Figure 11.8.

The characteristics of the considered system can be determined on the basis of the formulas presented in Section 11.2.2, assuming "soft" capacity for the uplink and the downlink direction (Equations (11.28) and (11.29)).

11.4 Cellular System with Hard and Soft Capacity Carrying a Mixture of Multi-Service Traffic Streams

This chapter presents models that can be used to determine traffic characteristics for the radio interface in mobile networks with both hard and soft capacity. The group of models considered here is based on the models with multi-rate traffic presented in Chapter 7. We will use a double-subscript notation, as Section 7.5. That is, stream parameters will be denoted by two subscripts $Parameter_{i,j}$, where the first subscript i indicates the type of the stream (subscript "1" corresponds to streams of PCT1, and subscript "2" to streams of PCT2), while j indicates the number of a class of a given type calls.

In the group of models considered here we will use the following notation:

$[P_n]_V$ – probability of n BBUs occupied in an interface;

M_i – number of traffic classes of type i offered to the radio interface of the cell ($i = 1$ to streams of PCT1, and $i = 2$ to streams of PCT2);
$E_{i,j}$ – blocking probability for calls of class j of type i offered to the radio interface of the cell;
$B_{i,j}$ – loss probability for calls of class j of type i offered to the radio interface of the cell;
L_{BBU} – value of the basic bandwidth unit, in percentage of the interface load;
$A_{i,j}$ – intensity of PCT1 traffic of class j of type i offered to the radio interface of the cell;
$L_{i,j}$ – load factor (in percentage) introduced by one call of class j of type i into the radio interface of the cell;
$t_{i,j}$ – number of BBUs required by call of class j call of type i to set up a connection by the radio interface of the cell;
$Y_{i,j}$ – intensity of traffic of class j of type i carried by the radio interface of the cell;
$R_{i,j}$ – intensity of traffic of class j of type i lost in the radio interface of the cell.

11.4.1 Model of the Radio Interface Servicing PCT1 Traffic Streams

The model concerns the radio interface in the cell that services a mixture of multi-service traffic streams generated by the number of users greater than the capacity of the interface (PCT1 traffic). We also assume that each traffic class requires a different number of BBUs to set up a connection.

A diagram of the cell model under consideration here is shown in Figure 11.9. All traffic characteristics – blocking probability, carried and offered traffic – result from the generalized Kaufman–Roberts distribution (Section 7.3.5):

- Basic bandwidth unit (Equation (11.13)):

$$L_{\text{BBU}} = \text{GCD}(L_{1,1}, L_{1,2}, \ldots, L_{1,M_1}) \tag{11.30}$$

- Equivalent capacity of the system in BBUs for the uplink (Equation (11.28)):

$$V = V_{\text{UL}} = \left\lfloor \frac{\eta}{L_{\text{BBU}}(1+\delta)} \right\rfloor \tag{11.31}$$

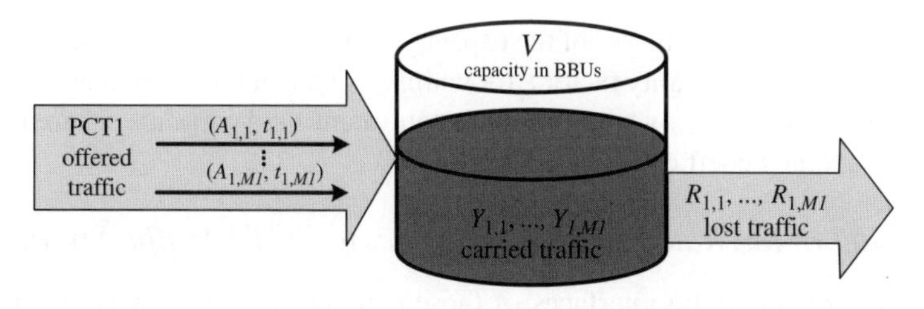

Figure 11.9 Model of the radio interface in the cell servicing multi-rate PCT1 traffic.

- Equivalent capacity of the system in BBUs for the downlink direction (Equation (11.29)):

$$V = V_{\mathrm{DL}} = \left\lfloor \frac{\eta}{L_{\mathrm{BBU}}(1 - \xi + \delta)} \right\rfloor \tag{11.32}$$

- Number of BBUs demanded by a class i call (Equation (11.15)):

$$t_{1,i} = \left\lceil \frac{L_{1,i}}{L_{\mathrm{BBU}}} \right\rceil \tag{11.33}$$

- Occupancy distribution (Section 7.3.5, Equation 7.39)):

$$n\,[P_n]_V = \sum_{i=1}^{M_1} A_{1,i} t_{1,i} \left[P_{n-t_{1,i}}\right]_V \tag{11.34}$$

where $[P_{n-t_{1,i}}]_V = 0$, if $n < t_i$, and the value $[P_0]_V$ results from the normalization condition:

$$\sum_{n=0}^{V} [P_n]_V = 1 \tag{11.35}$$

- Blocking (loss) probability for class i calls (Section 7.3.6, Equation (7.42)):

$$E_{1,i} = B_{1,i} = \sum_{n=V-t_{1,i}+1}^{V} [P_n]_V \tag{11.36}$$

- Carried traffic intensity of class i calls in the cell:

$$Y_{1,i} = A_{1,i}\,[1 - E_{1,i}] \tag{11.37}$$

- Intensity of the traffic carried by one BBU of the radio interface in the cell:

$$y_V = \sum_{i=1}^{M_1} Y_{1,i} \Big/ V \tag{11.38}$$

- Mean loss traffic intensity of class i calls (the difference between the intensity of the offered traffic and the intensity of the carried traffic):

$$R_{1,i} = A_{1,i} - Y_{1,i} = A_{1,i}\,E_{1,i} \tag{11.39}$$

In the above model, the "softness" of the capacity of the radio interface is determined by the coefficient δ and the orthogonality factor ξ (downlink). Thus, in this particular case, we model the system with soft capacity by a system with equivalent hard capacity, where this capacity depends on the adopted coefficients δ and ξ.

11.4.2 Model of the Radio Interface Servicing PCT2 Traffic Streams

The model concerns the radio interfaces of those cells in the mobile network in which the number of traffic sources of a given class N_i is comparable to, although higher than, the

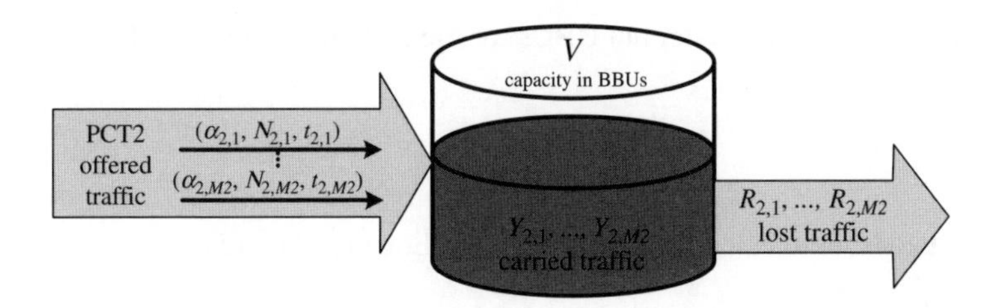

Figure 11.10 Model of the radio interface in the cell servicing PCT2 multi-rate traffic.

capacity of the cell expressed in the required numbers of BBUs of calls of this class:

$$\forall_{1 \leq i \leq M} \; N_{2,i} > \lfloor V/t_{2,i} \rfloor \tag{11.40}$$

We assume that a radio interface is offered traffic of the PCT2 type only. In the model of the interface we proceed exactly as in Section 11.4.1 – we treat the system with soft capacity like the system with equivalent hard capacity, in which this capacity depends on the values of the adopted coefficient δ and the orthogonality factor ξ.

In the model considered here we also used the parameter $\gamma_{2,i}$ which is the arrival rate of calls generated by one free source of class i and $\alpha_{2,i}$ defined as the intensity of the traffic offered by one free source of class i.

Figure 11.10 shows visually the considered model of the radio interface. All traffic characteristics – blocking probability, loss probability, carried and offered traffic – result from the approximate algorithm of the calculation of multi-service Erlang-Engset distribution (Section 7.5.3), referred to here as the algorithm. Since the radio interface is treated as a full-availability group servicing exclusively traffic streams of the type PCT2, we can thus assume in the algorithm:

$$\begin{cases} M_1 = 0 \\ M_2 \neq 0 \end{cases} \tag{11.41}$$

and

$$\forall_{i \in \{1,\ldots,M_2\}} \quad \forall_{0 \leq n \leq V - t_{2,i}} \sigma_{2,i}(n) = 1 \tag{11.42}$$

where conditional transition coefficient $\sigma_{2,i}(n)$ in the model of the full-availability group, which is the model of the radio interface, is equal to one.

With such assumptions, the appropriate characteristics of the system can be determined on the basis of the following formulas:

- Basic bandwidth unit (Equation (11.13)):

$$L_{\mathrm{BBU}} = \mathrm{GCD}(L_{2,1}, L_{2,2}, \ldots, L_{2,M_2}) \tag{11.43}$$

- Equivalent capacity of the system in BBUs for the uplink (Equation (11.28)):

$$V = V_{\mathrm{UL}} = \left\lfloor \frac{\eta}{L_{\mathrm{BBU}}(1+\delta)} \right\rfloor \tag{11.44}$$

- Equivalent capacity of the system in BBUs for the downlink direction (Equation (11.29)):

$$V = V_{\mathrm{DL}} = \left\lfloor \frac{\eta}{L_{\mathrm{BBU}}(1 - \xi + \delta)} \right\rfloor \tag{11.45}$$

- Number of BBUs, demanded by a call of class i (Equation (11.15):

$$t_{2,i} = \left\lceil \frac{L_{2,i}}{L_{\mathrm{BBU}}} \right\rceil \tag{11.46}$$

- Occupancy distribution:
  occupancy distribution $[P_n]_V$ is determined on the basis of item 3 of the algorithm (Section 7.5.3, Equation (7.79)):

$$n\,[P_n]_V = \sum_{i=1}^{M_2} A_{2,i}(n - t_{2,i})t_{2,i}\left[P_{n-t_{2,i}}\right]_V \tag{11.47}$$

- Blocking probability of calls of class i:
  blocking probability $E_{2,i}$ is determined on the basis of item 4 of the algorithm (Section 7.5.3, Equation (7.81)):

$$E_{2,i} = \sum_{n=V-t_{2,i}+1}^{V} [P_n]_V \tag{11.48}$$

- Loss probability of class i calls:
  Loss probability $B_{2,i}$ is determined on the basis of item 4 of the algorithm (Section 7.5.3, Equation (7.82)):

$$B_{2,i} = \frac{\sum\limits_{n=V-t_{2,i}+1}^{V} [P_n]_V\,[N_{2,i} - y_{2,i}(n)]\gamma_{2,i}}{\sum\limits_{n=0}^{V} [P_n]_V\left[N_{2,i} - y_{2,i}(n)\right]\gamma_{2,i}} \tag{11.49}$$

  where $y_{2,i}(n)$ is the number of serviced class i calls in state n, defined in item 2 of the algorithm (Section 7.3.6, Equation (7.78)).
- Intensity of class i traffic carried in the radio interface:
  The intensity of the carried traffic of the given class i is determined on the basis of the definition formulated in Section 5.3.10, according to which traffic intensity is equal to the average number of simultaneously occupied BBUs in a group with the capacity of V BBUs. Thus, we can write:

$$Y_{2,i} = \sum_{n=0}^{V} y_{2,i}(n)\,[P_n]_V \tag{11.50}$$

- Intensity of the traffic carried by one BBU of the radio interface in the cell:

$$y_V = \sum_{i=1}^{M2} Y_{2,i} \Big/ V \tag{11.51}$$

- The mean value of the intensity of lost traffic of class i in the radio interface is the difference between the intensity of the offered traffic and the intensity of the carried traffic:

$$R_{2,i} = A_{2,i} - Y_{2,i} \tag{11.52}$$

11.4.3 Model of the Radio Interface Servicing PCT1 and PCT2 Traffic Streams

This model concerns radio interface in a cell of a mobile network that services simultaneously traffic streams generated by infinite and finite numbers of traffic classes (PCT1 and PCT2 traffic).

In this model we proceed in exactly the same way as in Sections 11.4.1 and 11.4.2: we model the system with soft capacity using a system with equivalent hard capacity, in which this capacity depends on the values of the adopted coefficient δ and orthogonality factor ξ.

Figure 11.11 shows this model of the radio interface carrying a mixture of PCT1 and PCT2 traffic streams. All traffic characteristics – blocking probability, loss probability, carried and offered traffic – result from the approximate calculational algorithm of the multi-service Erlang-Engset distribution (Section 7.5.3), referred to here as the algorithm. The cell is a full-availability system, servicing simultaneously streams of PCT1 and PCT2, so we assume in the algorithm:

$$\begin{cases} M_1 \neq 0 \\ M_2 \neq 0 \end{cases} \tag{11.53}$$

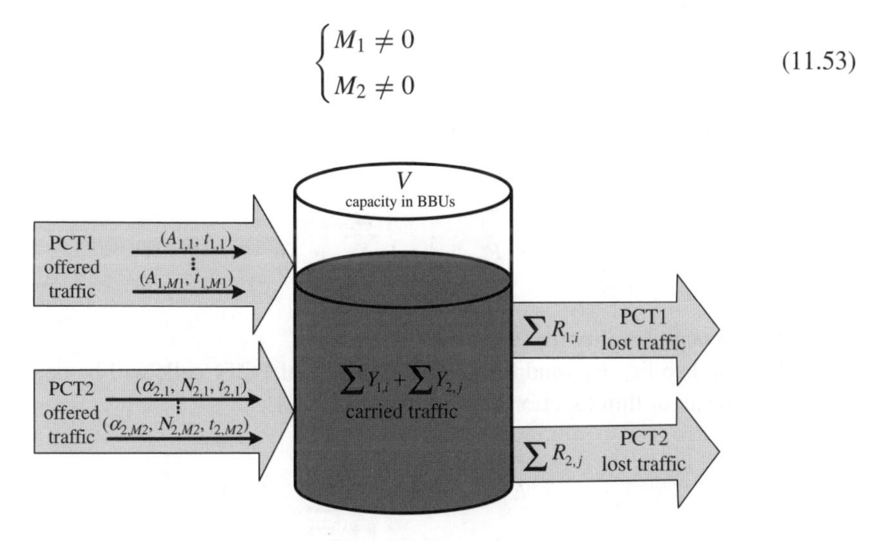

Figure 11.11 Model of the radio interface in the cell servicing multi-rate traffic of the type PCT1 and PCT2.

$$\forall_{i\in\{1,\ldots,M_1\}} \quad \forall_{0\leq n\leq V-t_{1,i}} \sigma_{1,i}(n) = 1 \tag{11.54}$$

and

$$\forall_{j\in\{1,\ldots,M_2\}} \quad \forall_{0\leq n\leq V-t_{2,j}} \sigma_{2,j}(n) = 1 \tag{11.55}$$

With such assumptions, the corresponding characteristics of the system can be determined on the basis of the following formulas:

- Basic bandwidth unit (Equation (11.13)):

$$L_{\text{BBU}} = \text{BBU}(L_{1,1}, \ldots, L_{1,M1}, \; L_{2,1}, \ldots, L_{2,M_2}) \tag{11.56}$$

- Equivalent capacity of the system in BBUs for the uplink (Equation (11.28)):

$$V = V_{\text{UL}} = \left\lfloor \frac{\eta}{L_{\text{BBU}}(1+\delta)} \right\rfloor \tag{11.57}$$

- Equivalent capacity of the system in BBUs for the downlink direction (Equation (11.29)):

$$V = V_{\text{DL}} = \left\lfloor \frac{\eta}{L_{\text{BBU}}(1-\xi+\delta)} \right\rfloor \tag{11.58}$$

- Number of BBUs demanded by a PCT1 call of class i (Equation (11.15)):

$$t_{1,i} = \lceil L_{1,i}/L_{\text{BBU}} \rceil \tag{11.59}$$

- Number of BBUs demanded by a PCT2 call of class j (Equation (11.15)):

$$t_{2,j} = \lceil L_{2,j}/L_{\text{BBU}} \rceil \tag{11.60}$$

- Occupancy distribution: $[P_n]_V$ is determined after step 3 of the algorithm (Section 7.5.3, Equation (7.79)):

$$n\,[P_n]_V = \sum_{i=1}^{M_1} A_{1,i} t_{1,i} \left[P_{n-t_{1,i}}\right]_V + \sum_{i=1}^{M_2} A_{2,i}(n - t_{2,i}) t_{2,i} \left[P_{n-t_{2,i}}\right]_V \tag{11.61}$$

- Blocking/loss probability for PCT1 calls of class i:
  Blocking probability $E_{1,i}$ and loss probability $B_{1,i}$ of PCT1 calls will be determined after step 4 of the algorithm (Section 7.5.3, Equation (7.81)):

$$E_{1,i} = B_{1,i} = \sum_{n=V-t_{1,i}+1}^{V} [P_n]_V \tag{11.62}$$

- Blocking probability of PCT2 calls of class j:

Blocking probability $E_{2,i}$ will be determined after step 4 of the algorithm (Section 7.5.3, Equation (7.81)):

$$E_{2,j} = \sum_{n=V-t_{2,j}+1}^{V} [P_n]_V \tag{11.63}$$

- Loss probability of PCT2 calls of class j:
Loss probability $B_{2,j}$ will be determined on the basis of step 4 of the algorithm (Section 7.5.3, Equation (7.82)):

$$B_{2,j} = \frac{\sum_{n=V-t_{2,j}+1}^{V} [P_n]_V \, [N_{2,j} - y_{2,j}(n)]\gamma_{2,j}}{\sum_{n=0}^{V} [P_n]_V \left[N_{2,j} - y_{2,j}(n)\right] \gamma_{2,j}} \tag{11.64}$$

where $y_{2,j}(n)$ is the number of serviced calls of class j in state n, determined in step 2 of the algorithm (Section 7.5.3, Equation (7.78)).

- PCT1 traffic intensity of class i, carried in the interface:

$$Y_{1,i} = A_{1,i} \, [1 - E_{1,i}] \tag{11.65}$$

- PCT2 traffic intensity of class j, carried by the radio interface:
The intensity of the carried traffic is determined on the basis of the definition (Section 5.3.10), according to which traffic intensity is equal to the mean number of simultaneously occupied BBUs in the cell with the capacity of V BBUs. Thus, we can write:

$$Y_{2,j} = \sum_{n=0}^{V} y_{2,j}(n) \, [P_n]_V \tag{11.66}$$

- Intensity of the traffic carried by one BBU of the radio interface in the cell:

$$y_V = \left(\sum_{i=1}^{M_1} Y_{1,i} + \sum_{j=1}^{M_2} Y_{2,j} \right) \bigg/ V \tag{11.67}$$

- Intensity of PCT1 lost traffic of class i in the interface:

$$R_{1,i} = A_{1,i} - Y_{1,i} \tag{11.68}$$

- Intensities of PCT2 lost traffic of class j in the interface:

$$R_{2,j} = A_{2,j} - Y_{2,j} \tag{11.69}$$

11.4.4 Threshold Model of the Radio Interface

The threshold model of a radio interface is a model that takes into consideration a change in QoS parameters depending on the load of a cell. To determine the characteristics of a cell in which an appropriate access algorithm has been applied – the one in which, along with an increase in the load, the number of BBUs allocated to the calls of particular classes decreases, and, at the same time, the mean service time of the calls can be increased – we will use the multi-threshold model presented in Section 7.8.2.

We proceed with the model under consideration here in exactly the same way as in the earlier chapters – we model the system with soft capacity using a system with equivalent hard capacity in which this capacity depends on the values of adopted coefficient δ and orthogonality factor ξ.

In the threshold model we will use a notation in which three subscripts are introduced, similar to notation proposed earlier in Sections 7.5 and 7.8.2: stream parameters will be denoted by three subscripts $Parameter_{i,j,k}$, where the first subscript i indicates the type of the stream (subscript "1" corresponds to streams of PCT1, and subscript "2" to streams of PCT2), parameter j indicates classes while k represents the number of post-threshold areas:

- $A_{i,j,k}$ – intensity of traffic of class j of type i offered to the radio interface of the cell being in the post-threshold area k;
- $L_{i,j,k}$ – load factor introduced by one call of class j of type i into the radio interface of the cell being in the post-threshold area k;
- $t_{i,j,k}$ – number of BBUs required by a call of class j of the type i to set up a connection in the radio interface of the cell being in in the post-threshold area k;
- $Y_{i,j,k}$ – intensity of traffic of class j of type i carried by the radio interface of the cell being in the post-threshold area k.

11.4.4.1 Assumptions for the Model

We assume that the radio interface in the model that we are considering here services M_1 classes of the PCT1 traffic and M_2 classes of the PCT2 traffic. In the radio interface of the cell, for each traffic class carried by the interface, a set of thresholds is introduced individually. For example, for the PCT1 traffic of class i, the set of adopted thresholds will be written as follows: $\{Q_{1,i,1}, Q_{1,i,2}, \ldots, Q_{1,i,qi}\}$, while for the PCT2 traffic of class j, the set of adopted thresholds will be written in the following way: $\{Q_{2,j,1}, Q_{2,j,2}, \ldots, Q_{2,j,qj}\}$. With the choice of thresholds, it is assumed that:

$$\{Q_{1,i,1} \leq Q_{1,i,2} \leq \ldots \leq Q_{1,i,q_i}\} \quad \text{and} \quad \{Q_{2,j,1} \leq Q_{2,j,2} \leq \ldots \leq Q_{2,j,q_j}\}$$

The set of states n, such that: $Q_{1,i,k} < n \leq Q_{1,i,k+1}$, is called the post-threshold area k of PCT1 traffic of class i, whereas the set of states n, such that: $Q_{2,j,k} < n \leq Q_{2,j,k+1}$ is called the post-threshold area k of PCT2 traffic of class j. All thresholds in the system are selected in such a way that the following condition is fulfilled:

$$\begin{cases} V - t_{1,i,k} \geq Q_{1,i,k+1} \\ V - t_{1,i,k} > Q_{1,i,k} \\ V - t_{2,j,k} \geq Q_{2,j,k+1} \\ V - t_{2,j,k} > Q_{2,j,k} \end{cases} \quad \text{for} \quad k \neq qi \quad \text{and} \quad k \neq qj \tag{11.70}$$

In considering the threshold model for a radio interface we assume that thresholds will be selected in such a way that the condition (11.70) should always be fulfilled. The fulfillment of the condition (11.70) means that each state of the post-threshold area, except the last

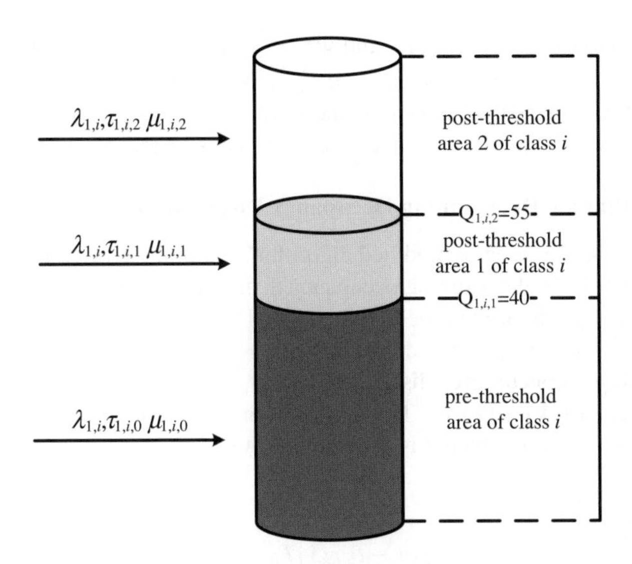

Figure 11.12 Resources of the cell with threshold mechanism for i class calls.

post-threshold area, can service appropriate calls. With such assumptions, blocking states in the interface will occur exclusively in the last post-threshold area. Such an approach can be justified practically: the reason behind the implementation of thresholds in the system is to minimize losses of calls with worst parameters that occur in the last post-threshold area. It should be emphasized here that only such systems have been hitherto considered in the literature on this subject [8–10].

Figure 11.12 shows the resources of the radio interface from the point of view of a traffic stream of class i. In the interface with a capacity of $V = 60$ BBUs two thresholds were introduced: $Q_{1,i,1} = 40$ and $Q_{1,i,2} = 55$. To simplify the considerations, it was assumed that only traffic of the type PCT1 was offered. In the pre-threshold area, each call of class i demands $t_{1,i,0} = 10$ BBUs, the post-threshold area 1 demands $t_{1,i,1} = 5$ BBUs, whereas post-threshold area 2 demands $t_{1,i,2} = 2$ BBUs. Note that the mutual positioning of thresholds and reference states presented in the figure meets the condition (11.70), because:

$$\begin{cases} V - t_{1,i,0} \geq Q_{1,i,1} & (50 > 40) \\ V - t_{1,i,0} > Q_{1,i,0} & (50 > 0\ ) \end{cases}$$

and

$$\begin{cases} V - t_{1,i,1} \geq Q_{1,i,2} & (55 = 55) \\ V - t_{1,i,1} > Q_{1,i,1} & (55 > 40) \end{cases}$$

The operation of the threshold access algorithm in the radio interface can be presented in the following way: in each post-threshold area k of the PCT1 traffic of class i, traffic stream of class i is offered, defined by a set of parameters $\{\lambda_{1,i}, t_{1,i,k}, \mu_{1,i,k}\}$, and when $t_{1,i,0} > t_{1,i,1} > \ldots > t_{1,i,k} > \ldots > t_{1,i,qi}$ and $\mu_{1,i,0}^{-1} < \mu_{1,i,1}^{-1} < \ldots < \mu_{1,i,k}^{-1} < \ldots < \mu_{1,i,qi}^{-1}$. Similarly, in each post-threshold area k of the PCT2 traffic of class j, traffic stream of class j is offered, defined by a set of parameters $\{\lambda_{2,j}, t_{2,j,k}, \mu_{2,j,k}\}$, and while $t_{2,j,0} > t_{2,j,1} > \ldots > t_{2,j,k} > \ldots > t_{2,j,qj}$ and $\mu_{2,j,0}^{-1} < \mu_{2,j,1}^{-1} < \ldots < \mu_{2,j,k}^{-1} < \ldots < \mu_{2,j,qj}^{-1}$. Such a choice of parameters implies that, along with an increase in the load of a group, the number of BBUs allocated to calls of particular classes decreases while, the mean service time of the calls increases.

11.4.4.2 Algorithm for Determining Occupancy Distribution

In this section we shall present a modified algorithm of a multi-service Erlang-Engset distribution (Section 7.5.3) for determining occupancy distribution in the threshold model of a cell servicing multi-rate traffic of the type PCT1 and PCT2. The algorithm under consideration – as in Section 7.5.3 – will be presented in four steps.

1. Determination of occupancy distribution $[P_n]_V$ with the assumption that the offered traffic streams are PCT1 streams. This distribution will be determined on the basis of the recursive formula (7.75), in which it is adopted that $A_{2,j,k} = N_{2,j}\alpha_{2,j,k}$:

$$n\,[P_n]_V = \sum_{i=1}^{M_1}\sum_{k=0}^{qi} A_{1,i,k}t_{1,i,k}\sigma_{1,i,k}(n - t_{1,i,k})\left[P_{n-t_{1,i,k}}\right]_V + \sum_{j=1}^{M_2}\sum_{k=0}^{qj} N_{2,j}\alpha_{2,j,k}t_{2,j,k}\sigma_{2,j,k}(n - t_{2,j,k})\left[P_{n-t_{2,j,k}}\right]_V \quad (11.71)$$

where:

$$A_{1,i,k} = \lambda_{1,i}/\mu_{1,i,k} \quad (11.72)$$

$$\alpha_{2,j,k} = \gamma_{2,j}/\mu_{2,j,k} \quad (11.73)$$

and for each of the pair $\{i, k\}$, such that: $1 \leq i \leq M_1$ and $0 \leq k \leq q_i$, we have:

$$\sigma_{1,i,k}(n) = \begin{cases} 1 & \text{for } Q_{1,i,k} < n \leq Q_{1,i,k+1} \\ 0 & \text{for remaining } n \end{cases} \quad (11.74)$$

and for each pair $\{j, k\}$, such that: $1 \leq j \leq M_2$ and $0 \leq k \leq q_j$, we have:

$$\sigma_{2,j,k}(n) = \begin{cases} 1 & \text{for } Q_{2,j,k} < n \leq Q_{2,j,k+1} \\ 0 & \text{for remaining } n \end{cases} \quad (11.75)$$

In Equations (11.74) and (11.75) we assume that $Q_{1,i,0} = 0$ and $Q_{2,j,0} = 0$.

2. Determination of the number of occupied BBUs by calls of Engset streams in a given state of the post-threshold area. Once all probabilities $[P_n]_V$ of macrostates have been determined, the number of calls of the Engset stream of class j in the macrostate n, i.e. $y_{2,j}(n)$, can be determined on the basis of Equation (7.121) (Section 7.8.1), which, in the case currently being considered, will be rewritten, with traffic $A_{2,j,k} = N_{2,j}\alpha_{2,j,k}$ taken into account, in the following way:

$$y_{2,j,k}(n) = N_{2,j}\alpha_{2,j,k}\sigma_{2,j,k}(n - t_{2,j,k})\left[P_{n-t_{2,j,k}}\right]_V / [P_n]_V$$
$$\text{for} \quad 0 \le k \le q_j \quad \text{and} \quad Q_{2,j,k} + t_{2,j,k} < n \le Q_{2,j,k+1} + t_{2,j,k} \quad (11.76)$$

For n from beyond the interval adopted in Equation (11.76), the value of the parameter $y_{2,j,k}$ is equal to zero (the explanation of the determination intervals used in Equation (11.76) is included in Section 7.8.2).

3. Determination of occupancy distribution of macrostates $[P_n]_V$ with the assumption that the offered streams are PCT1 and PCT2 streams. Knowing the parameters $y_{2,j,k}(n)$, we can determine once again the distribution of macrostates $[P_n]_V$, this time taking into consideration the state-dependable PCT2 traffic. The distribution will be determined on the basis of the recursive formula (7.75), in which it is assumed that:

$$A_{2,j,k}(n) = \left[N_{2,j} - \sum_{z=0}^{k} y_{2,j,z}(n)\right]\alpha_{2,j,k} \tag{11.77}$$

The expression within the square bracket, that is $[N_{2,j} - \sum_{z=0}^{k} y_{2,j,z}(n)]$ in Equation (11.77), determines the number of free traffic sources PCT2 of class j in state n belonging to the post-threshold area k. The sum in this expression indicates that, in a given post-threshold area, k calls of a given class from the preceding post-threshold areas can also be serviced. Therefore, taking into consideration (11.77) in (7.75), we obtain:

$$\begin{aligned} n\,[P_n]_V &= \sum_{i=1}^{M_1}\sum_{k=0}^{qi} A_{1,i,k}t_{1,i,k}\sigma_{1,i,k}(n - t_{1,i,k})\left[P_{n-t_{1,i,k}}\right]_V \\ &+ \sum_{j=1}^{M_2}\sum_{k=0}^{qj}\left[N_{2,j} - \sum_{z=0}^{k} y_{2,j,z}(n)\right]\alpha_{2,j,k}t_{2,j,k}\sigma_{2,j,k}(n - t_{2,j,k})\left[P_{n-t_{2,j,k}}\right]_V \end{aligned} \tag{11.78}$$

4. Calculation of blocking probabilities and loss probabilities in PCT1 and PCT2 call streams.

Erlang stream of class i:

$$E_{1,i} = B_{1,i} = \sum_{n=V-t_{1,i,qi}+1}^{V} [P_n]_V \tag{11.79}$$

Engset stream of class j:

$$E_{2,j} = \sum_{n=V-t_{2,j,qj}+1}^{V} [P_n]_V \tag{11.80}$$

In each state n, free PCT2 traffic sources of class j in a post-threshold area k generate a call stream with the intensity: $[N_{2,j} - \sum_{z=0}^{k} y_{2,j,z}(n)]\gamma_{2,j,k}$. The loss probability can therefore be written in the following way:

$$B_{2,j} = \frac{\sum\limits_{n=V-t_{2,j,qj}+1}^{V} [P_n]_V \left[N_{2,j} - \sum\limits_{z=0}^{qj} y_{2,j,z}(n) \right] \gamma_{2,j}}{\sum\limits_{k=0}^{qj} \sum\limits_{n=Q_{2,j,k}+1}^{Q_{2,j,k+1}} [P_n]_V \left[N_{2,j} - \sum\limits_{z=0}^{k} y_{2,j,z}(n) \right] \gamma_{2,j}} \tag{11.81}$$

where $Q_{2,j,0} = 0$ and $Q_{2,j,qj+1} = V$.

11.4.4.3 Characteristics of the Radio Interface

All traffic characteristics – the blocking probability, loss probability, carried and offered traffic – result from the approximate algorithm of the multi-service Erlang-Engset distribution presented in Section 11.4.4.2, hereafter referred to as the algorithm. Thus, the appropriate equations will be rewritten with the condition (11.15) and the notation used taken into consideration.

- Basic bandwidth unit (Equation (11.13)):

$$L_{\text{BBU}} = \text{GCD}(L_{1,1,0}, \ldots, L_{1,1,q_i}, \ldots, L_{1,M_1,0}, \ldots, L_{1,M1,q_{M_1}}, \ldots$$
$$\ldots, L_{2,1,0}, \ldots, L_{2,1,q_j}, \ldots, L_{2,M_2,0}, \ldots, L_{2,M_2,q_{M2}}) \tag{11.82}$$

- Equivalent capacity of the system in BBUs for the uplink (Equation (11.28)):

$$V = V_{\text{UL}} = \left\lfloor \frac{\eta}{L_{\text{BBU}}(1+\delta)} \right\rfloor \tag{11.83}$$

- Equivalent capacity of the system in BBUs for the downlink direction (Equation (11.29)):

$$V = V_{\text{DL}} = \left\lfloor \frac{\eta}{L_{\text{BBU}}(1-\xi+\delta)} \right\rfloor \tag{11.84}$$

- Number of BBUs demanded by a PCT1 call of class i in the post-threshold area k (Equation (11.15)):

$$t_{1,i,k} = \left\lceil \frac{L_{1,i,k}}{L_{\text{BBU}}} \right\rceil \tag{11.85}$$

- Number of BBUs demanded by a PCT2 call of class j in the post-threshold area k (Equation (11.15)):

$$t_{2,j,k} = \left\lceil \frac{L_{2,j,k}}{L_{\text{BBU}}} \right\rceil \tag{11.86}$$

- Occupancy distribution:
  Occupancy distribution $[P_n]_V$ is determined on the basis of step 3 of the algorithm (Equation (11.77)).
- The intensity of the PCT1 traffic of class i, offered to the interface in the post-threshold area k $(0 \leq k \leq q_i)$:

$$A_{1,i,k} = \lambda_{1,i}/\mu_{1,i,k} \tag{11.87}$$

- The intensity of the PCT2 traffic of class j, offered to the interface in the post-threshold area k $(0 \leq k \leq q_j)$:

$$\alpha_{2,j,k} = \gamma_{2,j}/\mu_{2,j,k} \tag{11.88}$$

- Blocking/loss probability of PCT1 calls of class i:
  Blocking probability $E_{1,i}$ and loss probability $B_{1,i}$ of PCT1 calls will be determined on the basis of step 4 of the algorithm (Equation (11.79)):

$$E_{1,i} = B_{1,i} = \sum_{n=V-t_{1,i,qi}+1}^{V} [P_n]_V \tag{11.89}$$

- Blocking probability of PCT2 calls of class j:
  Blocking probability of PCT2 calls $E_{2,j}$ is determined on the basis of step 4 of the algorithm (Equation (11.80):

$$E_{2,j} = \sum_{n=V-t_{2,j,qj}+1}^{V} [P_n]_V \tag{11.90}$$

- Loss probability for PCT2 calls of class j:
  Loss probability $B_{2,j}$ is determined on the basis of step 4 of the algorithm (Equation (11.81)):

$$B_{2,j} = \frac{\sum\limits_{n=V-t_{2,j,qj}+1}^{V} [P_n]_V \left[N_{2,j} - \sum\limits_{z=0}^{qj} y_{2,j,z}(n) \right] \gamma_{2,j}}{\sum\limits_{k=0}^{qj} \sum\limits_{n=Q_{2,j,k}+1}^{Q_{2,j,k+1}} [P_n]_V \left[N_{2,j} - \sum\limits_{z=0}^{k} y_{2,j,z}(n) \right] \gamma_{2,j}} \tag{11.91}$$

  where $Q_{2,j,0} = 0$ and $Q_{2,j,qj+1} = V$. The parameter $y_{2,j,z}(n)$ is the number of calls of class j serviced in state n of the post-threshold area z, determined in item 2 of the algorithm (Equation (11.76)).
- Intensity of the PCT1 traffic of class i carried in the interface in the post-threshold area k $(0 \leq k \leq qi)$:

$$Y_{1,i,k} = \sum_{z=0}^{k} \sum_{n=Q_{1,i,k}+1}^{\Phi_1} [P_n]_V \, y_{1,i,z}(n) \tag{11.92}$$

where:

$$\Phi_1 = \begin{cases} 0 & \text{for} \quad Q_{1,i,k} + 1 > Q_{1,i,z} + t_{1,i,z} \\ Q_{1,i,z} + t_{1,i,z} & \text{for} \quad Q_{1,i,k} + 1 \le Q_{1,i,z} + t_{1,i,z} < Q_{1,i,k+1} \\ Q_{1,i,k+1} & \text{for} \quad Q_{1,i,z} + t_{1,i,z} \ge Q_{1,i,k+1} \end{cases} \tag{11.93}$$

whereas, service streams are determined on the basis of the (7.121):

$$y_{1,i,k}(n) = A_{1,i,k}\,\sigma_{1,i,k}(n - t_{1,i,k})\left[P_{n-t_{1,i,k}}\right]_V / [P_n]_V$$
$$\text{for} \quad 0 \le k \le q \quad \text{and} \quad Q_{1,i,k} + t_{1,i,k} < n \le Q_{1,i,k+1} + t_{1,i,k} \tag{11.94}$$

- Total intensity of the PCT1 traffic of class i carried in the interface:

$$Y_{1,i} = \sum_{k=0}^{Q_{1,i,k}} Y_{1,i,k} \tag{11.95}$$

- Intensity of the PCT2 traffic of class j carried by the interface in the post-threshold area k $(0 \le k \le qj)$:

$$Y_{2,j,k} = \sum_{z=0}^{k} \sum_{n=Q_{2,j,k}+1}^{\Phi_2} [P_n]_V\, y_{2,j,z}(n) \tag{11.96}$$

where:

$$\Phi_2 = \begin{cases} 0 & \text{for} \quad Q_{2,j,k} + 1 > Q_{2,j,z} + t_{2,j,z}, \\ Q_{2,j,z} + t_{2,j,z} & \text{for} \quad Q_{2,j,k} + 1 \le Q_{2,j,z} + t_{2,j,z} < Q_{2,j,k+1} \\ Q_{2,j,k+1} & \text{for} \quad Q_{2,j,z} + t_{2,j,z} \ge Q_{2,j,k+1} \end{cases} \tag{11.97}$$

whereas service streams are determined on the basis of Equation (11.77):

$$y_{2,j,k}(n) = A_{2,j,k}\,\sigma_{2,j,k}(n - t_{2,j,k})\left[P_{n-t_{2,j,k}}\right]_V \Big/ [P_n]_V$$
$$\text{for} \quad 0 \le k \le q \quad \text{and} \quad Q_{2,j,k} + t_{2,j,k} < n \le Q_{2,j,k+1} + t_{2,j,k} \tag{11.98}$$

- Total intensity of the PCT2 traffic of class j carried by the interface:

$$Y_{2,j} = \sum_{k=0}^{Q_{2,j,k}} Y_{2,j,k} \tag{11.99}$$

- Intensity of the traffic carried by one BBU of the radio interface in the cell:

$$y_V = \left(\sum_{i=1}^{M_1} Y_{1,i} + \sum_{j=1}^{M_2} Y_{2,j} \right) \Big/ V \tag{11.100}$$

- Intensity of PCT1 lost traffic of class i in the interface:

$$R_{1,i} = A_{1,i} - Y_{1,i} \tag{11.101}$$

- Intensity of the PCT2 lost traffic of class i in the interface:

$$R_{2,j} = A_{2,j} - Y_{2,j} \tag{11.102}$$

11.4.5 Priorities in the Radio Interface

Each of the radio interfaces presented in (Sections 11.4.1, 11.4.2, 11.4.3 and 11.4.4) can service a mixture of multi-rate traffic streams that are characterized by different priorities. Priorities can be introduced both for services offered to a given cellular network and for a group of subscribers that use a given service in a cellular network with soft or hard capacity. Let us consider a radio interface that services a mixture of multi-service traffic generated by a Poisson call stream (PCT1 traffic).

11.4.5.1 Description of the Model

Let us assume that the radio interface of a given cell services M traffic classes offered to the interface by G different users group. Figure 11.13 presents the relation between a number of classes and the number of group of users. In the figure, the grey rectangle indicates which traffic class is represented in each group of users. The model also assumes:

- each traffic class is characterized by a different priority: the first class has the highest priority and the last class (M) has the lowest priority;
- each group of users is designated by a different priority: the first group has the highest priority and the last group (G) is characterized by the lowest priority;
- each traffic class in each group of users has a different priority: the higher group priority the higher class priority;
- each group of users does not have to consist of users representing all traffic classes; each group of users consist of users equipped with different types of terminal and only some of them are capable of servicing all traffic classes.

A result of the adopted assumptions each traffic class in each group forms a flat priority hierarchy – the first class has the highest priority in the first group of users, whereas class M

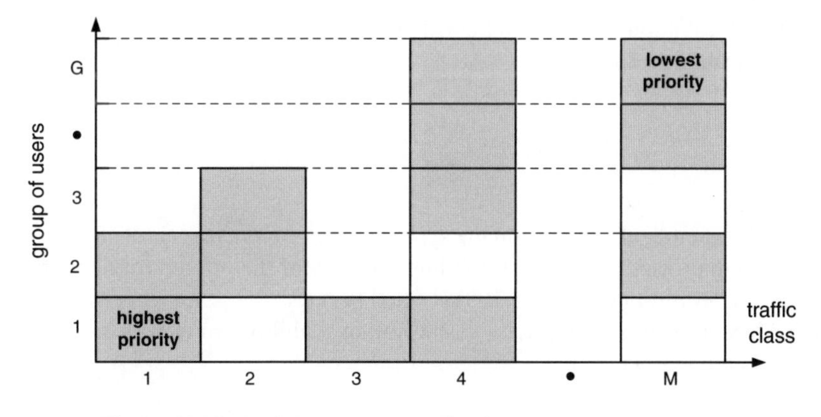

Figure 11.13 Relation between traffic classes and groups of users.

in the group G has the lowest priority (Figure 11.13). The traffic of class j ($j \in \{1, \ldots, M\}$) belonging to the group of users i ($i \in \{1, \ldots, G\}$) has priority z with respect to all classes and all group of users:

$$z = M(i-1) + j \tag{11.103}$$

Equation (11.103) transforms a two-dimensional description of priorities into a one-dimensional description, which is convenient in considerations of the model.

This phenomenon means that the radio interface of the cell can be modeled by the full-availability group carrying multi-rate traffic with priorities presented in Section and all traffic characteristics – blocking probability, carried and offered traffic – result from this model. In the algorithm the following formulas will be used:

- Basic bandwidth unit (Equation (11.13)):

$$L_{\mathrm{BBU}} = \mathrm{GCD}(L_{1,1}, L_{1,2}, \ldots, L_{1,M_1}) \tag{11.104}$$

- Equivalent capacity of the system in BBUs for the uplink (Equation (11.28)):

$$V = V_{\mathrm{UL}} = \left\lfloor \frac{\eta}{L_{\mathrm{BBU}}(1+\delta)} \right\rfloor \tag{11.105}$$

- Equivalent capacity of the system in BBUs for the downlink (Equation (11.29)):

$$V = V_{\mathrm{DL}} = \left\lfloor \frac{\eta}{L_{\mathrm{BBU}}(1-\xi+\delta)} \right\rfloor \tag{11.106}$$

- Number of BBUs demanded by a class i call (Equation (11.15)):

$$t_{1,i} = \left\lceil \frac{L_{1,i}}{L_{\mathrm{BBU}}} \right\rceil \tag{11.107}$$

- Occupancy distribution in the system without priorities and with z traffic classes (Section 7.10, Equation (7.146)):

$$n\,[P_n]_V = \sum_{i=1}^{z} A_{1,i} t_{1,i} \left[P_{n-t_{1,i}}\right]_V \tag{11.108}$$

  where z is the number of priorities in the system. When a particular class of calls j does not exist in i group of users, then corresponding intensity of the offered traffic with subscript $z = M_1(i-1) + j$ in Equation (11.108) is equal to zero.
- The blocking probability for class z in the full-availability group with priorities can be expressed as follows [11, 12]:

$$[E_{1,z}]_z = f((A_{1,1}, t_{1,1}), (A_{1,2}, t_{1,2}), \ldots, (A_{1,z}, t_{1,z})) \tag{11.109}$$

where $z = M_1(i-1) + j$, M_1 is the maximum number of traffic classes carried by the interface, i is the number of group of users and j is the sequential number of class in group i. Function (11.109) can be determined in accordance with the algorithm presented in the next chapter.

- Class i traffic intensity carried in the cell:

$$Y_{1,i} = A_{1,i}\left[1 - E_{1,i}\right] \tag{11.110}$$

- The mean load of one BBU in the cell:

$$y_V = \sum_{i=1}^{M_1} Y_{1,i} \Big/ V \tag{11.111}$$

11.4.5.2 Calculation Algorithm

In the full-availability group model considered here, the blocking probability for each traffic class in the system with z different priorities can be expressed as an $z-1$ step iterative calculation algorithm (Section 7.10):

1. Setting up the initial value of current calculation priority $z = MG$.
2. Determination of the occupancy distribution in the full-availability group with z traffic classes (Section 7.10, Equation (7.146)):

$$n\left[P_n\right]_V = \sum_{i=1}^{z} A_{1,i} t_{1,i} \left[P_{n-t_{1,i}}\right]_V \tag{11.112}$$

3. Calculation of the values of blocking probabilities in the full-availability group with z traffic classes (Section 7.10, Equation (7.146)):

$$\forall_{1 \le i \le z}\ [E_i]_z = \sum_{n=V-t_{1,i}+1}^{V} \left[P_n\right]_V \tag{11.113}$$

4. Decreasing the value of z: $z = z - 1$, and if $z > 0$, return to the step 2.
5. Designation of the values of the blocking probabilities for all traffic classes in the full-availability group with priorities (Section 7.10, Equation (7.161)):

$$[E_z]_z^P = \frac{A_{1,z} t_{1,z} [E_z]_z + \sum_{i=1}^{z-1} A_{1,i} t_{1,i}\left([E_i]_z - [E_i]_{z-1}\right)}{A_{1,z} t_{1,z}} \tag{11.114}$$

where $[E_z]_z^P$ is the blocking probability of class z in the system with priorities that services z traffic classes and $[E_z]_z$ is the blocking probability of class z in the system without priorities that also services z traffic classes (Equation (11.113)).

A system with two priorities is considered at each step of the algorithm. In the first step of the algorithm, the calls of the lowest priority are considered, under the assumption that the remaining classes of calls $(1, \ldots, z-1)$ have a higher priority and push out the calls of class z. In the successive steps of the algorithm, traffic of class $z-1$ has the lowest priority, while the remaining traffic classes $(1, \ldots, z-2)$ have a higher priority and can push out traffic of class $z-1$. The calculation process is repeated until, in the last step of the algorithm, the system services only two highest priority classes.

11.5 High-Speed Packet Data Transmission (HSPA) Traffic in the Radio Interface of the UMTS Network

High-speed packet data transmission (HSPA) traffic in the radio interface of the cell in the UMTS mobile network is serviced in the downlink (HSDPA) and in the uplink (HSUPA) direction (Section 2.8). In HSUPA each user has a dedicated logical channel, while HS-DPA users share common radio interface resources. In the mobile networks there are two main techniques used for introducing HSDPA to the radio interface. In the first method, R99 and HSDPA users are serviced by the common resources of the radio interface. The second method assumes that the R99 and HSDPA users are serviced by the separated resources of the radio interface.

11.5.1 Description of the Model

Analytical models for the HSUPA and HSDPA (in both types) are based on the model of the radio interface with soft capacity servicing multi-rate models presented in Section 11.4 for R99 traffic streams. In the calculation algorithm we discussed an example in which R99 and HSPA traffic classes are carried by common resources of the radio interface.

11.5.2 Calculation Algorithm

The model assumed a radio interface carrying a mixture of M_1 classes of the PCT1 traffic and M_2 classes of the PCT2 traffic offered by R99 and group of HSDPA users. The basic model for such a radio interface is a full-availability group servicing multi-rate traffic with the capacity determined by Equation (11.14). All traffic characteristics for R99 traffic – blocking probability, loss probability, carried and offered traffic – are obtained from the approximate calculational algorithm of the multi-service Erlang-Engset distribution (Section 7.5.3), referred to here as the algorithm. Since the cell is a full-availability system, servicing simultaneously streams of PCT1 and PCT2, we then assume in the algorithm:

$$\begin{cases} M_1 \neq 0 \\ M_2 \neq 0 \end{cases} \tag{11.115}$$

$$\forall_{i\in\{1,\ldots,M_1\}} \quad \forall_{0\leq n\leq V-t_{1,i}} \sigma_{1,i}(n) = 1 \tag{11.116}$$

and

$$\forall_{j\in\{1,\ldots,M_2\}} \quad \forall_{0\leq n\leq V-t_{2,j}} \sigma_{2,j}(n) = 1 \tag{11.117}$$

With such assumptions, the corresponding characteristics of the system can be determined on the basis of the following equations:

- Basic bandwidth unit (Equation (11.13)):

$$L_{\mathrm{BBU}} = \mathrm{BBU}(L_{1,1}, \ldots, L_{1,M1},\ L_{2,1}, \ldots, L_{2,M2}) \tag{11.118}$$

- Equivalent capacity of the system in BBUs for the uplink (Equation (11.28)):

$$V = V_{\mathrm{UL}} = \left\lfloor \frac{\eta}{L_{\mathrm{BBU}}(1+\delta)} \right\rfloor \tag{11.119}$$

- Equivalent capacity of the system in BBUs for the downlink direction (Equation (11.29)):

$$V = V_{\mathrm{DL}} = \left\lfloor \frac{\eta}{L_{\mathrm{BBU}}(1-\xi+\delta)} \right\rfloor \tag{11.120}$$

- Number of BBUs demanded by a PCT1 call of class i (Equation (11.15)):

$$t_{1,i} = \lceil L_{1,i}/L_{\mathrm{BBU}} \rceil \tag{11.121}$$

- Number of BBUs demanded by a PCT2 call of class j (Equation (11.15)):

$$t_{2,j} = \lceil L_{2,j}/L_{\mathrm{BBU}} \rceil \tag{11.122}$$

- Occupancy distribution $[P_n]_V$ is determined after step 3 of the algorithm (Equation (7.79)):

$$n\,[P_n]_V = \sum_{i=1}^{M_1} A_{1,i} t_{1,i} \left[P_{n-t_{1,i}}\right]_V + \sum_{i=1}^{M_2} A_{2,i}(n - t_{2,i}) t_{2,i} \left[P_{n-t_{2,i}}\right]_V \tag{11.123}$$

- Blocking/loss probability for PCT1 calls of class i:
  Blocking probability $E_{1,i}$ and loss probability $B_{1,i}$ of PCT1 calls will be determined after step 4 of the algorithm (Equation (7.81)):

$$E_{1,i} = B_{1,i} = \sum_{n=V-t_{1,i}+1}^{V} [P_n]_V \tag{11.124}$$

- Blocking probability of PCT2 calls of class j:
  Blocking probability of PCT2 calls $E_{2,i}$ will be determined after item 4 of the algorithm (Equation (7.81)):

$$E_{2,j} = \sum_{n=V-t_{2,j}+1}^{V} [P_n]_V \tag{11.125}$$

- Loss probability of PCT2 calls of class j:

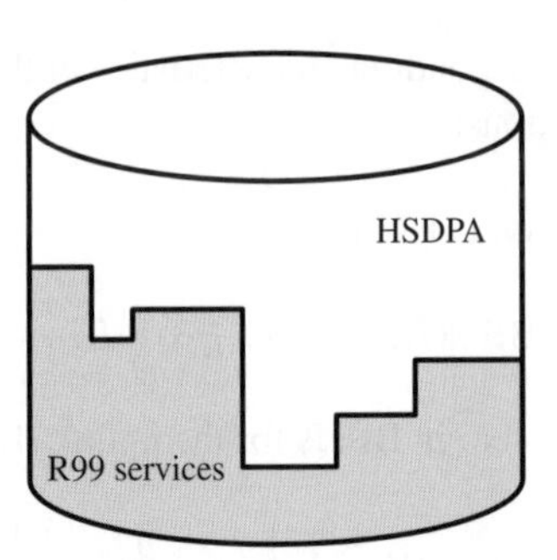

Figure 11.14 The capacity division in the HSDPA radio interface.

Loss probability $B_{2,j}$ will be determined on the basis of step 4 of the algorithm (Equation (7.82)):

$$B_{2,j} = \frac{\sum_{n=V-t_{2,j}+1}^{V} [P_n]_V \, [N_{2,j} - y_{2,j}(n)]\gamma_{2,j}}{\sum_{n=0}^{V} [P_n]_V \left[N_{2,j} - y_{2,j}(n)\right] \gamma_{2,j}} \tag{11.126}$$

where $y_{2,j}(n)$ is the number of serviced calls of class j in state n, determined in item 2 of the algorithm (Equation (7.78)).

If we assume that, in the downlink direction, R99 services have higher priority than HSDPA services and only those resources of the interface that are not used by R99 users can be offered to HSDPA users (Figure 11.14) than each new R99 call limits the bandwidth available for HSDPA users. On the basis of the occupancy distribution we can determine the mean value of resources available for HSDPA users. In the first step we calculate the mean number of serviced calls of individual traffic classes in state n (Section 7.2.10, Equation (7.27)):

$$y_{1,i}(n) = \frac{A_{1,i}[P_{n-t_{1,i}}]_V}{[P_n]_V} \tag{11.127}$$

$$y_{2,j}(n) = \frac{A_{2,j}[P_{n-t_{2,j}}]_V}{[P_n]_V} \tag{11.128}$$

Having the knowledge of the average number of calls $y_{1,i}(n)$ and $y_{2,j}(n)$ of each of the traffic classes, it is possible, for each state n, to determine the bandwidth (the number of available BBUs) that can be used by users of HSDPA as the difference between the total capacity of the radio interface and the number of BBUs occupied by calls. The mean value of resources (expressed in the number of BBUs) offered to HSDPA calls is equal [13, 14]:

$$T_x = \sum_{n=0}^{V} \left[V - \sum_{i=1}^{M_1} y_{1,i}(n)t_{1,i} - \sum_{j=1}^{M_2} y_{2,j}(n)t_{2,j} \right] [P_n]_V \tag{11.129}$$

The link capacity T_x can be available to one user only (then it will reach the highest transmission speed), or can be shared by many users of HSDPA.

11.6 Comments

Radio interface characteristics can be determined on the basis of the models proposed in this chapter. A choice of a particular model depends on the structure of the offered traffic (for example, single-service traffic / multi-service traffic; limited and/or unlimited number of subscribers) and on the technology employed by a system (GSM and/or UMTS). The technology of a mobile system also influences the choice of the model with hard or soft capacity. A detailed comparison of the models, along with the conditions of their applications is presented in Tables A.1 and A.2.

References

[1] Roberts, J.W. (ed.) (1992) *Performance Evaluation and Design of Multi-Service Networks, Final Report COST 224*, Commission of the European Communities.

[2] Roberts, J.W., Mocci, V., and Virtamo, I. (eds) (1996) *Broadband Network Teletraffic, Final Report of Action COST 242*, Commission of the European Communities, Springer.

[3] Holma, H. and Toskala, A. (2000) *WCDMA for UMTS. Radio Access for Third Generation Mobile Communications*, John Wiley & Sons, Ltd.

[4] Laiho, J., Wacker, A., Novosad, T. (2006) *Radio Network Planning and Optimization for UMTS*, John Wiley & Sons, Ltd.

[5] Stasiak, M., Wiśniewski, A., and Zwierzykowski, P. (2004) *Blocking Probability Calculation in the Uplink Direction for Cellular Systems with WCDMA Radio Interface*. Third Polish-German Teletraffic Symposium, Dresden, September VDE Verlag GMBH, Berlin.

[6] Engineering Services Group (2007) Aspects of HSUPA Network Planning. Technical Report No. 80-W1159-1, Revision B, Qualcomm Incorporated, San Diego.

[7] Faruqe, S. (1997) *Cellular Mobile Systems Engineering*, Artech House.

[8] Kaufman, J.S. (1992) Blocking with retrials in a completely shared resource environment. *Journal of Performance Evaluation*, **15**, 99–113.

[9] Moscholious, I.D., Logothetis, M.D., and Kokkinakis, G.K. (2002) Connection-dependent threshold model: a generalization of the Erlang multiple rate loss model. *Journal of Performance Evaluation*, **48** (1–4), 177–200.

[10] Moscholios, I.D., Logothetis, M.D., and Nikolaropoulos, P.I. (2005) Engset multi-rate state-dependent loss models. *Journal of Performance Evaluation*, **59** (2–3), 247–77.

[11] Stasiak, M., Wiewióra, J., and Zwierzykowski, P. (2009) *Analytical Model of the WCDMA Radio Interface in UMTS Network with User Differentiation*. Proceedings of 5th Advanced International Conference on Telecommunications, Venice, Italy, May IEEE, Venice.

[12] Stasiak, M., Wiewióra, J., and Zwierzykowski, P. (2009) *WCDMA Interface in UMTS Network Carrying a Mixture of Multi-Rate Traffic with Different Priorities*. Proceedings of International Conference on Telecommunications, Marrakech, Maroco, May IEEE, Marrakech.

[13] Głąbowski, M. Hanczewski, S. and Stasiak M. (2007) Calculation of Available Bandwidth for UMTS-HSDPA/HSUPA Users. *Proceedings of Eurocon 2007*, Warsaw, September IEEE, Warsaw.

[14] Głąbowski, M., Hanczewski, S., and Stasiak, M. (2007) *Available Bandwidth for HSDPA and HSUPA Services*. Proceedings of 14th IEEE International Conference On Telecommunications, Penang, May.

12

Modeling and Dimensioning of the Iub Interface

12.1 Introduction

Bearing in mind the time needed for network expansion and the huge costs involved, as well as the potential for cost savings, operators of cellular networks are inclined to implement technological solutions that optimize investments but, nevertheless, make it possible to retain the complexity and quality of service.

The dimensioning process for the third-generation Universal Mobile Telecommunications System (UMTS) should make it possible for individual elements of the system to secure the required grade of service (GoS) given the assumed load on the system.

One of the most important elements of the architecture of the UMTS system is the radio access network. The functions of UMTS Terrestrial Radio Access Network (UTRAN) are performed by two basic elements: by the Node B base station and by the Radio Network Controller (RNC) connected by the Iub interface (Figure 12.1). The present chapter proposes methods for dimensioning of the Iub interface.

12.2 Example Architecture of the Iub Interface

In the Iub interface the mobile operators used asynchronous transfer mode (ATM) technology or Internet protocol (IP). Currently, the most commonly used transport technology in the Iub interface is ATM (Section 2.2, Figure 2.4). The possibility of introducing the IP protocol in the Iub interface has also been considered in subsequent versions of the UMTS system (Section 2.2, Figure 2.6) and in the LTE system (Section 3.2). This chapter presents an exemplary architecture based on the application of the ATM technology in the Iub interface. Inverse Multiplexing for ATM (IMA) technology [1] has been widely used for this interface. This makes it possible to create several logical ATM paths on the basis of separate physical links.[1] This solution

[1]For example E1/T1 digital time division multiplexing (TDM) links.

Modeling and Dimensioning of Mobile Networks: From GSM to LTE
Maciej Stasiak, Mariusz Głąbowski, Arkadiusz Wiśniewski and Piotr Zwierzykowski

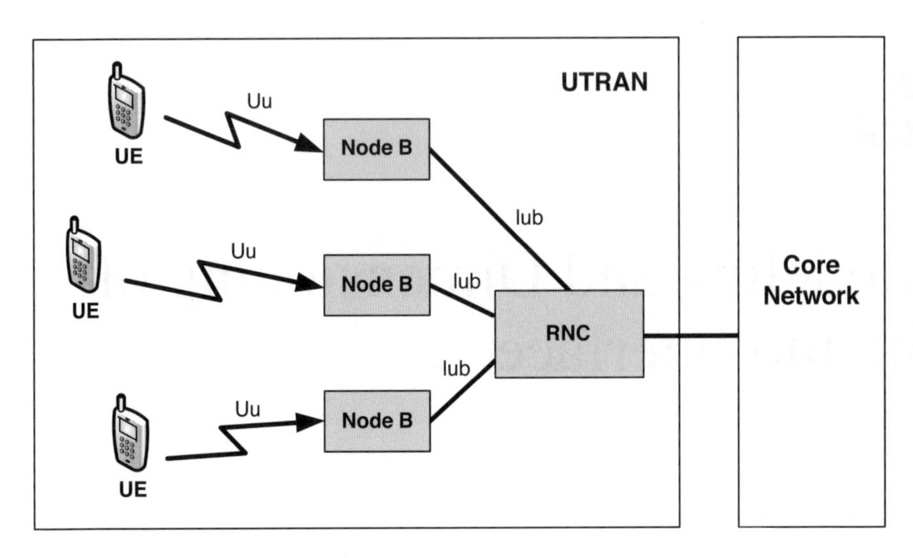

Figure 12.1 Access part of the UMTS network.

provides a more efficient use of resources (in accordance with the group conservation principle, Section 5.2.9). It is worth emphasizing that the use of IMA has also been a result of the introduction of HSPA technology to mobile, networks which require the capacity of link above 2 Mbps. So the example interface architecture has assumed a separation of E1[2] physical links on the Iub interface. The solution (IMA) enables the creation of one, or many, virtual paths (VPs), and, within each of them, one or many virtual channels (VCs). The decision on the number of paths and channels is solely that of the producers of the hardware and software and depends on the configuration principles and guidelines adopted by a given operator of the cellular network. Most frequently, one or two VPs are defined on the Iub interface. Where on VP is defined, virtual channels are further defined within this VP. Operators of mobile networks usually define at least two VCs for the so-called constant bit rate (CBR) data and the so-called unspecified bit rate (UBR) data.[3] The latter solution has been adopted in the case being considered here (Figure 12.2). It should be noticed, however, that it is not necessary to assign particular services to CBR and UBR data types. In particular cases, services that are not characterized by a constant packet generation rate (according to the definition of CBR) are treated by the operator as CBR services. This has been the case with the architecture presented in Figure 12.2.

Table 12.1 shows an example of UMTS packet switched (PS) and circuit switched (CS) services carried out by logical ATM paths dedicated to servicing respectively *unspecified bit rate* traffic and *constant bit rate* traffic (Figure 12.2).

The architecture of the Iub interface presented in this chapter assumes a fixed distribution of resources in the interface between *constant bit rate* (such as services related to Release 99 [2]) and *unspecified bit rate* (such as services related to HSPA [3]) traffic classes. In this case the algorithm presented in Section 12.3.1, can be used for dimensioning the Iub interface. The

[2]E1 is an interface at 2 Mbps in the plesiochronous digital hierarchy (PDH), which is in common use in Europe [4].

[3]Sometimes so-called unspecified bit rate plus data (UBR+) is specified which enables the operator to define the minimum guaranteed bit rate for a given type of service.

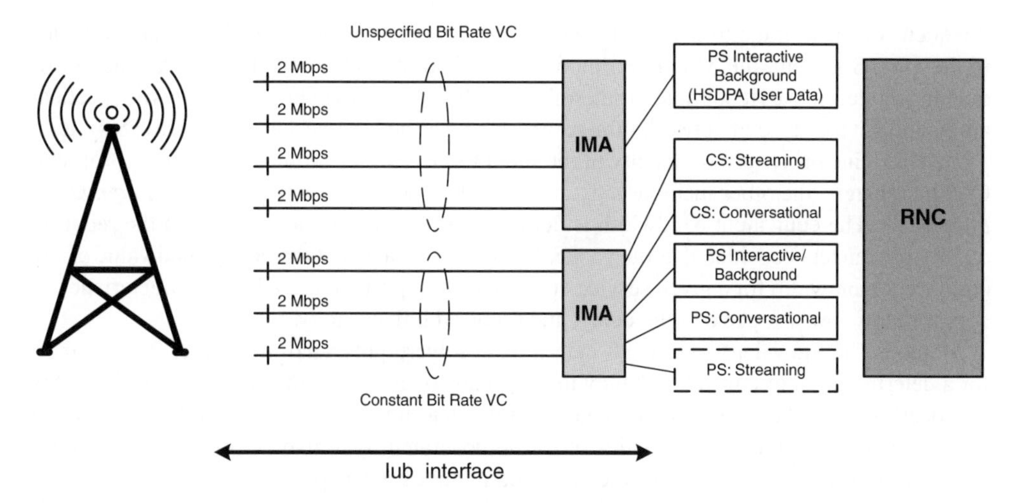

Figure 12.2 One of the most common methods for establishing a connection between the UMTS base station and Radio Network Controller is the one applying IMA technology.

Table 12.1 An example of service class mapping into ATM classes

ATM class of service	UMTS class of service	Example service
Unspecified bit rate	Interactive background (HSDPA user data)	WWW
Constant bit rate	CS: conversational	Speech
Constant bit rate	CS: streaming	Modem connection
Constant bit rate	PS: interactive/background	FTP, realtime gaming
Constant bit rate	PS: conversational	Speech (VoIP)
Constant bit rate	PS: streaming	Mobile TV

This kind of mapping is popular in ATM-based core of mobile networks.

algorithm makes it possible to dimension the constant bit rate VC and unspecified bit rate VC paths separately.

It should be noticed that the solutions typically adopted by hardware suppliers for the UMTS network also include architectures in which priorities for individual services are introduced (Section 12.3.2) and solutions in which Release 99 and HSPA-based services are serviced by shared resources of the interface (Section 12.3.3)[4] [5, 6].

12.3 Modeling of the Iub Interface

Two methods can be used to define the traffic characteristics of the Iub interface. The first uses queuing models to analyze and determine, for example, the average number of packets in

[4]In former technological solutions, the Iub interface also used dynamic distribution of resources between R99 and HSDPA-related services [7]. These solutions, however, have only historical relevance and are not therefore presented in this chapter.

the queue or the average time spent by a call in the queue. To determine these characteristics in the UMTS system, the M/G/R PS model, described in Section 10.10, is used. The use of queuing models makes it possible to determine the characteristics of buffers, but has no practical application at the stage of network dimensioning, or for the Iub interface in the UMTS system.

To determine the required capacity of the Iub interface, which would ensure the appropriate QoS parameters, the other methodology is used – the one based on the concept of *equivalent bandwidth*. The equivalent bandwidth is determined for each service class with the required quality parameters considered at the packet level. This means that the determined value of the equivalent bandwidth for a given service ensures such important operational characteristics as appropriately low lost packet rate, delay, and required link capacity.

Methods for equivalent bandwidth determination were presented in Chapter 9. A method for a determination of equivalent bandwidth depends on a version of the UMTS system. This approach assumes the application of a method for a determination of the equivalent bandwidth that corresponds to the characteristics of traffic generated by a given mobile network under consideration (for example, one of the proposed methods in Chapter 9).

The determined equivalent bandwidth values, expressed in kbps, enable us to apply the full-availability group model with multi-rate traffic to the process of modeling and dimensioning the Iub interface. The method proposed for modeling and dimensioning of the Iub interface presented in the chapter is based on this approach. The following sections propose algorithms which can be used in the process of dimensioning and optimizing the Iub interface in the UMTS network.

12.3.1 Basic Algorithm for Dimensioning of the Iub Interface

The defined values of the equivalent bandwidth, expressed in kbps, allow us to use the full-availability group with multi-rate traffic in the process of dimensioning the Iub interface. We will assume that the interface considered here will be treated as a system with hard capacity in which aggregated traffic streams are generated by an "unlimited" number of traffic sources. We will further assume that the border blocking probabilities for all serviced classes in the Iub interface are known. These incoming parameters will be designated by symbols $E_1^*, \ldots, E_M^*$. Following the above assumptions, the Iub interface dimensioning process can be expressed in the form of the following algorithm:

1. Determination of the equivalent bandwidth, expressed in kbps, for particular traffic classes (services), that is $C_{E,1}, C_{E,2}, \ldots, C_{E,M}$.[5]
2. Determination of the basic bandwidth unit (BBU) as the greatest common divisor of equivalent bandwidths:

$$C_{\text{BBU}} = \text{GCD}(C_{E,1}, C_{E,2}, \ldots, C_{E,M}) \tag{12.1}$$

3. The expression of equivalent bandwidths $C_{E,1}, C_{E,2}, \ldots, C_{E,M}$ of particular traffic classes in the number of BBUs – that is, the determination of the number of BBUs required by calls

[5]The choice of the method for the determination of the equivalent bandwidth depends on the types of services, for example taking into account whether they are prone to delay or packet losses. It is worth remembering here that the equivalent bandwidth can also be determined for the so-called self-similar traffic streams that are characterized by their extremely unfavorable influence on network systems, even with relatively small average values of the link capacity.

of particular traffic classes:

$$\forall_{i\in\{1,\ldots,M\}} t_i = \lceil C_{E,i}/C_{\mathrm{BBU}} \rceil \tag{12.2}$$

4. Determination of the initial value of the link capacity C in the first iteration:

$$C_0 = \sum_{i=1}^{M} C_{E,i} \quad \text{and} \quad C = C_0 \tag{12.3}$$

5. Setting up the capacity increment parameter ΔC.
A method for a determination of ΔC is related to the network dimensioning method adopted by the mobile operator.
6. A change of the current value of the capacity of the Iub interface C:

$$C = C + \Delta C \tag{12.4}$$

7. Determination of the equivalent capacity of the interface expressed in BBU:

$$V = \lfloor C/C_{\mathrm{BBU}} \rfloor \tag{12.5}$$

where C is the assumed interface capacity expressed in kbps.
8. Determination of the occupancy distribution in the system on the basis of the full-availability group model (Section 7.3.5, Equation (7.39)):

$$n[P_n]_V = \sum_{i=1}^{M} A_i t_i [P_{n-t_i}]_V \tag{12.6}$$

where A_i is the intensity of the offered traffic of class i.
9. Determination of the blocking and loss probability in the system on the basis of the state probabilities given by Equation (12.6):

$$E_i = \sum_{n=V-t_i+1}^{V} [P_n]_V \tag{12.7}$$

10. Repetition of steps 6–10 until we reach the minimum capacity of the Iub interface that guarantees the following condition to be fulfilled:

$$\forall_{1\le i\le M} \left(E_i \le E_i^*\right)$$

12.3.2 Dimensioning of the Iub Interface with Priorities

The Iub interface dimensioning algorithm presented in Section 12.3.1 can also be used for dimensioning the interface that services a mixture of different traffic classes with priorities. When this is the case, the method for determining the blocking probability for calls of individual classes will have to be changed.

Let us assume that the interface services a mixture of multi-rate traffic classes that are characterized by different priorities. This means that the Iub interface can be modeled by the full-availability group that services multi-rate traffic with priorities presented in Section 7.10.

In this model, the blocking probability for each traffic class in the system with z different priorities can be expressed as a $z-1$ step calculation algorithm that has replaced steps 8 and 9 in the basic algorithm (Section 12.3.1):

1. Setting up the initial value of z: $z = M$.
2. Determination of the occupancy distribution in the full-availability group for z traffic classes (Section 7.10, Equation (7.146)):

$$n\left[P_n\right]_V = \sum_{i=1}^{z} A_i t_i \left[P_{n-t_i}\right]_V \tag{12.8}$$

3. Determination of the values of blocking probabilities in the full-availability group for z traffic classes (Section 7.10, Equation (7.147)):

$$\forall_{1\le i\le z}\ \left[E_i\right]_z = \sum_{n=V-t_i+1}^{V} \left[P_n\right]_V \tag{12.9}$$

4. Decreasing the value of z: $z = z-1$, and if $z > 0$, return to the step 2.
5. Determination of the values of the blocking probabilities for all traffic classes in the full-availability group with priorities (Section 7.10, Equation (7.161)):

$$\left[E_z\right]_z^P = \frac{A_z t_z \left[E_z\right]_z + \sum_{i=1}^{h-1} A_i t_i \left(\left[E_i\right]_z - \left[E_i\right]_{z-1}\right)}{A_z t_z} \tag{12.10}$$

where $[E_z]_z^P$ is the blocking probability of class z in the system with priorities that services z traffic classes and $[E_z]_z$ is the blocking probability of class z in the system without priorities that also services z traffic classes (Equation (12.9)).

The system with two priorities is considered at each step of the algorithm. In the first step of the algorithm, class M calls with the lowest priority are considered, on the assumption that the remaining classes of calls $(1, \ldots, M-1)$ have a higher priority and push out the calls of class M. In subsequent steps of the algorithm, calls of class $M-1$ have the lowest priority, while the remaining call classes $(1, \ldots, M-2)$ have a higher priority and can push out calls of class $M-1$. The calculation process is repeated and in each step the total number of classes decreases. In the last step of the algorithm the system services only two classes, 1 and 2.

12.3.3 Dimensioning of the Iub Interface Carrying HSPA Traffic

Assume that the Iub interface carries both Release 99 and HSPA traffic streams. Also assume that there are traffic classes belonging to HSPA traffic with calls that can change occupied resources during service time. It is therefore assumed that the Iub interface services simultaneously a mixture of different multi-rate traffic classes, while these classes are divided into M_k classes whose calls can change requirements while being serviced, and M_{nk} classes, which do not change their demands in the service time.

The use of a given analytical model depends on particular mechanisms used in the solutions provided by vendors of equipment for the UMTS network. Let us consider a scenario in which the average bandwidth is assigned to all users unevenly. Let us further assume that the users in this network have different types of terminals and those subscribers that have newer terminals are capable of achieving higher maximum throughput. This scenario can be further considered using the model with uneven compression, presented in Section 7.9.

According to this model, the basic algorithm (Section 12.3.1) will undergo slight changes in the method for determining the occupancy distribution (step 9) and the blocking probability (step 9).

The occupancy distribution will be determined as in Section 7.9, Equation (7.133):

$$n\left[P_n\right]_V = \sum_{i=1}^{M_{nk}} A_i t_i \left[P_{n-t_i}\right]_V + \sum_{j=1}^{M_k} A_j t_{j,\min} \left[P_{n-t_{j,\min}}\right]_V \tag{12.11}$$

where $t_{j,\min}$ is the minimum number of BBUs demanded by a call of class j, which undergoes compression. The blocking probability will be calculated on the basis of the following formula (Section 7.9, Equation (7.134)):

$$E_i = \begin{cases} \sum\limits_{n=V-t_i+1}^{V} \left[P_n\right]_V & \text{if} \quad \text{class } i \text{ call belongs to } M_{nk} \text{ classes} \\ \\ \sum\limits_{n=V-t_{i,\min}+1}^{V} \left[P_n\right]_V & \text{if} \quad \text{class } i \text{ call belongs to } M_k \text{ classes} \end{cases} \tag{12.12}$$

After the dimensioning process, according to the basic algorithm, is complete, it is then possible to determine such link characteristics as the average throughput and available throughput (Section 12.3.1).

12.3.3.1 Average Throughput

Determination of the average throughput is only important for those classes of HSPA traffic[6] that can undergo compression.

In the first stage of the determination of the average throughput we determine the compression coefficient $\xi_i(n)$. The coefficient, following the dependence (7.142) (Section 7.9), takes on the following form:

$$\xi_i(n) = \begin{cases} K_{i,\max} & \text{for } \frac{V-Y_{nk}(n)}{n-Y_{nk}(n)} \geq K_{i,\max} \\ \\ \frac{V-Y_{nk}(n)}{n-Y_{nk}(n)} & \text{for } 1 \leq \frac{V-Y_{nk}(n)}{n-Y_{nk}(n)} < K_{i,\max} \end{cases} \tag{12.13}$$

[6]The analysis of the influence of HSPA traffic on modeling of the Iub interface is limited here to the downlink direction (HSDPA) [3] because in solutions provided by some vendors the uplink direction (HSUPA) is based on resources dedicated for R99 traffic services (*real-time* classes). In the case of different vendors, HSUPA resources can be determined in a similar way as for HSDPA resources.

where $Y_{nk}(n)$ parameter is expressed as follows (Section 7.9, Equation (7.137)):

$$Y_{nk}(n) = \sum_{i=1}^{M_{nk}} y_i(n)t_i \tag{12.14}$$

and $Y_k(n)$ can be determined based on Equation (7.140):

$$Y_k(n) = \xi_{(n)} \sum_{i=1}^{M_k} y_i(n)t_{i,\min} \tag{12.15}$$

In Equations (12.14) and (12.15) the average number of calls of class i, serviced in the occupancy state n (parameter $y_i(n)$) can be determined in the following way (Section 7.9, Equation (7.135)):

$$y_i(n) = \begin{cases} \frac{A_i\left[P_{n-t_i}\right]_V}{[P_n]_V} & \text{if} \quad \text{class } i \text{ call belongs to } M_{nk} \text{ classes} \\ \\ \frac{A_i\left[P_{n-t_{i,\min}}\right]_V}{[P_n]_V} & \text{if} \quad \text{class } i \text{ call belongs to } M_k \text{ classes} \end{cases} \tag{12.16}$$

In the next step we can evaluate the average resources occupied by calls of class i (average throughput) on the basis of the formula:

$$Y_{i,k} = \sum_{n=0}^{V} y_i(n)\left[\xi_i(n)t_{i,\min}\right][P_n]_V \tag{12.17}$$

12.3.3.2 Average Throughput Available for HSDPA Traffic

The average capacity of the Iub interface available for the HSDPA traffic T_x can be determined in a similar way as the available throughput for the WCDMA interface presented in Section 11.5 (Equation (11.129)):

$$T_x = \sum_{n=0}^{V}\left[V - \sum_{i=1}^{M_{nk}} y_{i,nk}(n)t_i\right][P_n]_V \tag{12.18}$$

where

$$y_{i,nk}(n) = \frac{A_i[P_{n-t_i}]_V}{[P_n]_V} \tag{12.19}$$

Equation (12.18) determines average resources for subscribers of HSDPA services as a difference between the total capacity of the system and average carried traffic for classes that do not undergo compression. The link capacity T_x can be available to HSDPA traffic. For example, this parameter can be used for, UBR virtual channel dimensioning.

12.4 Comments

The methods for modeling and dimensioning the Iub interface in the UMTS network presented in this chapter can be used for modeling and dimensioning all interfaces in the transmission (wire) part of different generations of mobile networks of different generations (2G, 3G and 4G). The applicability of the presented models is only limited by the need to determine the equivalent bandwidth for each class of service.

References

[1] Bannister, J., Mather, P., and Coope, S. (2004) *Convergence Technologies for 3G Networks: IP, UMTS, EGPRS and ATM*, John Wiley & Sons, Ltd.

[2] Holma, H. and Toskala, A. (2000) *WCDMA for UMTS. Radio Access For Third Generation Mobile Communications*, John Wiley & Sons, Ltd.

[3] Holma, H. and Toskala, A. (2006) *HSDPA/HSUPA for UMTS: High Speed Radio Access for Mobile Communications*, John Wiley & Sons, Ltd.

[4] International Telecommunication Union (2001) *Physical/electrical Characteristics of Hierarchical Digital Interfaces*. Recommendation G.703, Geneva.

[5] Stasiak, M. and Zwierzykowski, P. (2009) Analytical model of the Iub interface carrying HSDPA traffic in the UMTS network, in *Computer Performance Engineering*, volume 5787 of *Lecture Notes in Computer Science*, (ed. J. Bradley). Springer.

[6] Stasiak, M., Zwierzykowski, P., Wiewióra, J., and Parniewicz, D. (2009) Analytical model of traffic compression in the UMTS network, in *Computer Performance Engineering*, volume 5652 of *Lecture Notes in Computer Science*, (ed. J. Bradley), Springer.

[7] Stasiak, M., Zwierzykowski, P., and Wiewióra, J. (2009) Modeling and dimensioning of the Iub interface in the UMTS network. *Ubiquitous Computing and Communication Journal. Special Issue CSNDSP 2008*, September.

13

Application of Multi-Rate Models for Modeling UMTS Networks

13.1 Introduction

In this chapter we will present and discuss traffic issues related to the multi-service cellular network with soft capacity that can be analyzed with the "multi-rate traffic models" presented in Chapter 7. We analyze the fixed-point method for soft capacity modeling in a group of cells (Section 13.2), intercell and intracell overflow in single-service mobile networks (Section 13.3) and call transition mechanism in mobile networks with hard and soft capacity (Section 13.4).

13.2 Models of Group of Cells Carrying Multi-Rate Traffic

In this chapter we will present traffic issues that deal with the cellular system that can be analyzed with the application of the "multi-rate models" presented in Chapter 7, and the fixed-point method, presented in Section 13.2.1. The methods discussed earlier (Sections 11.3 and 11.4) addressed intercellular interference exclusively through coefficients and equivalent capacities. The methods to be presented in this chapter will include the phenomenon of "soft" capacity (Section 11.1.2) directly in the computational method using the fixed-point method.

13.2.1 Fixed-Point Method

The fixed-point method is an important element of the models presented in this chapter. It is an approximate method for modeling systems in which calls are serviced simultaneously by the same servers and it allows interdependence of service processes of the call in servers to be taken into account. The method is characterized by a large simplicity of calculation and high accuracy of final results [1–3]. The concept of the fixed-point method will be presented with an example of two neighboring cells z and h with a radio interface servicing multi-rate traffic (Figure 13.1).

Modeling and Dimensioning of Mobile Networks: From GSM to LTE
Maciej Stasiak, Mariusz Głąbowski, Arkadiusz Wiśniewski and Piotr Zwierzykowski

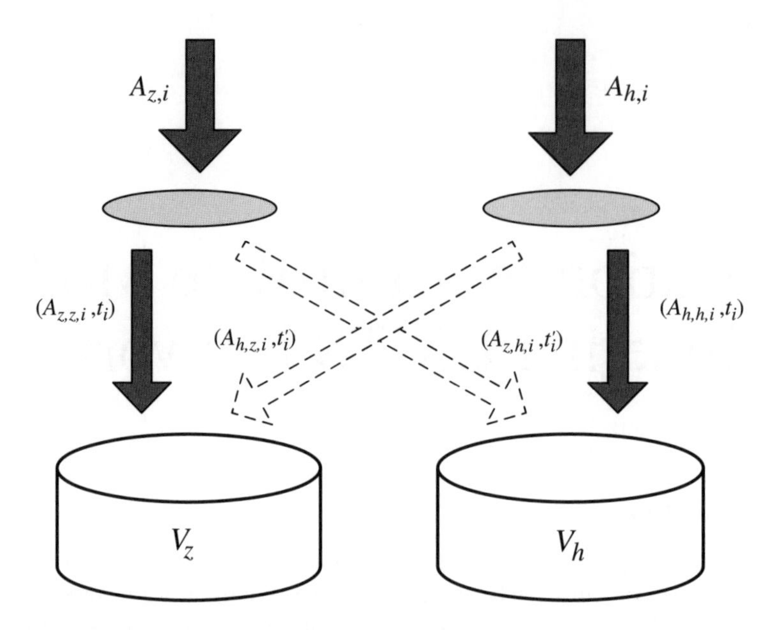

Figure 13.1 Concept of the application of the fixed-point method for intercell interference.

Assume that each call of class i offered to the cell z results in a load imposed on a radio interface of the neighboring cell, which can be interpreted as interference in the neighboring cell. This interference will be referred to further on in the section as *interference traffic*, while the load imposed by this interference will be called *interference load*. Interference traffic in a given cell is treated as subsequent traffic offered to this cell, which means that the blocking probability of interference traffic can be determined on the basis of the full-availability group model adopted for a given cell (Sections 11.4.1, 11.4.2 and 11.4.3). The influence of interference traffic upon blocking probability of calls in the mobile network can be included in the calculations using the fixed-point method.

According to the fixed-point method, each cell services the part of traffic offered to cells z or h ($A_{z,z,i}$ and $A_{h,h,i}$) that is not lost due to an excess of admissible level of load of the interference in the neighboring cells h and z, respectively. The total traffic offered to the cells z or h will be called *traffic* ($A_{z,i}$ and $A_{h,i}$), while the part of such traffic that is offered to the cell z or h, without blocking in the neighboring cell h (z) being taken into consideration, will be called *effective traffic* ($A_{z,z,i}$ and $A_{h,h,i}$). The traffic offered to cell z and h by calls originating from call h or z, respectively, will be called *effective interference traffic* ($A_{h,z,i}$ and $A_{z,h,i}$). Effective interference traffic has the same value in erlangs as traffic offered to a particular cell but the number of required BBUs in neighboring cell is much lower than the number of BBUs required in the reference cell i.e. $t_i' << t_i$.

The intensity of effective traffic of class i in the cell z can be determined by the formula.[1]

$$A_{z,z,i} = A_{z,i}(1 - E_{z,h,i}) \tag{13.1}$$

[1]For simplicity of presentation, in the description of the fixed-point method we will not divide traffic into types PCT1 and PCT2.

where $E_{z,h,i}$ is blocking probability of effective traffic $A_{z,h,i}$ generated in the cell z and offered in the cell h. For cell h Equation (13.1) takes on the following form:

$$A_{h,h,i} = A_{h,i}(1 - E_{h,z,i}) \tag{13.2}$$

The intensity of effective interference traffic of class i in cell h, originating from cell z, can be determined in the following way:

$$A_{z,h,i} = A_{z,i}(1 - E_{z,z,i}) \tag{13.3}$$

which, for the traffic originating from cell z, takes the following form:

$$A_{h,z,i} = A_{h,i}(1 - E_{h,h,i}) \tag{13.4}$$

With the fixed-point method we assume that effective interference traffic and effective traffic are responsible for any blocking of a call of a given class in each cell. Thus, whether a call is accepted or not, it is related to a need to provide a simultaneous service in the system secured for these both types of traffic. If we determine blocking probability $E_{z,z,i}$ of effective traffic of class i in cell z, and the blocking probability $E_{z,h,i}$ of effective interference traffic of class i in the neighboring cell h, then the blocking probability of traffic of class i in the system $E_{z,i}$, can be expressed with the following dependence:

$$E_{z,i} = 1 - (1 - E_{z,z,i})(1 - E_{z,h,i}) \tag{13.5}$$

Equation (13.5) for the cell h takes the following form:

$$E_{h,i} = 1 - (1 - E_{h,h,i})(1 - E_{h,z,i}) \tag{13.6}$$

Determination of particular probabilities is carried out using the iterative method. The iterative process terminates once the assigned level of the relative error of blocking probabilities of the traffic classes serviced in the system has been reached. Each next step requires a determination of new intensities of effective traffic and effective interference traffic (offered to a given cell) and the corresponding blocking probabilities.

A simplified calculation algorithm to calculate blocking with the application of the fixed-point method can be written as follows:

1. In the first step of the iteration we assume that blocking probabilities for all classes serviced by cell z and h are equal to zero:

$$\forall_{1 \le i \le M} \; E^{(0)}_{z,i} = E^{(0)}_{h,i} = E^{(0)}_{z,z,i} = E^{(0)}_{h,h,i} = E^{(0)}_{z,h,i} = E^{(0)}_{h,z,i} = 0 \tag{13.7}$$

2. Setting-up the incremental auxiliary variable: $k = k + 1$.
3. Determination of values of effective traffic and effective interference traffic in cell z and h (Equations (13.1)—(13.4)):

$$\forall_{1 \le i \le M} \; A^{(k)}_{z,z,i} = A_{z,i}(1 - E^{(k-1)}_{z,h,i}) \text{ and } \forall_{1 \le i \le M} \; A^{(k)}_{z,h,i} = A_{z,i}(1 - E^{(k-1)}_{z,z,i})$$

$$\forall_{1 \le i \le M} \; A^{(k)}_{h,h,i} = A_{h,i}(1 - E^{(k-1)}_{h,z,i}) \text{ and } \forall_{1 \le i \le M} \; A^{(k)}_{h,z,i} = A_{h,i}(1 - E^{(k-1)}_{h,h,i})$$

4. Determination of the occupancy distribution in cell z and h based on the model of the full-availability group servicing multi-rate traffic (Section 7.5.3, Equation (7.79)):

$$n\,[P_n]_{V_z}^{(k)} = \sum_{i=1}^{M}\left(A_{z,z,i}^{(k)} t_i \left[P_{n-t_i}\right]_{V_z} + A_{h,z,i}^{(k)} t_i' \left[P_{n-t_i'}\right]_{V_h}\right)$$

$$n\,[P_n]_{V_h}^{(k)} = \sum_{i=1}^{M}\left(A_{h,h,i}^{(k)} t_i \left[P_{n-t_i}\right]_{V_h} + A_{z,h,i}^{(k)} t_i' \left[P_{n-t_i'}\right]_{V_z}\right)$$

where t_i' is the number of required BBUs by the call of class i offered to a given cell in the neighboring cell[2] and the parameter V_z and V_h is the capacity of the radio interface in cells z and h, respectively.
5. Calculation of total blocking probability of calls of class i in the full-availability group, which correspond to cells z and h (Section 7.5.3, Equation (7.81)):

$$\forall_{1\le i\le M}\;\; E_{z,z,i}^{(k)} = \sum_{n=V_z-t_i+1}^{V_z} [P_n]_{V_z}{}^{(k)} \;\text{ and }\; \forall_{1\le i\le M}\;\; E_{h,z,i}^{(k)} = \sum_{n=V_z-t_i+1}^{V_z} [P_n]_{V_z}{}^{(k)}$$

$$\forall_{1\le i\le M}\;\; E_{h,h,i}^{(k)} = \sum_{n=V_h-t_i+1}^{V_h} [P_n]_{V_h}{}^{(k)} \;\text{ and }\; \forall_{1\le i\le M}\;\; E_{z,h,i}^{(k)} = \sum_{n=V_h-t_i+1}^{V_h} [P_n]_{V_h}{}^{(k)}$$

6. Calculation of blocking probabilities calls of class i in cells z and in cell h on the basis of Equations (13.5) and (13.6):

$$\forall_{1\le i\le M}\;\; E_{z,i}^{(k)} = 1-(1-E_{z,z,i}^{(k)})(1-E_{z,h,i}^{(k)})$$

$$\forall_{1\le i\le M}\;\; E_{h,i}^{(k)} = 1-(1-E_{h,h,i}^{(k)})(1-E_{h,z,i}^{(k)})$$

7. Repetition of steps 2–7 until the assumed accuracy ε of the iterative process is obtained:

$$\forall_{1\le i\le M}\left|\frac{E_{z,i}^{(k-1)}-E_{z,i}^{(k)}}{E_{z,i}^{(k)}}\right| \le \varepsilon \;\text{ and }\; \forall_{1\le i\le M}\left|\frac{E_{h,i}^{(k-1)}-E_{h,i}^{(k)}}{E_{h,i}^{(k)}}\right| \le \varepsilon \qquad (13.8)$$

The fixed-point method is an approximate method. It should be stressed, however, that it is characterized by great accuracy [2] and, hence, can be applied to blocking probability determination of connection paths and, by extension, to a study of traffic effectiveness of different algorithms for setting up connections in telecommunications networks. This method will be used in Section 13.2.2 to model a radio interface with intercellular interferences in the UMTS network.

[2]The value of t_i' is always smaller than t_i.

13.2.2 Model of the Group of Cells in the Uplink Direction

The model of a group of cells deals with a group of cells of the mobile network that jointly service a mixture of multi-rate traffic in the uplink direction. We assume that the system services streams of multi-rate traffic that can come from a finite (PCT2 type of traffic) or infinite (PCT1 type of traffic) number of traffic sources.[3]

In the model considered here we will use the following notation:

$[P_n]_V$ – occupancy probability of n BBUs in the radio interface of every cell in the group;

M – total number of traffic classes offered to the group of cells ($M = M_1 + M_2$);

$E^{\mathrm{UL}}_{z,j}$ – blocking probability for calls of class j in the uplink direction of cell z;

$E^{\mathrm{UL}}_{z,z,j}$ – blocking probability for calls of class j in the uplink direction of cell z, offered to cell z;

$E^{\mathrm{UL}}_{z,h_k,j}$ – blocking probability for calls of class j generated in cell z and offered to cell $h_k \in \mathbb{S}_z$, where set $\mathbb{S}_z = \{h_1, \ldots, h_f\}$ represents a set of cells neighboring cell z, whereas f is the number of elements in the set $\mathbb{S}_z$ ($|\mathbb{S}_z| = f$);

$E^{\mathrm{UL}}_{h_k,e_g,j}$ – blocking probability for calls of class j generated in cell h_k and offered to cell $e_g \in \mathbb{S}_{h_k}$, where set $\mathbb{S}_{h_k} = \{e_1, \ldots, e_u\}$ represents a set of cells neighboring cell h_k, whereas u is the number of elements in the set $\mathbb{S}_{h_k}$ ($\left|\mathbb{S}_{h_k}\right| = u$);

$A^{\mathrm{UL}}_{z,j}$ – intensity of traffic offered to the radio interface of the cell z by users of class j;

$A^{\mathrm{UL}}_{h_k,j}$ – intensity of traffic offered to the radio interface of the cell h_k by users of class j;

$A^{\mathrm{UL}}_{z,z,j}$ – intensity of effective traffic offered in cell z by users of class j in cell z (the difference between the parameters $A^{\mathrm{UL}}_{z,j}$ and $A^{\mathrm{UL}}_{z,z,j}$ will be explained in the next section);

$A^{\mathrm{UL}}_{z,h_k,j}$ – intensity of effective interference traffic offered in cell h_k by users of class j in cell z;

$A^{\mathrm{UL}}_{h_k,e_g,j}$ – intensity of effective interference traffic offered in cell e_g by users of class j in cell h_k;

η^{UL}_z – assumed capacity of the radio interface in cell z, in the uplink direction (as a percentage);

$\eta^{\mathrm{UL}}_{h_k}$ – assumed capacity of the radio interface in cell h_k, in the uplink direction (in percentage);

$L_{z,\mathrm{BBU}}$ – value of basic bandwidth unit, as a percentage of the interface load, for cell z;

$L_{z,z,j}$ – load factor (as a percentage) introduced by one call of class j originating from cell z into the radio interface of the cell z;

$L_{h_k,\mathrm{BBU}}$ – value of basic bandwidth unit, as a percentage of the interface load, for neighboring cell h_k;

$L'_{z,h_k,j}$ – load factor (as a percentage) introduced by one call of class j, originating from cell z, into the radio interface of the neighboring cell h_k;

$t_{z,z,j}$ – number of BBUs required by call of class j, originating from cell z, to set up a connection by the radio interface of the cell z;

---

[3]For simplicity of presentation, in the description of the fixed-point method we will not divide traffic into types PCT1 and PCT2.

$t'_{z,h_k,j}$ – number of BBUs required by call of class j, originating from cell z, to set up a connection by the radio interface of the neighboring cell h_k;

V_z^{UL} – capacity of the radio interface of cell z in the uplink direction, expressed in BBUs;

$V_{h_k}^{\text{UL}}$ – capacity of the radio interface of cell h_k in the uplink direction, expressed in BBUs.

All traffic characteristics – the blocking probability, loss probability, carried and offered traffic – result from the approximate calculation algorithm of the multi-service Erlang-Engset distribution (Section 7.5.3) and from the application of the fixed-point method (Section 13.2.1).

13.2.2.1 Generalized Model of the Group of Cells

The model can be applied to the uplink connections of any set of cells. Each of the cells of the system can be a reference cell, which means that calls can be generated in every one of them. A new call appearing in the system occupies resources in the reference cell and introduces interference load to neighboring cells (Figure 13.2). The model presents a situation in which traffic generated within the area of all cells is accompanied by considerable effective interference traffic, which influences the capacity of the whole system.

In this model we determine a set of neighboring cells for each of the cells. The sets of neighboring cells for an example group of seven cells (shown in Figure 13.2) are presented in Table 13.1. For the cell denoted by the symbol h_5 in Figure 13.2, the neighboring cells form the set $\mathbb{S}_{h_5} = \{z, h_4, h_6\}$, whereas cell z is surrounded by the following six cells $\mathbb{S}_z = \{h_1, h_2, h_3, h_4, h_5, h_6\}$. The model assumes that each of the cells can be a reference cell – that is, the cell in which calls appear. It was further assumed that a call of class j demands $t_{z,z,j}$ BBUs in cell z. At the same time, a call of class j generated in cell z demands $t'_{z,h_k,j}$ BBUs in cell h_k. The interference load of class j resulting from a call in cell z and imposed on each of the neighboring cells of the set $\mathbb{S}_z = \{h_1, \ldots, h_f\}$ is the same: $L'_{z,h_1,j} =, \ldots, = L'_{z,h_f,j}$. The interference load factor resulting from a call of class j in cell h_k imposed to each of the neighboring cells of the set $\mathbb{S}_{h_k} = \{e_1, \ldots, e_u\}$ is the same: $L'_{h_k,e_1,j} =, \ldots, = L'_{h_k,e_u,j}$.

On the basis of the adopted assumptions we determine first those parameters of the system that are dependent on the structure of serviced traffic [3]:

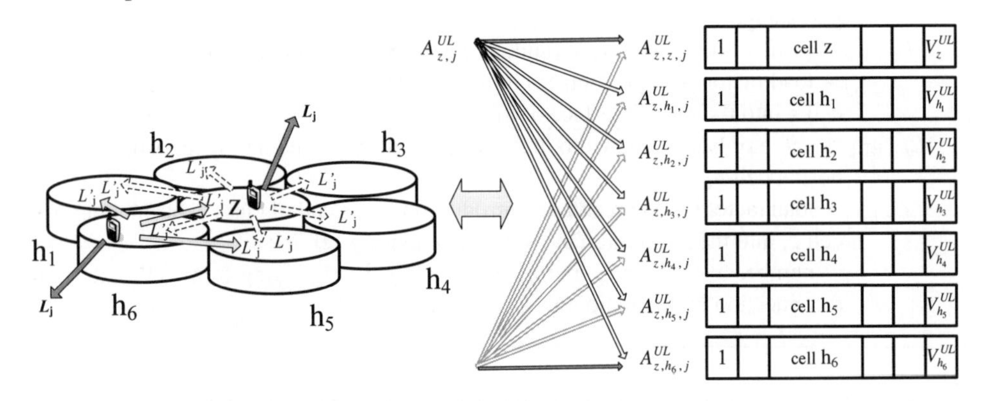

Figure 13.2 The generalized group of cells model.

Table 13.1 Sets of neighboring cells

Designation of the cell	Set of neighboring cells
z	$h_1, h_2, h_3, h_4, h_5, h_6$
h_1	z, h_2, h_6
h_2	z, h_1, h_3
h_3	z, h_2, h_4
h_4	z, h_3, h_5
h_5	z, h_4, h_6
h_6	z, h_1, h_5

Corresponding to Figure 13.2 [4].

- Basic bandwidth unit in cell z is determined on the basis of the loads imposed by streams generated in cell z as well as on the basis of the loads imposed by calls generated in neighboring cells. Therefore, we can write (Section 11.1.3, Equation (11.13)):

$$L_{z,\mathrm{BBU}} = \mathrm{GCD}\left\{\begin{matrix} L_{z,z,1}, & \cdots, & L_{z,z,j}, & \cdots, & L_{z,z,M}, \\ L'_{h_1,z,1}, & \cdots, & L'_{h_1,z,j}, & \cdots, & L'_{h_1,z,M} \\ . & \cdots & . & \cdots & . \\ L'_{h_k,z,1}, & \cdots, & L'_{h_k,z,j}, & \cdots, & L'_{h_k,z,M} \\ . & \cdots & . & \cdots & . \\ L'_{h_f,z,1}, & \cdots, & L'_{h_f,z,j}, & \cdots, & L'_{h_f,z,M} \end{matrix}\right\} \tag{13.9}$$

- Basic bandwidth unit in the neighboring cells (for the cell z), as in the previous case, defined in the following way (Section 11.1.3, Equation (11.13)):

$$L_{h_k,\mathrm{BBU}} = \mathrm{GCD}\left\{\begin{matrix} L_{h_k,h_k,1}, & \cdots, & L_{h_k,h_k,j}, & \cdots, & L_{h_k,h_k,M}, \\ L'_{e_1,h_k,1}, & \cdots, & L'_{e_1,h_k,j}, & \cdots, & L'_{e_1,h_k,M} \\ . & \cdots & . & \cdots & . \\ L'_{z,h_k,1}, & \cdots, & L'_{z,h_k,j}, & \cdots, & L'_{z,h_k,M} \\ . & \cdots & . & \cdots & . \\ L'_{e_u,h_k,1}, & \cdots, & L'_{e_u,h_k,j}, & \cdots, & L'_{e_u,h_k,M} \end{matrix}\right\} \tag{13.10}$$

- Knowing the value of BBU for each of the cells under consideration we can determine the equivalent capacity of the radio interface in the uplink direction of cell z, expressed in BBUs (Section 11.1.3, Equation (11.14)):

$$V_z^{\mathrm{UL}} = \left\lfloor \frac{\eta_z^{\mathrm{UL}}}{L_{z,\mathrm{BBU}}} \right\rfloor \tag{13.11}$$

- Similarly, the equivalent capacity of a neighboring cell h_k in relation to cell z ($h_k \in \mathbb{S}_z$) will be expressed by the following formula (Section 11.1.3, Equation (11.14)):

$$V_{h_k}^{\mathrm{UL}} = \left\lfloor \frac{\eta_{h_k}^{\mathrm{UL}}}{L_{h_k,\mathrm{BBU}}} \right\rfloor \tag{13.12}$$

- The values of BBU, determined by Equations (13.9) and (13.10) also make it possible to determine the number of BBUs required by a call of class j to set up a connection in cell z (Section 11.1.3, Equation (11.15)):

$$t_{z,z,j} = \left\lfloor \frac{L_{z,z,j}}{L_{z,\mathrm{BBU}}} \right\rfloor \tag{13.13}$$

- The number of BBUs carried in by a call of class j to neighboring cell h_k, generated by a user that remains in cell z, will be expressed by the equation (Section 11.1.3, Equation (11.15)):

$$t'_{z,h_k,j} = \left\lfloor \frac{L'_{z,h_k,j}}{L_{h_k,\mathrm{BBU}}} \right\rfloor \tag{13.14}$$

13.2.2.2 Calculation Algorithm

The iterative fixed-point method presented in Section 13.2.1 is used to obtain the values of appropriate parameters. Particular steps in the calculation algorithm therefore correspond to steps the fixed-point algorithm:

1. Setting up initial values of all blocking probabilities.

$$\forall_{1 \le j \le M} \; E_{z,j}^{\mathrm{UL}(0)} = E_{h_k,j}^{\mathrm{UL}(0)} = E_{z,z,j}^{\mathrm{UL}(0)} = E_{h_k,h_k,j}^{\mathrm{UL}(0)} = E_{z,h_k,j}^{\mathrm{UL}(0)} = E_{h_k,z,j}^{\mathrm{UL}(0)} = 0 \tag{13.15}$$

2. Increasing of the iteration step $k = k + 1$.
3. Determination of the values of effective traffic and interference effective traffic for cell z and cells $h_k \in \mathbb{S}_z$.
   Effective traffic $A_{z,z,j}^{UL(k)}$ of class j offered to cell z and generated in cell z can be determined on the basis of the following dependence:

$$A_{z,z,j}^{\mathrm{UL}(k)} = A_{z,j}^{\mathrm{UL}} \prod_{i=1}^{f} (1 - E_{z,h_i,j}^{\mathrm{UL}(k-1)}) \tag{13.16}$$

   where $A_{z,j}^{\mathrm{UL}}$ is the average traffic generated by users of class j in cell z, $E_{z,h_i,j}^{\mathrm{UL}(k)}$ is the blocking probability for calls of class j generated in cell z and offered to cell $h_i \in \mathbb{S}_z$.
   Effective interference traffic $A_{z,h_k,j}^{\mathrm{UL}(k)}$ can be determined as follows:

$$A_{z,h_k,j}^{\mathrm{UL}(k)} = \begin{cases} A_{z,j}^{\mathrm{UL}}(1 - E_{z,z,j}^{\mathrm{UL}(k-1)}) \prod\limits_{i=1, i \neq k}^{f} (1 - E_{z,h_i,j}^{\mathrm{UL}(k-1)}) & \text{for } h_i \in \mathbb{S}_z \\ 0 & \text{for } h_k \notin \mathbb{S}_z \end{cases} \tag{13.17}$$

where $E_{z,z,j}^{\mathrm{UL}(k)}$ is the blocking probability for calls of class j generated in cell z and offered to cell z.

Effective traffic $A_{h_k,h_k,j}^{\mathrm{UL}(k)}$ of class j offered to cell h_k and generated in cell h_k can be obtained in the following way:

$$A_{h_k,h_k,j}^{\mathrm{UL}(k)} = A_{h_k,j}^{\mathrm{UL}} \prod_{g=1}^{u} (1 - E_{h_k,e_g,j}^{\mathrm{UL}(k-1)}) \tag{13.18}$$

where $A_{h_k,j}^{\mathrm{UL}}$ is the average traffic generated by users of class j in cell h_k, whereas $E_{h_k,e_g,j}^{\mathrm{UL}(k)}$ is the blocking probability for calls of class j generated in cell h_k and offered to cell $e_g \in \mathbb{S}_{h_k}$.

Effective interference traffic $A_{h_k,z,j}^{\mathrm{UL}}$ can be determined as follows:

$$A_{h_k,z,j}^{\mathrm{UL}(k)} = \begin{cases} A_{h_k,j}^{\mathrm{UL}}(1 - E_{h_k,h_k,j}^{\mathrm{UL}(k-1)}) \prod\limits_{g=1,g\neq z}^{u} (1 - E_{h_k,e_g,j}^{\mathrm{UL}(k-1)}) & \text{for } e_g \in \mathbb{S}_{h_k} \\ 0 & \text{for } z \notin \mathbb{S}_{h_k} \end{cases} \tag{13.19}$$

where $E_{h_k,h_k,j}^{\mathrm{UL}(k)}$ is the blocking probability for calls of class j generated in cell h_k and offered to cell h_k.

4. Determination of the occupancy distribution in cell z and in each cell $h_k \in \mathbb{S}_z$.
   The occupancy distribution in cell z and in cell h_k can be determined on the basis of the following dependencies:

$$[P_n]_{V_z}^{(k)} = F \begin{Bmatrix} (A_{z,z,1}^{\mathrm{UL}(k)}, t_{z,z,1}), & \ldots, & (A_{z,z,M}^{\mathrm{UL}(k)}, t_{z,z,M}) \\ (A_{h_1,z,1}^{\mathrm{UL}(k)}, t'_{h_1,z,1}), & \ldots, & (A_{h_1,z,M}^{\mathrm{UL}(k)}, t'_{h_1,z,M}) \\ \ldots & \ldots & \ldots \\ (A_{h_k,z,1}^{\mathrm{UL}(k)}, t'_{h_k,z,1}), & \ldots, & (A_{h_k,z,M}^{\mathrm{UL}(k)}, t'_{h_k,z,M}) \\ \ldots & \ldots & \ldots \\ (A_{h_f,z,1}^{\mathrm{UL}(k)}, t'_{h_f,z,1}), & \ldots, & (A_{h_f,z,M}^{\mathrm{UL}(k)}, t'_{h_f,z,M}) \end{Bmatrix} \tag{13.20}$$

$$[P_n]_{V_{h_k}}^{(k)} = F \begin{Bmatrix} (A_{h_k,h_k,1}^{\mathrm{UL}(k)}, t_{h_k,h_k,1}), & \ldots, & (A_{h_k,h_k,M}^{\mathrm{UL}(k)}, t_{h_k,h_k,M}) \\ (A_{e_1,h_k,1}^{\mathrm{UL}(k)}, t'_{e_1,h_k,1}), & \ldots, & (A_{e_1,h_k,M}^{\mathrm{UL}(k)}, t'_{e_1,h_k,M}) \\ \ldots & \ldots & \ldots \\ (A_{z,h_k,1}^{\mathrm{UL}(k)}, t'_{z,h_k,1}), & \ldots, & (A_{z,h_k,M}^{\mathrm{UL}(k)}, t'_{z,h_k,M}) \\ \ldots & \ldots & \ldots \\ (A_{e_u,h_k,1}^{\mathrm{UL}(k)}, t'_{e_u,h_k,1}), & \ldots, & (A_{e_u,h_k,M}^{\mathrm{UL}(k)}, t'_{e_u,h_k,M}) \end{Bmatrix} \tag{13.21}$$

Function “F” can be obtained based on the model of the full-availability group servicing multi-rate traffic (Section 7.3.5, Equation (7.39)). Occupancy distribution in cell z for a set of variables in the Function (13.20), can be rewritten in the following way:

$$n\left[P_n\right]_{V_z}^{(k)} = \sum_{j=1}^{M} A_{z,z,i}^{\mathrm{UL}(k)}\, t_{z,z,j}\left[P_{n-t_{z,z,j}}\right]_{V_z} + \sum_{j=1}^{M}\sum_{k=1}^{f} A_{h_k,z,j}^{\mathrm{UL}(k)}\, t'_{h_k,z,j}\left[P_{n-t'_{h_k,z,j}}\right]_{V_z} \tag{13.22}$$

The occupancy distribution in cell $h_k \in \mathbb{S}_z$ for a set of variables in Equation (13.21) can be shown as follows:

$$n\left[P_n\right]_{V_{h_k}}^{(k)} = \sum_{j=1}^{M} A_{h_k,h_k,j}^{\mathrm{UL}(k)} t_{h_k,h_k,j}\left[P_{n-t_{h_k,h_k,j}}\right]_{V_{h_k}} + \sum_{j=1}^{M}\sum_{g=1}^{u} A_{e_g,h_k,j}^{\mathrm{UL}(k)} t'_{e_g,h_k,j}\left[P_{n-t'_{e_g,h_k,j}}\right]_{V_{h_k}} \tag{13.23}$$

5. Calculation of the value of the blocking probability for calls of class j in the cell z and in each cell $h_k \in \mathbb{S}_z$.
   The blocking probabilities $E_{z,z,j}^{\mathrm{UL}(k)}$, $E_{z,h_k,j}^{\mathrm{UL}(k)}$, $E_{h_k,h_k,j}^{\mathrm{UL}(k)}$ and $E_{h_k,z,j}^{\mathrm{UL}(k)}$ can be determined on the basis of the full-availability group servicing multi-rate traffic:

$$\forall_{1\le j\le M}\; E_{z,z,j}^{\mathrm{UL}(k)} = \sum_{n=V_z-t_{z,z,j}+1}^{V_z} \left[P_n\right]_{V_z}^{(k)} \tag{13.24}$$

$$\forall_{1\le j\le M}\; E_{h,z,j}^{\mathrm{UL}(k)} = \sum_{n=V_z-t'_{h_k,z,j}+1}^{V_z} \left[P_n\right]_{V_z}^{(k)}$$

$$\forall_{1\le j\le M}\; E_{h_k,h_k,j}^{\mathrm{UL}(k)} = \sum_{n=V_{h_k}-t_{h_k,h_k,j}+1}^{V_{h_k}} \left[P_n\right]_{V_{h_k}}^{(k)}$$

$$\forall_{1\le j\le M}\; E_{z,h_k,j}^{\mathrm{UL}(k)} = \sum_{n=V_{h_k}-t'_{z,h_k,j}+1}^{V_{h_k}} \left[P_n\right]_{V_{h_k}}^{(k)}$$

6. Determination of the value of total blocking probability for calls of class j in cell z and in each cell $h_k \in \mathbb{S}_z$.
   Given the blocking probability for calls of class j in cell z and in cell $h_k \in \mathbb{S}_z$, we can determine the total blocking probability $E_{z,j}^{\mathrm{UL}}$ for calls of class j in the reference cell z:

$$E_{z,j}^{\mathrm{UL}(k)} = 1 - (1 - E_{z,z,j}^{\mathrm{UL}(k)})\prod_{k=1}^{f}(1 - E_{z,h_k,j}^{\mathrm{UL}(k)}) \tag{13.25}$$

Equation (13.25) for cell $h_k \in \mathbb{S}_{h_k}$ can be rewritten as follows:

$$E_{h_k,j}^{\mathrm{UL}(k)} = 1 - (1 - E_{h_k,h_k,j}^{\mathrm{UL}(k)}) \prod_{g=1}^{u} (1 - E_{z,e_g,j}^{\mathrm{UL}(k)}) \tag{13.26}$$

7. Verification of the accuracy of calculations for cell z and cells $h_k \in \mathbb{S}_z$.
   Repetition of steps 2–7 until the assumed accuracy ε of the iterative process is obtained:

$$\forall_{1\le j\le M} \left| \frac{E_{z,j}^{\mathrm{UL}(k-1)} - E_{z,j}^{\mathrm{UL}(k)}}{E_{z,j}^{\mathrm{UL}(k)}} \right| \le \varepsilon \quad \text{and} \quad \forall_{1\le j\le M} \left| \frac{E_{h_k,j}^{\mathrm{UL}(k-1)} - E_{h_k,j}^{\mathrm{UL}(k)}}{E_{h_k,j}^{\mathrm{UL}(k)}} \right| \le \varepsilon \tag{13.27}$$

where $E_{z,j}^{\mathrm{UL}(k)}$, $E_{h_k,j}^{\mathrm{UL}(k)}$, $E_{z,j}^{\mathrm{UL}(k-1)}$ and $E_{h_k,j}^{\mathrm{UL}(k-1)}$ are the values of total blocking probability for class j calls, in cell z and cell $h_k \in \mathbb{S}_z$ in the uplink direction, in iteration k and $k-1$, respectively.

Conversely, if the approximation error is lower, or equal than the accuracy ε , then the iteration process terminates. The iteration process is a convergent process, and the number of steps of the iteration depends on the accuracy of calculations ε. In analyzed cases the iteration process terminated after a dozen of steps. The problem of accuracy and the convergence of the iterative process related to the fixed-point method is analyzed in detail in [5].

13.2.2.3 A Simplified Model of the Group of Cells – Macro and Micro Cell Consideration

Let us consider a model for one-sided influence of interference – that is, a model in which a centrally located cell services a large amount of traffic that influences neighboring cells, while the neighboring cells carry a relatively small amount of traffic concentrated near base stations [1, 6]. One can therefore assume that the influence of effective interference traffic on the central cell is insignificant. Thus, in the model presented here, the access cell influences the neighboring cells, whereas the neighboring cells have no influence on the access cell. A good example of the above system is an internal installation in a shopping center (a macro base station generating a large amount of traffic). The shopping center is surrounded by microcells that generate small traffic.

In the model it was assumed that intensities of traffic offered to all cells are the same and equal to the maximum intensity of traffic offered to the cell belonging to the group of cells under consideration here:

$$\forall_{1\le i\le M} \; A_h = \max(A_z, A_{h_1}, A_{h_2}, A_{h_3}, A_{h_4}, A_{h_5}, A_{h_6}) \tag{13.28}$$

Such an assumption can also be justified by the method of operation of the call admission control system: a new call is rejected if in at least one of the neighboring cells the load will be higher than the admissible limit. This means that the application of this model can be limited to the analysis of two cells: the reference cell and the neighboring cell with the lowest capacity (Figure 13.4):

$$V_h^{\mathrm{UL}} = \min\left\{ V_{h_1}^{\mathrm{UL}}, V_{h_2}^{\mathrm{UL}}, V_{h_3}^{\mathrm{UL}}, V_{h_4}^{\mathrm{UL}}, V_{h_5}^{\mathrm{UL}}, V_{h_6}^{\mathrm{UL}}, V_{h_7}^{\mathrm{UL}} \right\} \tag{13.29}$$

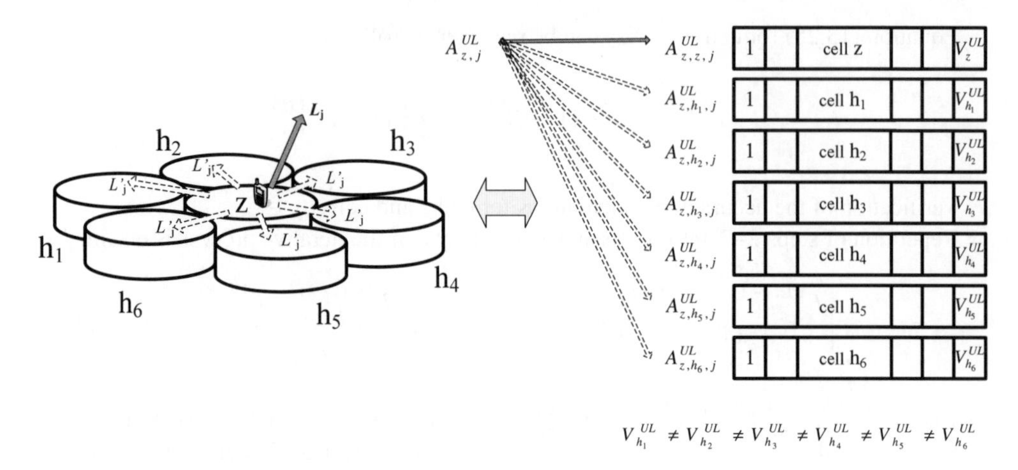

Figure 13.3 Simplified group of cells.

Further on in the chapter, we will denote the reference cell with the symbol z and the cell with the lowest capacity with the symbol h. Figure 13.4 shows the one-sided influence of interference in the simplified group of cells. Traffic $A^{\text{UL}}_{z,j}$ influences cell h (traffic $A^{\text{UL}}_{z,h,j}$), while traffic $A^{UL}_{h,j}$ has no influence upon the cell z (traffic $A_{h,z,j} = 0$).

Let us consider the model of the central access cell z surrounded by six cells, presented in Figure13.3. The model assumes that the cells can have different capacities, but a microcell with the heaviest load is taken into consideration in the model. Since we consider a model of two cells, therefore, for a determination of the traffic characteristics of such a one-sided system, the fixed-point algorithm can be used (Section 13.2.1), which will be rewritten in the following form:

1. In the first step of the iteration we assume that blocking probabilities for all classes serviced by cell z and h are equal to zero:

$$\forall_{1\leq i\leq M}\ E^{\text{UL}(0)}_{z,i} = E^{\text{UL}(0)}_{h,i} = E^{\text{UL}(0)}_{z,z,i} = E^{\text{UL}(0)}_{h,h,i} = E^{\text{UL}(0)}_{z,h,i} = 0 \tag{13.30}$$

2. Increase the value of the parameter $k = k + 1$.

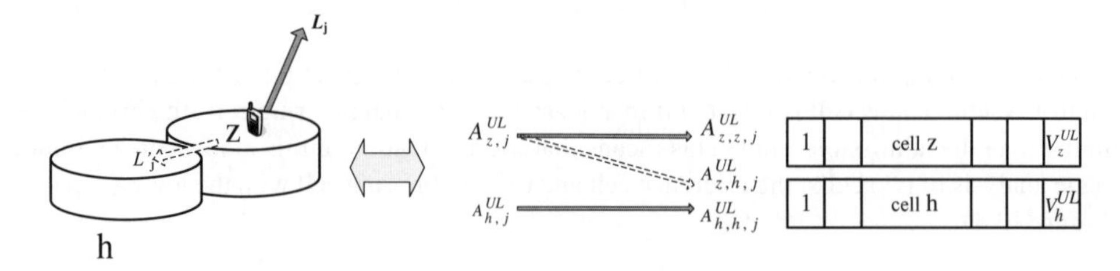

Figure 13.4 Two-cell simplified model.

3. Determination of values of effective traffic and effective interference traffic in cell z and h:

$$\forall_{1\le i\le M}\ A_{z,z,i}^{\mathrm{UL}(k)} = A_{z,i}(1 - E_{z,h,i}^{\mathrm{UL}(k-1)}) \text{ and } \forall_{1\le i\le M}\ A_{z,h,i}^{\mathrm{UL}(k)} = A_{z,i}(1 - E_{z,z,i}^{\mathrm{UL}(k-1)})$$

$$\forall_{1\le i\le M}\ A_{h,h,i}^{\mathrm{UL}(k)} = A_{h,i} \qquad \text{and } \forall_{1\le i\le M}\ A_{h,z,i}^{\mathrm{UL}(k)} = 0$$

4. Determination of the occupancy distribution in cell z and h based on the model of the full-availability group servicing multi-rate traffic (Section 7.5.3, Equation (7.79)):

$$n\,[P_n]_{V_z}^{(k)} = \sum_{i=1}^{M} A_{z,z,i}^{\mathrm{UL}(k)} t_{z,z,i} \left[P_{n-t_{z,z,i}}\right]_{V_z}$$

$$n\,[P_n]_{V_h}^{(k)} = \sum_{i=1}^{M} \left(A_{h,h,i}^{\mathrm{UL}(k)} t_{h,h,i} \left[P_{n-t_{h,h,i}}\right]_{V_h} + A_{z,h,i}^{\mathrm{UL}(k)\,\prime} t_{z,h,i}' \left[P_{n-t_{z,h,i}'}\right]_{V_z} \right)$$

where $t_{z,h,i}'$ is the number of BBUs required by a call of class i generated in cell z and offered to cell h. The parameter V_z and V_h is the capacity of the radio interface in the cell z and h, respectively.

5. Calculation of the total blocking probability for calls of class i in the full-availability group that correspond to cells z and h Section 7.5.3, Equation (7.81)):

$$\forall_{1\le i\le M}\ E_{z,z,i}^{\mathrm{UL}(k)} = \sum_{n=V_z-t_{z,z,i}+1}^{V_z} [P_n]_{V_z}{}^{(k)} \text{ and } E_{h,z,i}^{\mathrm{UL}(k)} = 0$$

$$\forall_{1\le i\le M}\ E_{h,h,i}^{\mathrm{UL}(k)} = \sum_{n=V_h-t_{h,h,i}+1}^{V_h} [P_n]_{V_h}{}^{(k)} \text{ and } E_{z,h,i}^{\mathrm{UL}(k)} = \sum_{n=V_h-t_{z,h,i}'+1}^{V_h} [P_n]_{V_h}{}^{(k)}$$

6. Calculation of blocking probabilities for calls of class i in cells z and in cell h on the basis of Equations (13.5) and (13.6):

$$\forall_{1\le i\le M}\ E_{z,i}^{\mathrm{UL}(k)} = 1 - (1 - E_{z,z,i}^{\mathrm{UL}(k)})(1 - E_{z,h,i}^{\mathrm{UL}(k)})$$

$$\forall_{1\le i\le M}\ E_{h,i}^{\mathrm{UL}(k)} = E_{h,h,i}^{\mathrm{UL}(k)}$$

7. Repetition of steps 2–7 until the assumed accuracy ε of the iterative process is obtained:

$$\forall_{1\le i\le M} \left| \frac{E_{z,i}^{\mathrm{UL}(k-1)} - E_{z,i}^{\mathrm{UL}(k)}}{E_{z,i}^{\mathrm{UL}(k)}} \right| \le \varepsilon \text{ and } \forall_{1\le i\le M} \left| \frac{E_{h,i}^{\mathrm{UL}(k-1)} - E_{h,i}^{\mathrm{UL}(k)}}{E_{h,i}^{\mathrm{UL}(k)}} \right| \le \varepsilon \qquad (13.31)$$

13.2.2.4 Comments

This model of a set of cells is based on the assumption that a new call in the reference cell is accompanied by interference in neighboring cells, and that the degree of interference has some influence upon the blocking probability of the reference cell. If we assume that interference generated in neighboring cells does not influence the blocking probability of the call in the reference cell, then the computational process can be simplified, to the algorithm presented above. If the blocking probability of effective interference traffic $E_{z,h,i}^{\mathrm{UL}}$ in cell h is small (for example, of an order of magnitude lower than the probability $E^{\mathrm{UL}}_{z,z,i}$), then the fixed-point method can be omitted. When this is the case, the calculation process is not iterative and the values of probabilities are determined directly on the basis of the full-availability group that services multi-rate traffic streams of a given type.

13.2.3 Model of the Group of Cell in the Downlink Direction

This model deals with a group of cells of a mobile network that jointly service a mixture of multi-rate traffic in the downlink direction. We assume, for simplicity, that the system services streams of multi-rate traffic that may come from an infinite number of traffic sources (PCT1-type traffic).

The model represents the downlink direction in a given group of neighboring cells. It is assumed that the group consists of seven cells. In each of the cells, calls of different traffic classes can be generated. A further assumption is that each of the cells will be treated separately, which means that traffic generated in each of the cells comes exclusively from the calls that occur in that particular cell (Figure 13.5). Such an assumption is justified due to the fact that orthogonal OVSF codes are used in the downlink direction. Any interference and the power used in the downlink direction by the common pilot channel (CPICH) can be taken into consideration in the appropriately reduced capacity of the system assumed in the calculations. The model of the downlink direction also assumes that each cell in the system is a reference cell, whereas the capacity of individual cells can be different. The following notation has been adopted in the considerations:

- $E^{\mathrm{DL}}_{z,j}$ – blocking probability for calls of class j in the downlink direction of the cell z;
- $A^{\mathrm{DL}}_{z,j}$ – intensity of traffic offered to the radio interface of the cell z by users of class j;
- $L_{z,\mathrm{BBU}}$ – the value of the basic bandwidth unit for the radio interface of the cell z, expressed as a percentage of the interface load;
- $t_{z,j}$ – number of BBUs required by call of class j to set up a connection by the radio interface of the cell z;
- η^{DL}_{z} – assumed capacity if the radio interface in the cell z in the downlink direction, expressed as a percentage;

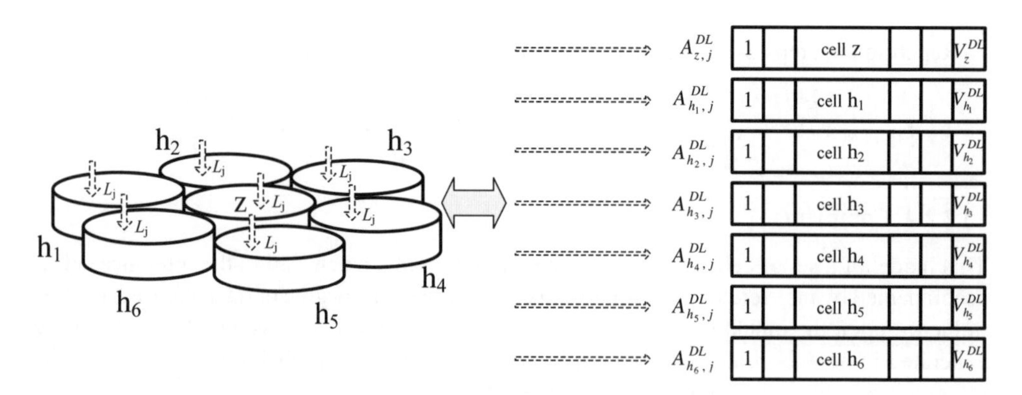

Figure 13.5 Model of the cell in the downlink direction.

V_z^{DL} – capacity of the radio interface of cell z in the downlink direction, expressed in BBUs.

The following parameters can be defined:

- Basic Bandwidth Unit:

$$L_{z,\text{BBU}} = \text{GCD}\{L_{z,1}, \ldots, L_{z,j}, \ldots, L_{z,M}\} \tag{13.32}$$

- Equivalent capacity of cell z expressed in BBUs:

$$V_z^{\text{DL}} = \left\lfloor \frac{\eta_{z,\text{DL}}}{L_{z,\text{BBU}}} \right\rfloor \tag{13.33}$$

- Number of BBUs demanded by a call of class j in cell z:

$$t_{z,j} = \left\lfloor \frac{L_{z,j}}{L_{z,\text{BBU}}} \right\rfloor \tag{13.34}$$

The blocking probability of calls of class j in cell z in the downlink direction can be expressed by the following function:

$$E_{z,j}^{\text{DL}} = f\left\{(A_{z,j}^{\text{DL}}, t_{z,j}), \ldots, (A_{z,M}^{\text{DL}}, t_{z,M})\right\} \tag{13.35}$$

The blocking probability $E_{z,j}^{\text{DL}}$ (Equation (13.35)) can be determined on the basis of the full-availability group model (Section 7.3.5, Equation (7.39)).

13.2.4 Models of Group of Cells in the Uplink and Downlink Directions

In a cellular network with a WCDMA radio interface, many services require resource allocation both in the uplink direction and in the downlink direction. The service processes of calls in both directions are mutually dependent processes and this dependence can also be taken into account with the fixed-point method (Section 13.2.1), which has also been used for a determination of the blocking probability in the uplink direction (Section 13.2.2). The calculation process can be presented in the form of the following algorithm:

1. Setting up initial values of all blocking probabilities:

$$\forall_{1 \le j \le M} \; E_{z,j}^{\text{UL}(0)} = E_{z,j}^{\text{DL}(0)} = 0 \tag{13.36}$$

2. Increasing the iteration step $k = k + 1$.
3. Determination of the values of effective traffic in the uplink and downlink directions. Effective traffic $A_{z,j}^{\text{UL}(k)}$ of class j offered to cell z can be determined on the basis of the following dependence:

$$A_{z,j}^{\text{UL}(k)} = A_{z,j}(1 - E_{z,j}^{\text{DL}(k-1)}) \tag{13.37}$$

where $A_{z,j}$ is the average traffic generated by users of class j in cell z and $E_{z,j}^{\text{DL}(k-1)}$ is the blocking probability for calls of class j in cell z in the downlink direction.

Effective traffic $A_{z,j}^{\mathrm{DL}(k)}$ can be determined as follows:

$$A_{z,j}^{\mathrm{DL}(k)} = A_{z,j}(1 - E_{z,j}^{\mathrm{UL}(k-1)}) \tag{13.38}$$

where $E_{z,j}^{\mathrm{UL}(k-1)}$ is the blocking probability for calls of class j in cell z in the uplink direction.

4. Determination of the occupancy distribution in cell z in the uplink and in the downlink direction.
   The occupancy distribution $[P_n]_{V_z}^{\mathrm{UL}(k)}$ in cell z in the uplink direction can be calculated as the previous sentence on the basis of the algorithm presented in Section 13.2.1 for the uplink direction (step 4).
   The occupancy distribution $[P_n]_{V_z}^{\mathrm{DL}(k)}$ in cell z in the uplink direction can be calculated on the basis of Equation (13.35).
5. Calculation of the value of blocking probability for calls of class j in cell z in the uplink and in the downlink direction.
   The blocking probabilities $E_{z,j}^{\mathrm{DL}(k)}$ can be determined on the basis of the downlink model (Section 13.2.2).
6. Determination of the value of total blocking probability for calls of class j.
   From the knowledge of the blocking probability for calls of class j in the uplink and in the downlink direction, we can determine the total blocking probability $E_{z,j}^{(k)}$ for calls of class j in the reference cell z using the dependence:

$$E_{z,j}^{(k)} = 1 - (1 - E_{z,j}^{\mathrm{UL}(k)})(1 - E_{z,j}^{\mathrm{DL}(k)}) \tag{13.39}$$

7. Verification of the accuracy of calculations.
   Repetition of steps 2–7 until the assumed accuracy ε of the iterative process is obtained:

$$\forall_{1\leq j\leq M} \left| \frac{E_{z,j}^{(k-1)} - E_{z,j}^{(k)}}{E_{z,j}^{(k)}} \right| \leq \varepsilon \tag{13.40}$$

where $E_{z,j}^{(k)}$ and $E_{z,j}^{(k-1)}$ are the value of total blocking probability in the cell z, in iteration k and $k - 1$, respectively.

Conversely, if the approximation error is lower than or equal to the assumed accuracy ε, then the iteration process terminates.

13.3 Models of Traffic Overflow

This chapter discusses two possibilities of the application of the overflow traffic models described in Chapters 6 and 8. An intercell traffic overflow model and a model for traffic overflow between macro and microcells are proposed.

13.3.1 Model of Intercell Overflow of Single-Rate Traffic

The model concerns those cells in which there is a possibility of overflowing unserviced traffic to other cells. We assume that all cells service traffic of the PCT1 type and, according to the accepted standards, that each one demands 1 BBU to set up a connection. The model of intercell

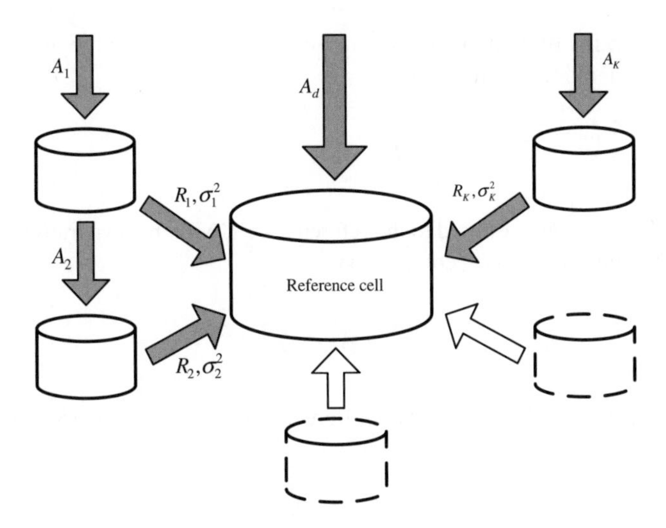

Figure 13.6 Model of intercell overflow.

overflow is shown in Figure 13.6. The cell under consideration – further on referred to as the reference cell – absorbs the overflow traffic from K neighboring cells.

The following notation is adopted for the considered model:

A_d – PCT1 traffic intensity offered to the reference cell;
A_i – PCT1 traffic intensity, offered in the neighboring cell i;
V – capacity of the reference cell;
V_i – capacity of the cell i;
R_i – mean value of the overflow traffic directed to the reference cell from the cell i;
σ_i^2 – variance of the traffic that overflows to the reference cell from the cell i;
R – mean value of the total traffic offered to the reference cell;
σ^2 – variance of the total traffic offered to the reference cell;
Z – peakedness coefficient of traffic in the reference cell;
E_d – blocking probability in the reference cell.

To determine the characteristics of the system, either the ERT method (Section 6.4.2) or the Fredericks–Hayward method (Section 6.4.5) can be used. In the present model, the simpler Fredericks–Hayward method is used, on the basis of which the following characteristics of the system can be obtained:

- Mean value of the overflow traffic from cell i to the reference cell (Section 6.3.3, Riordan formulas):

$$R_i = A_i E_{V_i}(A_i) \tag{13.41}$$

- Variance of the overflow traffic from cell i to the reference cell (Section 6.3.3, Riordan formulas):

$$\sigma_i^2 = R_i \left(\frac{A_i}{V_i + 1 - A_i + R_i} + 1 - R_i \right) \tag{13.42}$$

- Mean value of traffic offered to the reference cell with the own traffic of the reference cell taken into consideration (Section 6.3.4):

$$R = A_d + \sum_{i=1}^{K} R_i \tag{13.43}$$

- Variance of the total traffic offered to the reference cell, with the own traffic of the reference cell taken into consideration (Section 6.3.4):

$$\sigma^2 = A_d + \sum_{i=1}^{K} \sigma_i^2 \tag{13.44}$$

- Peakedness coefficient of the total traffic offered to the reference cell (Section 6.3.4):

$$Z = \frac{\sigma^2}{R} \tag{13.45}$$

- Blocking probability of the reference cell (Section 6.4.5, Hayward formula):

$$E_d = E_{\frac{V}{Z}}\left(\frac{R}{Z}\right) \tag{13.46}$$

If the value V/Z is a noninteger, then to obtain the value E_d we use the interpolation method or the integral form of Erlang B-formula [7]:

$$E_V(A) = \left[A \int_0^{\infty} \mathrm{e}^{-At}(t+1)^V \mathrm{d}t\right]^{-1} \tag{13.47}$$

13.3.2 Model of Single-Rate Traffic Overflow between Macro and Microcells

In this model we consider a macrocell divided into M microcells. Each microcell services the PCT1 traffic and has some assigned resources of channels at its disposal. A certain number of channels, shared by all the microcells, has been allocated in the macrocell to service traffic that is not serviced in microcells. The system is then characterized by a particular hierarchy – traffic overflows only in the direction: microcell $\rightarrow$ macrocell. Overflow in the opposite direction is not possible. We assume conventionally that each call demands one BBU to set up a connection. The considered model of overflow between microcells and macrocell is presented visually in Figure 13.7. Traffic from K microcells overflows to the macrocell.

In this model, the following notation is used:

A_i – PCT1 traffic intensity offered to the microcell with the serial number i;
V – capacity of the shared part of the macrocell;
V_i – capacity of the microcell i;
R_i – mean value of traffic overflowing to shared BBUs of the macrocell from the microcell i;
σ_i^2 – variance of traffic overflowing to shared BBUs of the macrocell from the microcell with the serial number i;
R – mean value of traffic offered to shared BBUs of the macrocell;
σ^2 – variance of traffic offered to shared BBUs of the macrocell;

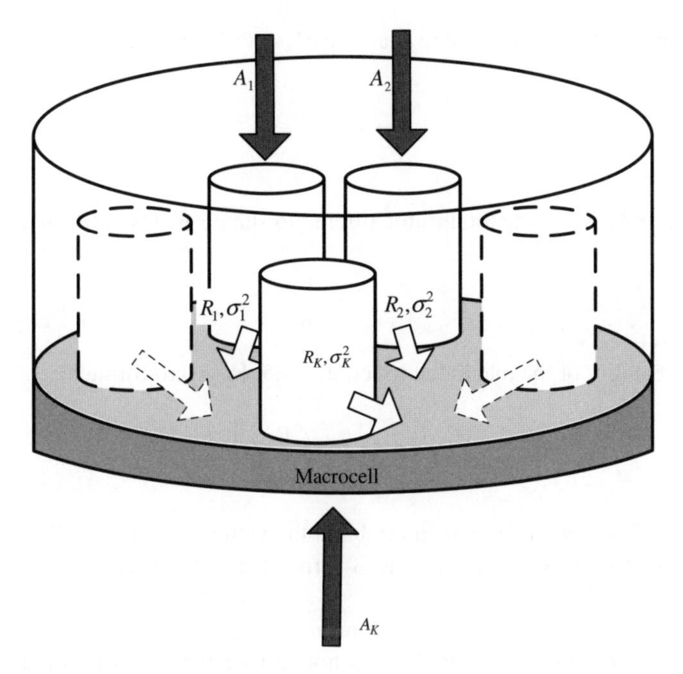

Figure 13.7 Model of overflow between microcells and macrocell.

Z – peakedness coefficient of traffic offered to shared BBUs of the macrocell;
E – blocking probability of the macrocell.

With the above notation at hand, we can write mathematical expressions determining the characteristics of the system:

- Mean value of traffic overflow, from microcell i, to shared BBUs of the macrocell (Section 6.3.3):

$$R_i = A_i E_{V_i}(A_i) \tag{13.48}$$

- Variance of traffic overflow, from microcell i to shared BBUs of the macrocell (Section 6.3.3):

$$\sigma_i^2 = R_i \left(\frac{A_i}{V_i + 1 - A_i + R_i} + 1 - R_i \right) \tag{13.49}$$

- Mean value of traffic offered to shared channels of the macrocell (Section 6.3.4):

$$R = \sum_{i=1}^{K} R_i \tag{13.50}$$

- Variance of the total offered traffic to shared BBUs of the macrocell (Section 6.3.4):

$$\sigma^2 = \sum_{i=1}^{K} \sigma_i^2 \tag{13.51}$$

- Peakedness coefficient of the total traffic offered to shared BBUs of the macrocell:

$$Z = \frac{\sigma^2}{R} \tag{13.52}$$

- Blocking probability of the macrocell (Section 6.4.5, Hayward formula):

$$E = E_{\frac{V}{Z}}\left(\frac{R}{Z}\right) \tag{13.53}$$

If the value V/Z is a noninteger, then to obtain the value E we use the linear interpolation method or the integral form of the Erlang B-formula (Equation (13.47)).

This model differs from the previous one, presented in Section 13.3.1 – in that the resources to which traffic overflows (shared BBUs) do not service their own PCT1 traffic (compare Equations (13.43) and (13.44) with Equations (13.50) and (13.51)). If we were to assume that the shared BBUs of the macrocell do service some of their own traffic, then the models would be identical.

In the case of modeling systems with single-rate overflow traffic generated by a finite number of traffic sources (PCT2 traffic), one can also use the calculation process presented in Sections 13.3.1 and 13.3.2. This process should be preceded according to the method presented in Section 6.5.

13.3.3 Model of Intercell Overflow of Multi-Rate Traffic

This model refers to those cells of the UMTS system that service a mixture of multi-rate traffic from a finite number of traffic sources (PCT1 streams).[4] We also assume that in the cellular system there is a possibility of traffic that is not serviced to overflow to other cells [8]. The scheme of system with intercell overflow multi-rate traffic is presented in Figure 13.8. Traffic from K neighboring cells overflows to the considered cell, which, further on in the text, will be referred to as the reference cell.

The following notation is used in the model:

$[P_n]_{V_f}$ – probability of the occupancy of n BBUs in cell f;
M_f – number of offered traffic classes in cell f;
$E_{i,f}$ – blocking probability of class i calls in cell f;
L_{BBU} – value of the basic bandwidth unit, in percentage of interference load;
V_f – cell interface capacity expressed in the number of available BBUs (equivalent hard capacity of the cell);

[4] For simplicity we assume that we will not divide traffic into types PCT1 and PCT2.

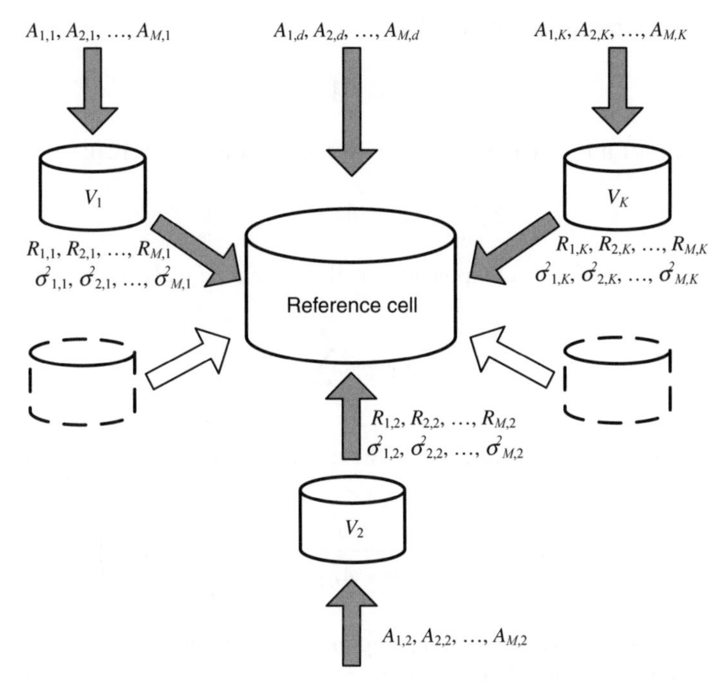

Figure 13.8 Model of a system with intercell overflow of multi-rate traffic.

$V_{f,i}$ – the capacity of the fictitious component of the cell with capacity f, calculated for calls of class i traffic stream, expressed in the number of BBUs (equivalent hard capacity of the cell);
$A_{i,f}$ – intensity of traffic of class i offered to cell f;
L_i – load factor (in percentage) introduced by one call of class i;
t_i – number of BBUs required by call of class i to set up a connection;
$Y_{i,f}$ – intensity of class i traffic carried in cell f;
$R_{i,f}$ – intensity of class i traffic that overflows from cell f;
$\sigma^2_{i,f}$ – variance of class i traffic that overflows from cell f;
$Z_{i,f}$ – peakedness coefficient of class i traffic that overflows from cell f.

The modified Hayword method, presented in Chapter 8, is used to determine characteristics of the system being considered. In order to determine the particular characteristics of individual cells the following formulas can be used:

- Occupancy distribution $[P_n]_{V_f}$ in cell f (Equation (7.79)).
- Blocking probability $E_{i,f}$ of class i calls in cell f (Equation (7.80)).

$$E_{i,f} = \sum_{n=V_f-t_i+1}^{V_f} [P_n]_{V_f} \tag{13.54}$$

- Intensity of class i traffic carried in cell f:

$$Y_{i,f} = A_{i,f}[1 - E_{i,f}] \tag{13.55}$$

- Intensities of class i traffic that overflows from cell f (Equation (8.12)):

$$R_{i,f} = A_{i,f}E_{i,f} \tag{13.56}$$

- Capacity $V_{i,f}$ of the fictitious component of cell f for class i calls (Equation (8.14)):

$$V_{i,f} = V_f - \sum_{l=1;l \neq i}^{M_1} Y_{l,f}t_l \tag{13.57}$$

- Variance of class i traffic that overflows to the reference cell from cell f (Equation (8.15)):

$$\sigma_{i,f}^2 = R_{i,f}\left(\frac{A_{i,f}}{\frac{V_{i,f}}{t_i} + 1 - A_{i,f} + R_{i,f}} + 1 - R_{i,f}\right) \tag{13.58}$$

The occupancy distribution and the blocking probability in the reference cell servicing its own traffic and traffic that overflows from other cells can be determined on the basis of the formulas derived in Section 8.4. We will denote the reference (overflow) cell by index d.

- The mean value of class i traffic offered to the access cell with the own traffic of the cell taken into consideration (Equation (8.15)):

$$R_{i,d} = A_{i,d} + \sum_{f=1;f \neq d}^{K} R_{i,f} \tag{13.59}$$

- Variance of the total class i traffic offered to the reference cell, with the own traffic of the cell taken into consideration (Equation (8.17)):

$$\sigma_{i,d}^2 = A_{i,d} + \sum_{f=1;f \neq d}^{K} \sigma_{i,f}^2 \tag{13.60}$$

- Peakedness coefficient of the total class i traffic offered to the reference cell:

$$Z_{i,d} = \frac{\sigma_{i,d}^2}{R_{i,d}} \tag{13.61}$$

- Overall peakedness coefficient Z (Equation (8.5)):

$$Z = \sum_{i=1}^{M} Z_{i,d}k_i \tag{13.62}$$

where

$$k_i = \frac{R_{i,d} t_i}{\sum_{l=1}^{M} R_{l,d} t_l} \tag{13.63}$$

- Occupancy distribution in the reference cell (Equation (8.3)):

$$n\,[P_n]_{V/Z} = \sum_{i=1}^{M} \frac{R_{i,d}}{Z_{i,d}} t_i \left[P_{n-t_i}\right]_{V_d/Z} \tag{13.64}$$

- Blocking probability of class i traffic in the reference cell (Equation (8.4)):

$$E_{i,d} = \sum_{n=\frac{V}{Z}-t_i+1}^{\frac{V}{Z}} [P_n]_{V/Z} \tag{13.65}$$

If the value V/Z is a non-integer number, then to obtain the value $E_{i,d}$ we use the interpolation method or the integral form of the Erlang B-formula (Equation (13.47)).

13.3.4 Comments

The overflow models presented in the chapter have been used for the case of single-service traffic. Overflows between sectors, cells and layers of the cellular network can also be executed for multi-rate traffic. In such cases models discussed in Chapter 8 can be also used.

With modeling systems in which PCT2 traffic streams are considered, the method for determination of the blocking probability $E_{i,f}$ has to be changed – this probability is determined on the basis of Equation (7.81). The value of the blocking probability $E_{i,f}$ enables us to determine the value of offered traffic $R_{i,f}$ on the basis of Equation (8.12) and makes it possible to perform a decomposition of the primary cell f (with the capacity of V_f BBUs), servicing M traffic classes, into M components, each having the capacity of $V_{i,f}$, described by Equation (8.14). Subsequently, we can perform a conversion of a PCT2 traffic stream to the equivalent PCT1 traffic stream according to the method described in Section 6.5, and, ultimately, determine the remaining parameters of the cellular system with multi-rate PCT2 overflow traffic according to Equations (13.59)–(13.65), presented above.

13.4 Handover Mechanisms

To secure both mobility for subscribers of a telephone cellular network and the increase in traffic capacity of the system, it is necessary to create mechanisms enabling continuity of the connection while moving the mobile station between different cells of the system. In the cellular system, such a mechanism is secured by the so-called handover. This occurs when a subscriber is moving out of the service range designated for a cell and, for some reason, it is convenient for the connection to be broken in the original cell and to be re-connected in a successor, neighboring cell. Hard handover of connections can occur between base stations or sectors of the same station within different frequency channels. Hard handover can be applied between 2G and 3G networks.

Another way of connection handover is the so-called soft handover, which can occur in a number of types: *soft handover*, *softer handover*, *soft-softer handover* (Section 2.7.2). In the case of a soft call connection transition of the type *soft handover*, the mobile station communicates with two or more sectors that belong to different base stations. The transmission between the mobile station and several base stations that occurs in this type of transition employs the diversity phenomenon. In the case of transitions of the type *softer handover*, the mobile station communicates with two or more sectors of a given cell. The advantage of this type of transition is the possibility of receiving and "assembling" signals that come from several sectors of the base station. Transition of the type *soft-softer handover* makes up a combination of the above-mentioned handoff methods and occurs when the base station communicates with several sectors of a given cell and, at least, one sector of another cell.

The last group of connection handover methods includes mechanisms that optimize the use of wireless network resources as a result of transferring users between neighboring cells such as directed retry [9], directed handover [10] and Virtual Channel Borrowing (VCB) [11, 12]. In the directed retry method, an attempt is made to admit a new call in sequentially chosen neighboring cells. With the second (directed handover), a cell in which a new call arrives attempts to transfer those calls that are already serviced to neighboring cells (with the instance of insufficient resources for the purpose of servicing the new call). The VBC method, in turn, transfers calls preventitatively to cells with the lowest actual traffic load in order to secure even load balance in cells and to decrease loss probabilities of calls.

13.4.1 The Model of the System Optimizing the Arrangement of Connections

In this chapter we will present the model of a group of cells that takes into consideration the phenomenon of connection transition between cells with the aim of increasing the traffic capacity of the system [13].

A group of cells in which the handover mechanism (directed retry, directed handover, VCB) has been introduced, can be treated as a system that optimizes the allocation of connections in particular cells to secure the best possible usage of the resources in the system. Figure 13.9 shows a group of seven selected cells. Each cell has a corresponding schematically presented

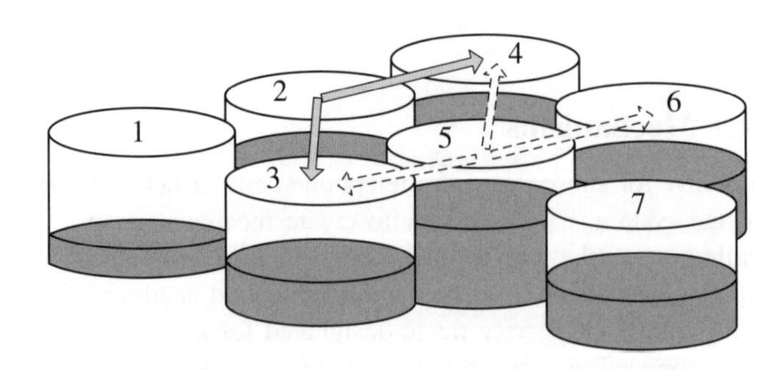

Figure 13.9 Connection handover in the group of cells.

cylindrical vessel, in which the level of the liquid corresponds to the degree of the radio interface load. Hence, cells 2 and 5 are those with most heavy loads. The handover mechanisms allow new connections that cannot be serviced in cells 2 and 5 to be directed to the neighboring cells, for example, from cell 2 to cells 3 and 4, and from cell 5 to the cells 3, 4, 6. In Figure 13.9, a new connection can be interpreted as an increase in the level of liquid in the vessel which corresponds to a given cell, then the excess of the liquid will overflow to the other neighboring vessels that still have enough space to accommodate the excess.

Such operation of the system resembles the operation of the limited-availability group. If a connection in the group cannot be set up in a given subgroup (cell), and other subgroups have sufficient resources to service it, then the connection is transferred to one of such subgroups.

13.4.2 Assumptions for the Model

Let us consider a group of cells in which each cell can have a different capacity. Let us assume that the term "group of cells" will designate a set of all cells within a given area, and the term "assembly of cells" will designate a set of cells directly adjacent to one another. Let us further assume that a group of cells is offered a multi-rate traffic,[5] in other words M independent call streams with the intensity: $\lambda_1, \lambda_2, \ldots, \lambda_M$. The demanded number of BBUs for the calls of particular classes will be designated, as earlier, by the symbols: $t_1, t_2, \ldots, t_M$. Service times of the calls of all classes have exponential distribution with the parameters: $\mu_1, \mu_2, \ldots, \mu_M$. The group of cells services a call only when it can be serviced entirely by one of the cells. Note that a call occurring in a given cell will be admitted if the cell, or any of the cells in its neighborhood (a particular assembly of cells), has a sufficient number of resources.

The following notation has been adopted for the considered model:

- $[P_n]_V$ – probability of n BBUs being in the group of cells;
- M – number of classes offered to the system;
- E_i – blocking probability of calls of class i in the group of cells;
- B_i – loss probability of calls of class i in the group of cells;
- η – interface capacity expressed in percentage;
- δ – ratio of the interference from other cells to the own cell interference in the (uplink/downlink) direction;
- ξ_j – orthogonality factor (downlink);
- L_i – load factor (in percentage) introduced by one call of class i;
- L_{BBU} – value of basic bandwidth unit, in percentage of interface load;
- q – number of cell types in the group, the type of the cell is identifiable by its capacity;
- k_s – number of cells of the type s;
- f_s – type s cell interface capacity expressed in the number of available BBUs;
- A_i – intensity of class i traffic, offered to the group of cells;
- t_i – number of BBUs required by a call of class i to set up a connection;
- Y_i – intensity of class i traffic carried in the group of cells;
- R_i – intensity of class i traffic lost in the group of cells;
- y_V – intensity of the traffic carried by one BBU of the radio interface in the cell.

[5]For simplicity of presentation, in the description of the method we will not divide traffic into types PCT1 and PCT2.

13.4.2.1 Characteristics of the Group of Cells with Multi-Rate Traffic

On the basis of the dependencies presented above, let us determine now the traffic characteristics of the group of cells with the included hard handoff and retry handoff mechanism. These characteristics result from the generalized model of the full-availability group that, will be used to model the cell system.

Let us consider the following method for modeling a group of cells servicing jointly (through the connection hand-off mechanism) multi-rate traffic streams. For each cell, we will first determine the neighboring cells and this assembly of cells will be then modeled by a limited-availability group. For instance, for the group shown in Figure 13.10, we will study seven assemblies of neighboring cells:

- cells: 1, 2, 3, 4, 5, 6, 7;
- cells: 2, 3, 1, 7;
- cells: 3, 4, 1, 2;
- cells: 4, 5, 1, 3;
- cells: 5, 6, 1, 4;
- cells: 6, 7, 1, 5;
- cells: 7, 2, 1, 6.

We determine the structure (capacities of individual cells) for each of the assemblies of cells, and the value of traffic offered to the group of cells. Then, for each of the assemblies, we determine the occupancy distribution and blocking (loss) probability. These blocking probability values form the input data in the determination of blocking probability in the whole area.

This textbook proposes a method that enables the determination of the blocking probability in the whole area (in the group of cells) on the basis of instances of blocking in individual assemblies of cells. This probability is determined as a weighted average, where the weights are the values of traffic offered to individual assemblies of cells.

A list of the formulas that define a proper determination of traffic characteristics of the considered system is presented below:

- Basic bandwidth unit (Section 11.1.3, Equation (11.13)):

$$L_{\mathrm{BBU}} = \mathrm{GCD}(L_1, L_2, \ldots, L_M) \tag{13.66}$$

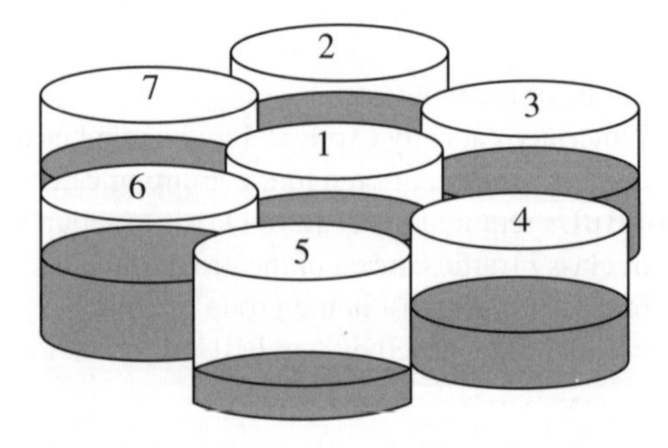

Figure 13.10 The formation of assemblies of neighboring cells.

- Equivalent capacity of the cell of the type s, expressed in BBUs for the uplink (Section 11.1.3):

$$f_s = f_{s,\text{UL}} = \left\lfloor \frac{\eta_s}{L_{\text{BBU}}(1+\delta)} \right\rfloor \tag{13.67}$$

- Equivalent capacity of the cell of the type s, expressed in BBUs for the downlink direction (Section 11.1.3):

$$f_s = f_{s,\text{DL}} = \left\lfloor \frac{\eta_s}{L_{\text{BBU}}(1-\xi+\delta)} \right\rfloor \tag{13.68}$$

- Number of BBUs demanded by a call of class i (Section 11.1.3, Equation (11.13)):

$$t_i = \lceil L_i / L_{\text{BBU}} \rceil \tag{13.69}$$

- Total capacity of the group of cells:

$$V = \sum_{s=1}^{q} k_s f_s \tag{13.70}$$

- Class i traffic offered to the group of cells:

$$A_i = \lambda_i / \mu_i \tag{13.71}$$

- Occupancy distribution in the assembly of cells No. j:

$$n\,[P_n]_{V_j} = \sum\nolimits_{i=1}^{M} A_{i,j} t_i \sigma_{i,j}(n-t_i)\left[P_{n-t_i}\right]_{V_j} \tag{13.72}$$

where V_j denotes the capacity of the assembly of cells No. j, $\sigma_{i,j}$ – conditional transition probability for class i calls in the assembly j, $A_{i,j}$ – value of traffic class i offered to the assembly No. j. Conditional transition probability $\sigma_{i,j}(n)$ for class i stream in the cell assembly j is determined by Equation (7.93) (Section 7.6.2):

$$\sigma_{i,j}(n) = 1 - \frac{F\{(V-n,(k_1 \ldots k_q),(t_i-1,\ldots,t_i-1),(0\ldots 0)\}}{\{F(V-n,(k_1 \ldots k_q),(f_1 \ldots f_q),(0\ldots 0)\}} \tag{13.73}$$

in which the value of the combinatorial function $F(\cdot)$ is determined on the basis of Equation (7.92) (Section 7.6.2):

$$F\{(x,(k_1,k_2,\ldots,k_q),(f_1,f_2,\ldots,f_q),(0,0,\ldots,0)\}$$

$$= \sum_{x_1=0}^{x} \cdots \sum_{x_{q-1}=0}^{x-\sum_{r=1}^{q-2} x_r} \left\{ \prod_{z=1}^{q-1} F(x_z,k_z,f_z,0) F(x - \sum_{r=1}^{q-1} x_r, k_q, f_q, 0) \right\} \tag{13.74}$$

where the parameter $F(x, k, f, 0)$ – that is, the number of possible arrangements of x free BBUs in k cells, each of which has a capacity equal to f BBUs, can be determined on the basis of Equation (7.83):

$$F(x,k,f,0) = \sum_{i=0}^{\left\lfloor \frac{x}{f+1} \right\rfloor} (-1)^i \binom{k}{i} \binom{x+k-1-i(f+1)}{k-1} \tag{13.75}$$

- Blocking (loss) probability for class i calls in the cell assembly No. j:

$$E_{i,j} = \sum\nolimits_{n=0}^{V_j} [P_n]_{V_j} [1 - \sigma_{i,j}(n)] \tag{13.76}$$

- Blocking (loss) probability for class i calls in the group of cells:

$$E_i = \sum\nolimits_{j=1}^{J} E_{i,j} \cdot w_{i,j} \tag{13.77}$$

where $w_{i,j}$ denotes a part of the total traffic of class i offered to the assembly j, whereas J denotes the number of cell assemblies in a given group of cells (within the considered area):

$$w_{i,j} = \frac{A_{i,j}}{\sum\limits_{i=1}^{M} \sum\limits_{j=1}^{J} A_{i,j}} \tag{13.78}$$

- Intensity of traffic class i, carried in the group of cells:

$$Y_i = A_i[1 - E_i] \tag{13.79}$$

- Intensity of the traffic carried by one BBU of the radio interface in the cell:

$$y_V = \sum_{i=1}^{M} Y_i \Big/ V \tag{13.80}$$

- Mean value of the lost traffic intensity of class i in the group of cells:

$$R_i = A_i - Y_i = A_i E_i \tag{13.81}$$

13.4.3 Group of Cells with Soft Handover Mechanism

In this section we will present a model of a group of cells that takes into consideration the phenomenon of soft handover between cells with the aim of increasing the traffic capacity of the system [14, 15]. A group of cells in which the soft handover mechanism has been introduced, can be also treated as a system that optimizes usage of the radio resources of the cellular system with soft capacity. Figure 13.11a shows a group of seven cells equipped with omnidirectional antennas. Each cell has a schematically presented cylindrical vessel. The soft handover mechanisms allow new connections that cannot be completely serviced by the radio interface resources in any of two cells (z and h_4) or three cells (z, h_5 and h_6) to be simultaneously serviced by all the above cells – that is, one connection can be serviced by cells z and h_4 and another connection can be carried out by radio resources in cells z, h_5 and h_6. It should be stressed that the greater the number of cells concurrently involved in a given

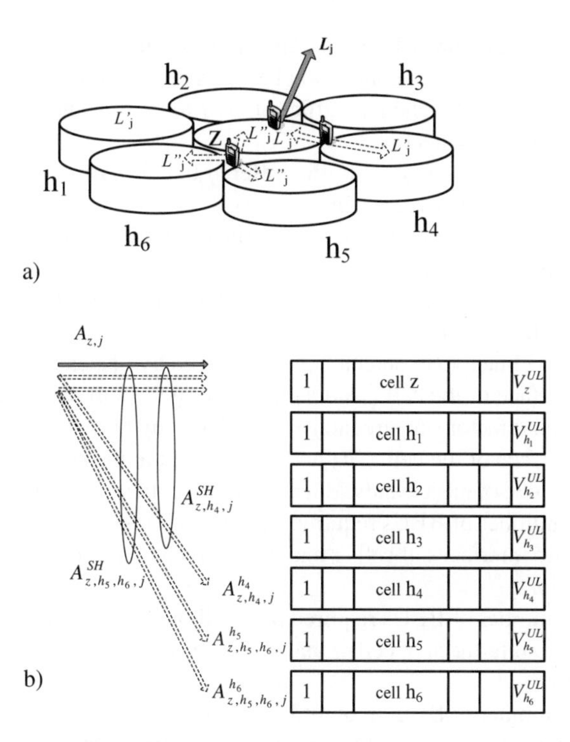

Figure 13.11 Fragment of the mobile network with soft handover mechanism.

connection, the lower the number of resources required for service of this connection in each of the cells.

Such operation of the system can be taken into considerations by the application of the fixed-point method presented Section 13.2.1.

Each of the radio interface in the group of cells can be treated as a full-availability group servicing a mixture of multi-rate traffic streams. All traffic characteristics – the blocking probability, loss probability, carried and offered traffic – result from the approximate calculation algorithm of the multi-service Erlang-Engset distribution (Section 7.5.3),[6] and from the application of the fixed-point method (Section 13.2.1).

13.4.3.1 Assumptions

Let us consider the system presented in Figure 13.11. Each cell in the group has a different capacity and services a mixture of multi-rate traffic. Assume that the cell in which new calls arrive is the cell denoted by z. Cell z is surrounded by six neighboring cells (Table 13.1).

Figure 13.11b shows a diagram of the distribution of traffic for the system under investigation. In the model we have adopted the notation introduced in the previous chapters:

$E^{\mathrm{SH}}_{z,h_4,j}$ – total blocking probability of calls of class j, involved in the soft handover connection between cells z and h_4;

[6]For simplicity of presentation, we will not divide traffic into types PCT1 and PCT2.

$E^{x}_{z,h_4,j}$ – blocking probability of calls of class j in cell x ($x \in \{z, h_4\}$), involved in the soft handover connection between cells z and h_4;

$A^{\mathrm{SH}}_{z,h_4,j}$ – intensity of class j traffic, offered to cells z and h_4 involved in the soft handover connection;

$A^{x}_{z,h_4,j}$ – intensity of class j traffic, offered to cell x involved in the soft handover connection between cells z and h_4;

$A_{y,j}$ – intensity of class j traffic, offered to cell y, where $y \in \{z, h_1, h_2, h_3, h_4, h_5, h_6\}$;

$L_{y,j}$ – load factor (in percentage) introduced by one call of class j into the radio interface of the cell y;

$L'_{z,h_4,j}$ or L'_j – load factor (in percentage) introduced by one call of class j, into the radio interface of the cell z and h_4 involved in the soft handover connection;

$L''_{z,h_5,h_6,j}$ or L''_j – load factor (in percentage) introduced by one call of class j, into the radio interface of the cell z, h_5 and h_6 involved in the soft handover connection;

$L_{y,\mathrm{BBU}}$ – value of basic bandwidth unit, in percentage of the interface load, for cell y;

$t_{z,h_4,j}$ or t'_j – number of BBUs required by a call of class j to set up a connection in the radio interfaces of cells z and h_4 involved in the soft handover connection, in cell h_4;

$t'_{h_4,z,j}$ or t'_j – number of BBUs required by a call of class j to set up a connection in the radio interfaces of cells z and h_4 involved in the soft handover connection, in cell z;

$t''_{z,h_5,h_6,j}$ or t''_j – number of BBUs required by a call of class j to set up a connection in the radio interfaces of cells z, h_5 and h_6 involved in the soft handover connection;

$t_{y,j}$ or t_j – number of BBUs required by call of class j to set up a connection by the radio interface of the cell y;

V^{UL}_{y} – equivalent hard capacity in the uplink direction of the cell y.

The proposed model assumes that a new call that employs a soft handover will be rejected if the increase in load of the interfaces of any of the cells that are to be involved in this connection exceeds the admissible load level. It should be stressed that the bigger the number of cells concurrently involved in a given connection, the lower the number of resources required for a service process of this connection in each of the cells. A call demands t_j BBUs in the reference cell and t'_j or t''_j BBUs in neighboring cells (for calls in soft handover), while $t''_j < t'_j < t_j$ (Figure 13.11).

13.4.3.2 Characteristics of the Group of Cells with Soft Handover Traffic

We first determine those parameters of the system that are dependent on the structure of serviced traffic. We will subsequently discuss the model from the point of view of soft handover connection between cells z and h_4. A list of the formulas that define the proper determination of traffic characteristics of the considered system is presented below:

- Determination of basic bandwidth unit (Section 11.1.3, Equation (11.13))
  $L_{z,\mathrm{BBU}}$ for cell z can be determined on the basis of the following dependency:

$$L_{z,\mathrm{BBU}} = \mathrm{GCD}(L_{z,1}, \ldots, L_{z,M}, L'_{h_4,z,1}, \ldots, L'_{h_4,z,M}) \tag{13.82}$$

and $L_{h4,\mathrm{BBU}}$ for cell h_4 is determined by the formula:

$$L_{h4,\mathrm{BBU}} = \mathrm{GCD}(L_{h4,1}, \ldots, L_{h4,M}, L'_{z,h_4,1}, \ldots, L'_{z,h_4,M}) \tag{13.83}$$

- Determination of number of BBUs demanded by a call of class i (Section 11.1.3, Equation (11.13)).
  Number of BBUs required by call of class j to set up a connection by the radio interface of the cell z and h_4:

$$t_{z,j} = \lfloor L_{z,j}/L_{z,\mathrm{BBU}} \rfloor \quad \text{and} \quad t_{h4,j} = \lfloor L_{h4,j}/L_{h4,\mathrm{BBU}} \rfloor \tag{13.84}$$

  The parameters $t'_{h_4,z,1}, \ldots, t'_{h_4,z,M}$ and $t'_{z,h_4,1}, \ldots, t'_{z,h_4,M}$ define the number of resources demanded in neighboring cells by a call of class i in soft handover:

$$t'_{z,h_4,j} = \left\lfloor L'_{z,h_4,j}/L_{h4,\mathrm{BBU}} \right\rfloor \quad \text{and} \quad t'_{h_4,z,j} = \left\lfloor L'_{h_4,z,j}/L_{z,\mathrm{BBU}} \right\rfloor \tag{13.85}$$

- Knowing the value of BBU for each of the cells under consideration, we can determine the equivalent capacity of the radio interface in the uplink direction of cell z and h_4, expressed in BBUs (Section 11.1.3, Equation (11.14)):

$$V_z^{\mathrm{UL}} = \lfloor \eta_z^{\mathrm{UL}}/L_{z,\mathrm{BBU}} \rfloor \quad \text{and} \quad V_{h4}^{\mathrm{UL}} = \lfloor \eta_{h4}^{\mathrm{UL}}/L_{h4,\mathrm{BBU}} \rfloor \tag{13.86}$$

13.4.3.3 Calculation Algorithm

In the calculation algorithm of the value of blocking probability for all traffic classes, the intensity of traffic offered to group of cells by calls serviced in soft handover connection, is needed. In the model we assume that the dependency in the service process of each call offered to the system by the soft handover traffic can be taken into account by fixed-point algorithm. In order to find values of appropriate parameters the iterative fixed-point method, presented in Section 13.2.1, is therefore used. This algorithm can be rewritten in the following form:

1. Setting up initial values of all blocking probabilities:

$$\forall_{1 \le j \le M} \; E_{z,j}^{(0)} = E_{h4,j}^{(0)} = E_{z,z,j}^{(0)} = E_{h4,h4,j}^{(0)} = E_{z,h4,j}^{(0)} = E_{h4,z,j}^{(0)} = 0 \tag{13.87}$$

2. Increasing of the iteration step $k = k + 1$.
3. Determination of the values of effective traffic and interference effective traffic for cell z and cells h_4.
   Effective traffic $A_{z,h_4,j}^{z(k)}$ of class j offered to cells z involved in soft handover connection with cell h_4 can be determined on the basis of the following dependency:

$$A_{z,h_4,j}^{z(k)} = A_{z,h_4,j}^{\mathrm{SH}}(1 - E_{z,h_4,j}^{h_4(k-1)}) \tag{13.88}$$

   Effective interference traffic $A_{z,h_4,j}^{h_4(k-1)}$ can be determined as follows:

$$A_{z,h_4,j}^{h_4(k)} = A_{z,h_4,j}^{\mathrm{SH}}(1 - E_{z,h_4,j}^{z(k-1)}) \tag{13.89}$$

4. Determination of the occupancy distribution in the cell z and h_4.
   The occupancy distribution in the cell z and in the cell h_4 can be determined on the basis of the following dependencies:

$$[P_n]_{V_z}^{(k)} = F\left\{\begin{array}{l}(A_{z,1}, t_{z,1}), \ldots, (A_{z,M}, t_{z,M}), \\ (A_{z,h_4,1}^{z(k)}, t'_{h_4,z,1}), \ldots, (A_{z,h_4,M}^{z(k)}, t'_{h_4,z,M})\end{array}\right\} \tag{13.90}$$

$$[P_n]_{V_{h_4}}^{(k)} = F\left\{\begin{array}{l}(A_{h_4,1}, t_{h_4,1}), \ldots, (A_{h_4,M}, t_{h_4,M}), \\ (A_{z,h_4,1}^{h_4(k)}, t'_{z,h_4,1}), \ldots, (A_{z,h_4,M}^{h_4(k)}, t'_{z,h_4,M})\end{array}\right\} \tag{13.91}$$

Function "F" can be obtained based on the model of the full-availability group servicing multi-rate traffic (Section 7.3.5, Equation (7.39)), which can be for Equation (13.90) rewritten in the following way:

$$n\,[P_n]_{V_z}^{(k)} = \sum_{j=1}^{M} A_{z,j}^{(k)} t_{z,j} \left[P_{n-t_{z,j}}\right]_{V_z} + \sum_{j=1}^{M} A_{z,h_4,j}^{z(k)} t'_{h_4,z,j} \left[P_{n-t'_{h_4,z,j}}\right]_{V_z} \tag{13.92}$$

and for Equation (13.91):

$$n\,[P_n]_{V_{h_4}}^{(k)} = \sum_{j=1}^{M} A_{h_4,j}^{(k)} t_{h_4,j} \left[P_{n-t_{h_4,j}}\right]_{V_{h_4}} + \sum_{j=1}^{M} A_{z,h_4,j}^{h_4(k)} t'_{z,h_4,j} \left[P_{n-t'_{z,h_4,j}}\right]_{V_{h_4}} \tag{13.93}$$

5. Calculation of the value of blocking probability for calls of class j in cell z and in each cell h_4.
   The blocking probabilities $E_{z,z,j}^{(k)}$, $E_{z,h_4,j}^{(k)}$, $E_{h_4,h_4,j}^{(k)}$ and $E_{h_4,z,j}^{(k)}$ can be determined on the basis of the full-availability group servicing multi-rate traffic:

$$\forall_{1\le j\le M} \; E_{z,j}^{(k)} = \sum\nolimits_{n=V_z-t_{z,j}+1}^{V_z} [P_n]_{V_z} \tag{13.94}$$

$$\forall_{1\le j\le M} \; E_{z,h_4,j}^{z(k)} = \sum\nolimits_{n=V_z-t'_{h_4,z,j}+1}^{V_z} [P_n]_{V_z}$$

$$\forall_{1\le j\le M} \; E_{h_4,j}^{(k)} = \sum\nolimits_{n=V_{h_4}-t_{h_4,h_4,j}+1}^{V_{h_4}} [P_n]_{V_{h_4}}$$

$$\forall_{1\le j\le M} \; E_{z,h_4,j}^{h_4(k)} = \sum\nolimits_{n=V_{h_4}-t'_{z,h_4,j}+1}^{V_{h_4}} [P_n]_{V_{h_4}}$$

6. Determination of the value of total blocking probability for calls of class j in cell z and h_4 involved in soft handover connection.

From the knowledge of the blocking probability for calls of class j in cell z and cell h_4 we can determine the blocking probability $E^{\mathrm{SH}}_{z,h_4,j}$ for calls of class j, involved in the soft handover connection, using the dependence:

$$E^{\mathrm{SH}(k)}_{z,h4,j} = 1 - (1 - E^{z(k)}_{z,h_4,j})(1 - E^{h_4(k)}_{z,h_4,j}) \tag{13.95}$$

7. Verification of the accuracy of calculations for cell z and cells h_4.
   Repetition of steps 2–7 until the assumed accuracy ε of the iterative process is obtained:

$$\forall_{1 \le j \le M} \left| \frac{E^{\mathrm{SH}(k-1)}_{z,h_4,j} - E^{\mathrm{SH}(k)}_{z,h_4,j}}{E^{\mathrm{SH}(k)}_{z,h_4,j}} \right| \le \varepsilon \tag{13.96}$$

where $E^{\mathrm{SH}(k)}_{z,h_4,j}$ and $E^{\mathrm{SH}(k-1)}_{z,h_4,j}$ are the value of blocking probability for class j calls participating in the soft handover connection with cells z and h_4, in iterations k and $k-1$, respectively.

Conversely, if the approximation error is lower than or equal to the adopted accuracy ε, then the iteration process terminates. The iteration process is a convergent process, and the number of steps of the iteration depends on the adopted accuracy of calculations ε. In the analyzed instances the iteration process terminated after a dozen steps.

Figure 13.11 also shows an example of a connection in which three cells are involved in soft handover. With the algorithm presented here, it also possible, in this case, to determine the blocking probability in a relatively easy way. Assuming that the cells that are involved in the connections are cells z, h_5 and h_6, we obtain:

$$E^{\mathrm{SH}}_{z,h_5,h_6,i} = 1 - (1 - E^{z}_{z,h_5,h_6,i})(1 - E^{h_5}_{z,h_5,h_6,i})(1 - E^{h_6}_{z,h_5,h_6,i}) \tag{13.97}$$

13.5 Comments

In the models presented in this chapter, only traffic of the type PCT1 is considered. This is because the intention is to facilitate comprehension of the dependencies presented. However, the proposed model can also be applied to a more general case, when a system services multi-rate traffic of types PCT1 and PCT2. In the generalized case, the change will only affect the way of determining occupancy distribution (7.79) and blocking (loss) probability, following the algorithm proposed in Section 7.5.3.

References

[1] Głąbowski, M., Stasiak, M., Wiśniewski, A., and Zwierzykowski, P. (2009) Uplink blocking probability calculation for cellular systems with WCDMA radio interface and finite source population, in *Information Science and Technology, vol. 2* (ed. D.D. Kouvatsos). River Publishers.

[2] Kelly, F.P. (1991) Loss networks. *The Annals of Applied Probability*, **1** (3), 319–78.

[3] Stasiak, M., Wiśniewski, A., Zwierzykowski, P., and Głąbowski, M. (2009) Blocking probability calculation for cellular systems with WCDMA radio interface servicing PCT1 and PCT2 multi-rate traffic. *IEICE Transactions on Communications*, **E920B**(4), 1156–65.

[4] Stasiak, M., Wiśniewski, A., and Zwierzykowski, P. (2004) *Blocking Probability Calculation in the Uplink Direction for Cellular Systems with WCDMA Radio Interface*. Third Polish-German Teletraffic Symposium, Dresden, September. VDE Verlag GMBH, Berlin.

[5] Wiśniewski, A. (2009) Modelling of cellular networks with the WCDMA radio interface. PhD thesis, Poznan University of Technology, Poznan, Poland, September.
[6] Stasiak, M., Wiśniewski, A. and Zwierzykowski, P. (2005) *Uplink Blocking Probability for a Cell with WCDMA Radio Interface and Differently Loaded Neighbouring Cells*. Service Assurance with Partial and Intermittent Resources Conference (SAPIR), Lisbon, June. IEEE Computer Society, Lisbon.
[7] Syski, R. (1986) *Introduction to Congestion Theory in Telephone Systems*. North Holland.
[8] M. Głąbowski, M. (2008) Modeling systems with multi-service overflow Erlang and Engset traffic streams. *International Journal On Advances in Telecommunications*, **1** (1), 14–26.
[9] Everitt, D. (1990) Traffic capacity of cellular mobile communications systems. *Computer Networks and ISDN Systems*, **20** (1), 447–54.
[10] Eklundh, E. (1986) Channel utilisation and blocking probability in a cellular mobile telephone system with directed retry. *IEEE Transactions on Communications*, **34** (4), 329–37.
[11] Al Agha, K., Pujolle, G., and Zeghlache, D. (2002) VCB: An efficient resource sharing scheme for cellular mobile systems. *Telecommunication Systems*, **19** (1), 101–10.
[12] Al-Meshhadany, T. and Al Agha, K. (2002) VCB by Means of Soft2hard Handover in WCDMA. Fourth IEEE Conference on Mobile and Wireless Communications Networks (MWCN), Stockholm, 9–11 September. IEEE, Stockholm. DOI: 10.1109/MWCN.2002.1045813.
[13] Głąbowski, M., Sobieraj, M., and Stasiak, M. (2007) *Blocking Probability Calculation in UMTS Networks with Bandwidth Reservation, Handoff Mechanism and Finite Source Population*. Proceedings of 7th International Symposium on Communications and Information Technologies, Sydney, October.
[14] Stasiak, M., Zwierzykowski, P., and Parniewicz, D. (2009) *Modelling of the WCDMA Interface in the UMTS Network with Soft Handoff Mechanism. Proceedings of IEEE Global Communications Conference*, Honolulu, December IEEE Communication Society, Honolulu.
[15] Stasiak, M., Zwierzykowski, P., and Parniewicz, D. (2009) Analytical model of the soft handoff mechanism in the UMTS network, in *Computer Performance Engineering*, volume 5652 of *Lecture Notes in Computer Science*, (ed. J. Bradley), Springer.

Conclusion

The development of present-day telecommunication networks is closely related to the increased tendency to combine services provided by traditional telecommunication networks and computer networks. This ongoing convergence of the network is accompanied by a growing need for a further development of methods for managing multi-service traffic.

Modern mobile telecommunication networks consist of many parallel systems operating in different frequency bands (GSM900, GSM 1800, different frequencies in UTRAN), and are built in varied technologies (GSM, UMTS Release 1999, HSPA).

The service of traffic in networks is also involved in business solutions that allow other clients to use network services provided by one company (a good example of the above is roaming across different group networks within one country). Such a complex level of development of mobile networks indicates how important an efficient formulation of a traffic allocation strategy in these systems is. Technological solutions allow us to execute these tasks precisely, both at the level of different frequency ranges of the same system, and at the level of Inter Radio Access Technology (InterRAT) the idea of transmitting signals and data. It should not be forgotten, though, that this kind of parametrization of network systems should be preceded by a strategy worked out by an operator, and then by the precise physical design of a network following this strategy. The changable market situation, combined with a reduction in fees for telecommunication services, makes operators pursue opportunities to increase the effectiveness of the network, maintaining, at the same time, the quality of services offered.

Design and optimization work (and consideration of the financial implications of the introduction of new services) requires a precise network dimensioning procedure. Unfortunately, the level of complexity of systems combined with the need for clients to possess particular capabilities, and the varying capabilities of popular technologies has led to a situation in which optimization of the traffic management system has become an enormous challenge not only from the engineering perspective but also scientifically. The use of models of network systems that are too simplified may result in an inadequate evaluation of the capabilities of network systems and this may lead to a substantial decrease in the competitiveness of a company.

The present book meets the needs for a manual for engineers involved in optimization of complex telecommunication networks and presents the reader with methods for network

Modeling and Dimensioning of Mobile Networks: From GSM to LTE
Maciej Stasiak, Mariusz Głąbowski, Arkadiusz Wiśniewski and Piotr Zwierzykowski

dimensioning that take into consideration, for example, so-called hard and soft capacity of the radio interface, the number of users serviced in a given cell or a group of cells, traffic overflow phenomena for single-service and multi-service traffic between cells (sectors) or layers of a mobile network, service prioritization in the radio interface and in the core network, resource differentiation for HSPA users and hard handover and soft handover mechanisms in the radio interface.

Appendix A

This appendix presents a tabular comparison of the models used in Part Three of the book for modeling of selected interfaces in the cellular network.

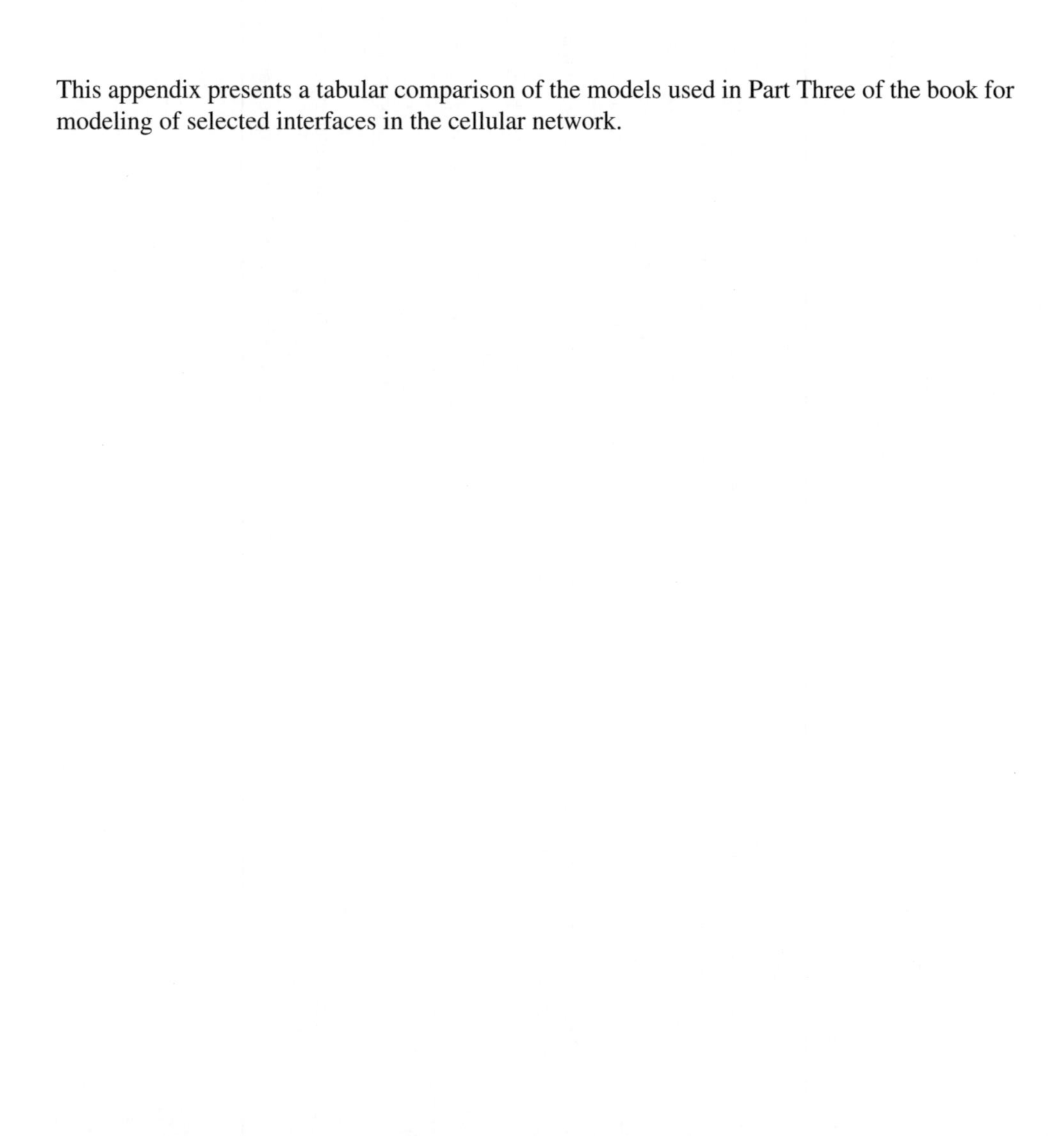

Modeling and Dimensioning of Mobile Networks: From GSM to LTE
Maciej Stasiak, Mariusz Głąbowski, Arkadiusz Wiśniewski and Piotr Zwierzykowski

Table A.1 Modeling of the radio interface in the single cell carrying single-service traffic

Model	Application	Assumptions	Section
Erlang	Determination of traffic characteristics of the cell of the GSM system or the cell of the single-service UMTS system with infinite number of traffic sources	• The number of subscribers substantially (>10) exceeds the radio interface capacity • A system with soft capacity is modeled by a system with equivalent hard capacity • Each call requires one BBU to set up a connection	11.2.1 and 11.3.1
Engset	Determination of traffic characteristics of a cell of the GSM system or a cell of single-service UMTS system with finite number of traffic sources	• Number of subscribers comparable (≥) to the radio interface capacity • System with soft capacity is modeled by a system with equivalent hard capacity • Each call requires one BBU to set up a connection	11.2.2 and 11.3.2
Erlang/Engset	Determination of traffic characteristics of a cell of the GSM or UMTS systems with two traffic streams, i.e. from an infinite number of traffic sources (PCT1 stream) and from a finite number of traffic sources (PCT2 stream)	• There are two traffic streams in the system, one of the type PCT1, the other of the type PCT2 (Engset stream) • System with soft capacity is modeled by a system with equivalent hard capacity • Each call requires one BBU to set up a connection	11.4.3

Table A.2 Modeling of the radio interface in the single cell carrying multi-service traffic

Model	Application	Assumptions	Section
Model of a cell with multi-rate traffic of the type PCT1	Determination of traffic characteristics of the radio interface of the UMTS/LTE system servicing a mixture of different traffic classes of PCT1 traffic	• The number of subscribers of a given class substantially exceeds (>10) the radio interface capacity • System with soft capacity is modeled by a system with equivalent hard capacity • Each traffic class demands a different number of BBUs to set up a connection	11.4.1
Model of a cell with multi-rate traffic of the type PCT2	Determination of traffic characteristics of the radio interface of the UMTS/LTE system servicing a mixture of different traffic classes of PCT2 traffic	• Number of traffic classes of a given class is comparable to, though greater than, the capacity of the radio interface • System with soft capacity is modeled by a system with equivalent hard capacity • Each traffic class demands a different number of BBUs to set up a connection	11.4.2
Model of a cell with multi-rate traffic of the type PCT1 and PCT2	Determination of traffic characteristics of the radio interface of the UMTS/LTE system servicing a mixture of traffic of the type PCT1 and PCT2 from a finite and a infinite number of traffic sources	• The number of PCT1 traffic sources substantially exceeds (>10) the capacity of the radio interface • The number of PCT2 traffic sources of a given class is comparable, though always higher, to the capacity of the radio interface • System with soft capacity is modeled by a system with equivalent hard capacity • Each traffic class demands a different number of BBUs to set up a connection	11.4.3

Table A.3 Modeling of the group of cells carrying multi-service traffic

Model	Application	Assumptions	Section
Generalized model of a group of cells with multi-rate traffic	Determination of traffic characteristics of the radio interface the UMTS system servicing a mixture of different traffic classes with the influence of neighboring cells taken into consideration in the uplink direction	• Group of cells consist of similarly overloaded cells • System with soft capacity is modeled by a system with equivalent hard capacity • Each traffic class requires a different number of BBUs to set up a connection	13.2.2
Simplified model of a group of seven cells with multi-rate traffic	Determination of traffic characteristics of the radio interface of the UMTS system servicing a mixture of different traffic classes with the influence from the neighboring cells taken into consideration in the uplink direction	• Group of cells consist of one huge cell and a group of cell with limited capacity • System with soft capacity is modeled by a system with equivalent hard capacity • Each traffic class requires a different number of BBUs to set up a connection	13.2.2
Model of a group of cells servicing multi-rate traffic in uplink and in downlink directions	Determination of traffic characteristics of the radio interface of the UMTS system servicing a mixture of different traffic classes with the influence from the neighboring cells taken into consideration in the uplink and in the downlink direction	• Group of cells consist of one similarly loaded group of cell • System with soft capacity is modeled by a system with equivalent hard capacity • Each traffic class requires a different number of BBUs to set up a connection	13.2.4

Table A.4 Modeling of the Iub interface in the UMTS network carrying multi-service traffic

Model	Application	Assumptions	Section
Basic model of the Iub interface with multi-rate traffic of the type PCT1	Determination of traffic characteristics of the Iub interface of the UMTS system servicing a mixture of different traffic classes of PCT1 traffic	• The number of subscribers of a given class substantially exceeds (>10) the Iub interface capacity • Each traffic class requires a different number of BBUs to set up a connection	12.3.1
Model of the Iub interface servicing multi-rate traffic of the type PCT1 with priorities	Determination of traffic characteristics of the Iub interface of the UMTS system servicing a mixture of different traffic classes with priorities	• Number of subscribers of a given class substantially exceeds (>10) the Iub interface capacity • Each traffic class requires a different number of BBUs to set up a connection • Each traffic class is characterized by different priority	12.3.2
Model of the Iub interface servicing a mixture of multi-rate traffic with and without compression	Determination of traffic characteristics of the Iub interface of the UMTS system servicing a mixture of different traffic classes with and without compression	• The number of traffic sources of PCT1 substantially exceeds (>10) the radio interface capacity • Each traffic class requires a different number of BBUs to set up a connection • Interface services a mixture of R99 and HSPA traffic classes	12.3.3

Table A.5 Modeling of a group of cells carrying multi-service with call handover mechanism

Model	Application	Assumptions	Section
Model of a set of cells with multi-rate traffic and hard handover mechanism	Determination of traffic characteristics of the group of cells in the mobile network servicing a mixture of different PCT1 traffic classes with the influence of neighboring cells taken into consideration In the network, handover methods which include mechanisms that optimize the use of wireless network resources as the result of the transferring users between neighboring cells such as directed retry, directed handover and Virtual Channel Borrowing were applied	• The number of subscribers of a given class substantially exceeds (>10) the radio interface capacity • System with soft capacity is modeled by a system with equivalent hard capacity • Each traffic class requires a different number of BBUs to set up a connection • Handover mechanism is modelled by the limited-availability group	13.4.1
Model of a set of cells with multi-rate traffic and soft handover mechanism	Determination of traffic characteristics of the group of cells in the mobile network servicing a mixture of different traffic classes of PCT1 traffic with the influence of neighboring cells taken into consideration. In the network soft handover mechanism is used	• Number of subscribers of a given class substantially exceeds (>10) the radio interface capacity • System with soft capacity is modeled by a system with equivalent hard capacity • Each traffic class requires a different number of BBUs to set up a connection • Soft handover mechanism is modeled by the fixed-point method	13.4.3

Table A.6 Modeling of the group of cells carrying overflow traffic

Model	Application	Assumptions	Section
Model of a set of cells servicing single-rate traffic of the type PCT1 with intercell overflow	Determination of traffic characteristics of the group of cells servicing a single-service PCT1 traffic with the influence of intercell overflow taken into consideration	• Number of subscribers of a given class substantially exceeds (>10) the radio interface capacity • System carry single-service overflow traffic • Overflow mechanism allows traffic overflow between cells in one layer of the mobile network to be executed	13.3.1
Model of a set of cells servicing single-rate traffic of the type PCT1 with traffic overflow between macro and microcells	Determination of traffic characteristics of the group of macro and microcells servicing a single-service PCT1 traffic with the influence of intercell overflow taken into consideration	• Number of subscribers of a given class substantially exceeds (>10) the radio interface capacity • System carry single-service overflow traffic • Overflow mechanism allows on traffic overflows only in the direction: microcell → *macrocell*	13.3.2
Model of a set of cells servicing multi-rate traffic of the type PCT1 with overflow	Determination of traffic characteristics of the group of cells servicing a multi-service PCT1 traffic with the influence of overflow taken into consideration	• Number of subscribers of a given class substantially exceeds (>10) the radio interface capacity • System carry multi-service overflow traffic	13.3.3

Table A.7 Modeling of the HSPA traffic in the UMTS network

Model	Application	Assumptions	Section
Model of a radio interface servicing HSDPA traffic	Determination of traffic characteristics of the radio interface in the UMTS system servicing a HSDPA traffic in the downlink direction	• The number of subscribers of a given class substantially exceeds (>10) the radio interface capacity • System with soft capacity is modeled by a system with equivalent hard capacity • Each traffic class requires a different number of BBUs to set up a connection	11.5
Model of the Iub interface servicing HSPA traffic	Determination of traffic characteristics of the Iub interface in the UMTS system servicing a mixture of different traffic streams on common resources – traffic classes with and without compression, which correspond to R99 and HSPA traffic classes. In the basic version of the model it was assumed that system services PCT1 traffic	• Number of subscribers of a given class substantially exceeds (>10) the radio interface capacity • System with soft capacity is modeled by a system with equivalent hard capacity • Each traffic class requires a different number of BBUs to set up a connection • Interface services classes that have compression	12.3.3

Index

Modeling and Dimensioning of Mobile Networks: From GSM to LTE
Maciej Stasiak, Mariusz Głąbowski, Arkadiusz Wiśniewski and Piotr Zwierzykowski